# NEUTRON INTERFEROMETRY

# NEUTRON INTERFEROMETRY

Proceedings of an International Workshop
held 5–7 June 1978 at the
Institut Max von Laue–Paul Langevin
Grenoble

Edited by
U. BONSE and H. RAUCH

Clarendon Press. Oxford
1979

*Oxford University Press, Walton Street, Oxford* OX2 6DP

OXFORD LONDON GLASGOW
NEW YORK TORONTO MELBOURNE WELLINGTON
KUALA LUMPUR SINGAPORE JAKARTA HONG KONG TOKYO
DELHI BOMBAY CALCUTTA MADRAS KARACHI
NAIROBI DAR ES SALAAM CAPE TOWN

ISBN 0 19 851947 8



Published in the United States by
Oxford University Press, New York



**British Library Cataloguing in Publication Data**

Neutron interferometry.
1. Neutron interferometry—Congresses
I. Bonse, U  II. Rauch, H
539.7'7  QC793.5.N462  79-41494

ISBN 0-19-851947-8

Printed in Great Britain by
Thomson Litho Ltd, East Kilbride, Scotland

# PREFACE

With the new interferometer at the Institut Max von Laue–Paul Langevin approaching its completion as a users instrument the Institute initiated a Workshop on Neutron Interferometry held from June 5 to June 7, 1978. It was attended by about 70 scientists from eight different countries. Most laboratories active in ångström range interferometry were represented. The workshop thus succeeded very well in its principal aim of establishing active discussion among leading interferometrists working in all wavelengths. The program included 11 invited guideline and 25 contributed papers, and also a final discussion on future trends of neutron interferometry. The subjects treated by the lectures and in the discussions included a variety of topics.

An essential part of the neutron interferometric activity at the high flux reactor at Grenoble is the precise determination of the index of refraction which establishes a new method for the measurement of coherent scattering lengths. Accurate values help in the separation of the nuclear contribution from electromagnetic and internal contributions. Bulk samples can be used which avoid errors due to surface effects. Applications in the field of solid state physics arise as soon as samples with an inhomogenous refractive index are inserted into the inferometer. Here the effects of magnetic domain structure, precipitates and long-range fluctuations near transition points are discussed. In this field neutron interferometry may compete with small-angle scattering as an investigative technique. Neutron phase topography may supplement diffraction topography of magnetic structure in crystals. The variation of the refractive index near to a reciprocal lattice point due to the dispersion relation was demonstrated with perfect crystal samples.

Neutron interferometers different from the LLL type, which so far have been employed exclusively, were discussed in some detail. Other possible versions treated include triple Bragg case (BBB) interferometers, two-component interferometers (LL or BB), four-component interferometers (LLLL), and mixed Bragg–Laue interferometers (LBBL). Some contributions showed the specific characteristics of dynamic diffraction of neutrons, namely the almost complete lack of absorption on one hand and the additional magnetic scattering effect with magnetic crystals, samples, or simply fields on the other.

A rather extended session of the workshop dealt with magnetic effects. The $4\pi$ periodicity of a spinor wavefunction shown in neutron interferometry and various spin superpositions enable a direct test of quantum-mechanical laws. Further experiments with spin rotations within the interferometer and three-dimensional polarization analysis were proposed. Neutron interferometry with polarized neutrons also has advantages for the separation of various weak neutron interaction effects from the larger nuclear and magnetic parts, e.g. electromagnetic interaction and neutron polarizibility. For most of these experiments large interferometers are needed to extend the neutron path length within the interferometer and therefore interferometers other than the LLL type will be used in the future.

The influence of gravity on the neutron phase is a further subject of research. The interferometer group of Columbia University is investigating the separation of the earth rotation effect. Various theoretical contributions deal with relativistic effects, view of symmetry, and coupling of the spin to the space—time curavture. All these effects are rather small and great progress in experimental techniques is needed to observe them.

The participants gained stimulation for new experiments in the field of neutron interferometry from the impressively high level of research in the more conventional interferometer techniques, like optical, electron, and X-ray interferometry, and also NMR interferometry. Many of the related experiments can be adapted to neutron interferometry where the special aspects of this radiation, like heavy particle mass, low velocity, and Fermi character, give additional effects.

The workshop closed with a panel discussion of future trends, where many suggestions for work were given to the audience. In addition to the perfect crystal interferometer, which was most discussed at the workshop, the biprism interferometer and interferometers with ultracold neutrons will increase the capability of neutron interferometry. Many investigations carried out to date may be classified as textbook experiments of quantum mechanics; future projects in the field of fundamental, nuclear, and solid state physics and crystallography will demonstrate the significance of this research technique.

The organizers would like to thank all participants for their very valuable contributions during the workshop and for taking the trouble of preparing excellent manuscripts. Special thanks are due to Dr. J.W. White and Prof. T. Springer, directors of the Institut Laue–Langevin (ILL), for their stimulating interest. The workshop profited very much from the scientific atmosphere of the ILL.

We experienced very essential and most welcomed help from Dr. B. Maier and his staff, for which we would like to thank them cordially. Financial support for the workshop by the ILL is also gratefully acknowledged.

*Dortmund*
*Vienna*

U. Bonse
H. Rauch

# PARTICIPANTS

D. Bader, Vienna
G. Badurek, Grenoble/Vienna
E. Balcar, Vienna
W. Bauspiess, Grenoble/Dortmund
P. Becker, Braunschweig
H. Bernstein, College Park, Md.
U. Bonse, Dortmund
J. Byrne, Grenoble/Brighton
R. Colella, Lafayette, Ind./Genoa
R.D. Deslattes, Washington, D.C.
K. Dorenwendt, Braunschweig
T. von Egidy, Grenoble
M. Françon, Paris
H. Friedrich, Braunschweig
R. Gähler, Garching
R. Golub, Grenoble/Brighton
J.P. Guigay, Grenoble
W. Graeff, Grenoble/Dortmund
D. Greenberger, New York
E. Granzer, Vienna
J. Hammer, Vienna
M. Hart, London
J. Hayter, Grenoble
M.A. Horne, Cambridge, Mass.
J. Joffrin, Grenoble
T.J.L. Jones, Harwell
H. Kaiser, Vienna
J. Kalus, Bayreuth
S. Kikuta, Tokyo
A.G. Klein, Melbourne
L. Koester, Garching
K. König, Garching
E. Krüger, Braunschweig
R. Lauer, Braunschweig
M.C. Li, Blacksburg, Va.
S.S. Malik, Kingston, R.I.
C. Malgrange, Paris
J.C. Marmeggi, Grenoble
G. Materlik, Hamburg
M. Mehring, Dortmund
F. Mezei, Grenoble/Budapest
G. Möllenstedt, Tübingen
G. Neilson, Bristol
L.A. Page, Pittsburgh, Pa.
T.J. Parker, London
J. Penfold, Grenoble/Harwell
D. Petraschek, Linz
H. Rauch, Vienna
M. Sauvage, Paris
O. Schärpf, Braunschweig
M. Schlenker, Grenoble
H.H. Schmidt, Garching
C.G. Shull, Cambridge, Mass.
I. Sosnowska, Warsaw
T. Springer, Grenoble
G.L. Squires, Cambridge, Mass./Cambridge, U.K.
J.L. Staudenmann, Columbia, Mo./Zurich
L. Stodolsky, Munich
W. Treimer, Berlin
S.A. Werner, Columbia, Mo.
J.W. White, Grenoble
W.G. Williams, Harwell
W. Yelon, Columbia, Mo./Grenoble
A. Zeilinger, Cambridge, Mass./Vienna

Scientific Secretary: B. Maier, Grenoble

## CONTENTS

# PART I METHODS AND INSTRUMENTATION IN NEUTRON INTERFEROMETRY

# 1. PRINCIPLES AND METHODS OF NEUTRON INTERFEROMETRY

U. BONSE

*Institut für Physik, University of Dortmund, Federal Republic of Germany*

## 1. General aspects of ångström range interferometry

Suppose in a scattering experiment the state $\psi$ of the probing (incident) beam is altered by some interaction $\hat{W}$ to $\hat{W}\psi$. A typical task then would be to determine $\hat{W}$ which in principle contains all the relevant properties of the scatterer. Since neither $\psi$ nor $\hat{W}\psi$ are physical observables themselves, the determination of $\hat{W}$ is not trivial.

Of special interest is the case of zero absorption within the sample because then the intensity of the incident beam is not altered in the scattering process and nothing about $W$ can be learned from simple intensity measurements of the forward-scattered beam with and without interaction $\hat{W}$. In more specific language, zero absorption corresponds to $\hat{W}$ being a unitary operator which can be written as $\hat{W} = \exp(i\hat{\phi})$ with $\hat{\phi}$ hermitian. Hence

$$I \equiv \int (\hat{W}\psi) * (\hat{W}\psi)d\tau = \int \psi * \psi d\tau = 1. \tag{1}$$

The essential idea in interferometry is then to circumvent the omission of $\hat{W}$ from the measurement by a coherent superposition of $\hat{W}\psi$ and $\psi$ (or some coherent wave derivative of $\psi$) and then to measure the intensity of the composed wave(s):

$$I_{\pm} \equiv \tfrac{1}{2} \int (\hat{W}\psi \pm \psi) * (\hat{W}\psi \pm \psi)d\tau = 1 \pm \langle \cos \hat{\phi} \rangle \tag{2}$$

The expectation values of $\hat{\phi}$ contain the properties of $\hat{W}$.

The following are examples of unitary interactions $\hat{W}$:

(a) the interaction of a plane parallel plate of thickness $D$ of homogenous material of refractive index $n$ with a neutron wave of wavelength $\lambda$

$$\left.\begin{aligned} W &= \exp(\mathrm{i}\varphi) \\ \varphi &= 2\pi D(n-1)/\lambda; \end{aligned}\right\} \qquad (3)$$

(b) the spinor rotation of a neutron when it passes through a homogenous magnetic field of strength $\mathbf{B}$ over a flight time $\tau$

$$\begin{aligned} \hat{W} &= \exp(-\mathrm{i}\boldsymbol{\sigma}\cdot\boldsymbol{\alpha}/2) \\ \boldsymbol{\alpha} &= \gamma\tau\mathbf{B} \end{aligned} \qquad (4)$$

where $\boldsymbol{\alpha}$ is the Pauli spin vector and $\gamma$ the gyromagnetic ratio.

Interferometry with light waves is an old and very well established art. With the advent of optical lasers (Maiman 1960) light interferometry expanded to holography (Gabor 1948, Leith and Upatnieks 1962), which is now a well-known technique, in which *deviated* scattered waves are superimposed with coherent reference waves. The essential feature in including deviated scattered waves is that in the hologram the total information about the *spatial structure* of the scatterer is recorded. Thus not only are bulk properties of homogenous samples measured, but images are formed of locally varying properties of structured objects. It is clearly very desirable to extend these techniques down to the ångström range of wavelengths, i.e. to electrons, X-rays, and thermal neutrons. Firstly, the study of the various typical interactions of ångström waves with solids, fluids, molecules, atoms, and possibly subatomic particles is augmented by the well-known accuracy and extreme sensitivity of interferometric methods. Secondly, if holography were to become feasible at ångström wavelengths, the problem of phase determination in structure investigations would in general be solved.

With neutrons (as with X-rays) the difficulties of beam handling, i.e. coherent splitting and recombining, arise from the well-known fact that for these radiations the refractive indexes of all materials are very close to unity. Nevertheless the first neutron interferometer (Maier-Leibnitz and Springer 1962) successfully employed a biprism for beam recombination. However, the biprism neutron interferometer suffered from too small a separation of the interfering beams (of the order of 60 μm) so that when samples were placed in the beams phase measurements were difficult, if not impossible (Landkammer 1966).

Beam handling is far less difficult if Bragg diffraction by perfect crystals is used, a technique first employed with X-rays (Bonse and Hart 1965a, b). With the simplest design, the so-called LLL type, Laue case diffraction by three crystal wafers behind each other (Fig. 1) provides beam splitting (S), beam bending (M), and beam superposition (A). However, Bragg type (BBB) and mixed Bragg–Laue type (LBBL) interferometers have also been operated with X-rays (Bonse and Hart 1966, 1968).

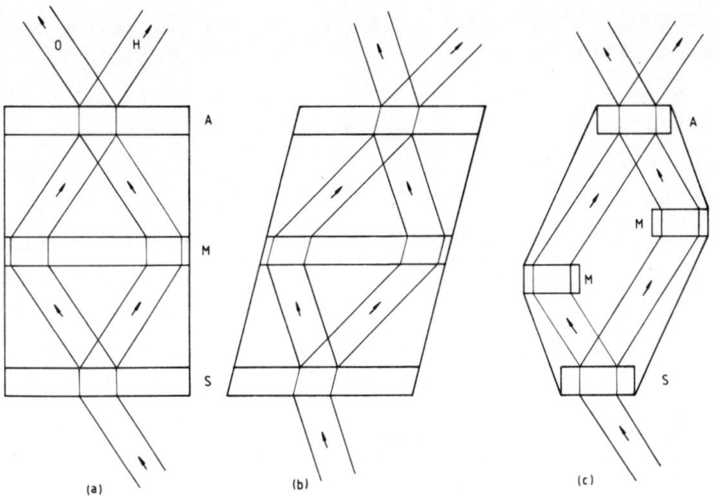


FIG. 1. Beam paths in different LLL neutron interferometers (beam smearing by Borrmann fans is neglected): (a) symmetric, (b) asymmetric, (c) skew symmetric. S, splitter; M, mirror; A, analyser crystal. S, M, and A stand on a common base and are integral parts of an extremely perfect single crystal. Stability is easier to maintain with such monolithic designs. Polylithic interferometers have been operated with X-rays.

In the first perfect-crystal neutron interferometer (Rauch, Treimer, and Bonse 1974, Bauspiess *et al.* 1974) the experience obtained with LLL X-ray interferometers was transferred to the neutron case. Here we restrict ourselves to LLL neutron interferometers. Aspects of other crystal neutron interferometers are discussed in paper I.2.

Further details on ångström interferometry in general can be found in the five reviews which have been written on the subject to date (Bonse 1969, 1975, Hart 1971, 1975, Bonse and Graeff 1977).

There are at least three variations of the LLL interferometer (Fig. 1): (a) the symmetric, (b) the asymmetric, and (c) the skew symmetric versions. Depending on the particular task to be solved, each of the three types may have advantages over the other two.

## 2. Wavelength range of crystal interferometers

### 2.1. *Geometrical restrictions*

Bragg diffraction by perfect crystals is welcomed in that it facilitates the coherent splitting of ångström waves and their deviation about large angles. At the same time, however, it introduces a mostly unwanted but principally unavoidable correlation $\theta(\lambda)$ between the angle of deviation $2\theta$ and the wavelength $\lambda$ which is expressed by Bragg's law $m\lambda = 2d \sin \theta$, where $d$ is the Bragg plane spacing of the particular reflection employed.

Realization of the ångström wave analog of an optical (semitransparent) mirror which is capable of reflecting *any* wavelength at a given angle $2\theta$ would be extremely valuable for interferometry and holography. However, as is well known, at present this is possible only near grazing incidence, implying values of $2\theta$ angles of a few degrees at the most. Ångström interferometers based on the use of grazing incidence mirrors are thus bound to have limited separation of interfering beams, similar to the biprism version mentioned above.

The restriction imposed by Bragg's law has (at least) the following consequences:

(a) intensity loss with ordinary $\theta(\lambda)$ uncorrelated (reactor beam hole) or unsuitably $\theta(\lambda)$ correlated (neutron guide tube mosaic fore-crystal) incident beams;

(b) $\lambda$ dispersion of beam geometries which with certain interferometer types results in 'defocusing' and loss of interference contrast, at least as long as sources with extreme time and space coherence (lasers) do not exist for ångström wavelengths;

(c) limitation of the accessible $\lambda$ range.

The last point is closely linked with the lattice parameter of available raw crystals of sufficient perfection. To date all neutron interferometers have been cut from silicon.

The perfection of crystals ordinarily supplied by crystal growers can be judged with high-order double-crystal topography or, more directly, by making a very short LLL sample interferometer of the crystal rod under consideration (Bonse, Graeff, and Materlik 1976). Sufficiently perfect crystals of up to 8 cm diameter which usually have a $\langle 100 \rangle$ growth direction can be purchased from Wacker-Chemitronic, Munich.

With silicon crystals of this size and orientation the following limits on the usuable $\lambda$ range can be estimated.

*Maximum* $\lambda$: Using the largest possible Bragg spacing of the silicon lattice (111 with $d$ = 3.135 Å) and assuming a maximum Bragg angle $\theta$ of 73° we obtain $\lambda \leqslant 6$ Å or a neutron energy $E \geqslant 2.3$ meV. The actual dimensions of such an interferometer with $\theta = 73°$ are given in Fig. 2(a). The assumed symmetry facilitates the measurement of a pair of samples of reasonable size, as indicated in Fig. 2(a). A *skew* symmetric version at $\theta = 73°$ is shown in Fig. 2(b). *Asymmetry* of ray paths generally appears to be disadvantageous with large $\theta$.

*Minimum* $\lambda$: The angular acceptance of the crystal

$$\Delta\theta = \frac{2}{\pi}\frac{b_c N\lambda^2}{\sin 2\theta}\, e^{-M} \tag{5}$$

decreases with $\lambda^2$ and hence results in too low an intensity below a particular value of $\lambda$. $b_c$ is the coherent scattering length, $e^{-M}$ the Debye–Waller factor, and $N$ the number of atoms per unit volume. To some extent the decrease of $\Delta\theta$ with $\lambda$ can be compensated by making $\theta$ small, i.e. by choosing reflection orders which are not too high. Similarly, as can be seen from Table 1 where $e^{-M}$ is listed for some reflections $hkl$, $e^{-M}$ also becomes small at high orders. On the other hand, in order to have good separation of interfering beams, $\theta$ should not be less than, say, about 6°. As a compromise for the limit we choose $\theta = 6°$ and the 880 reflection. This gives $\lambda \gtrsim 0.1$ Å or neutron energy $E \lesssim 8.18$ eV.

The actual dimensions of an interferometer with $\theta = 6°$ are

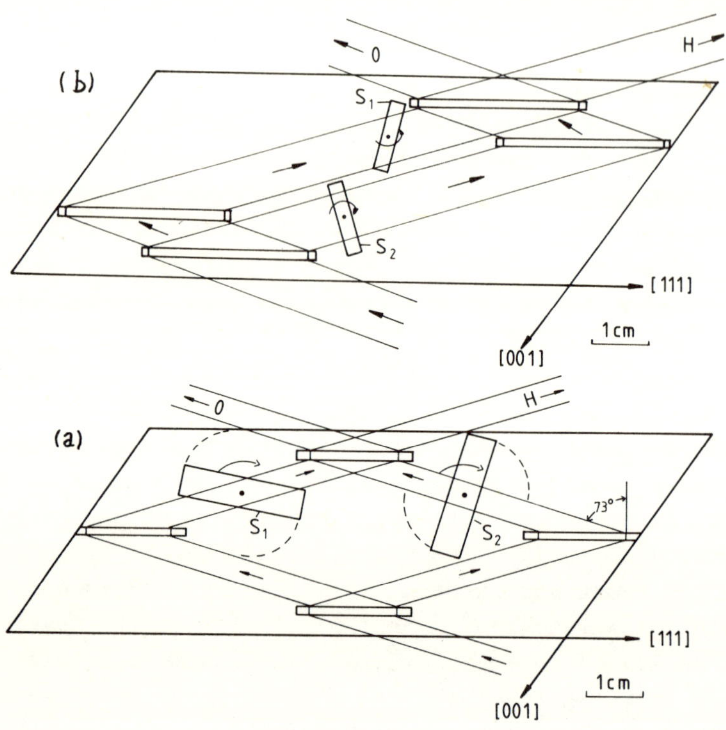


FIG. 2. Possible neutron interferometers for $E$ = 2. 3 meV (θ = 6 Å), θ = 73° and 111 reflection cut from a silicon crystal 8 cm in diameter with ⟨001⟩ growth direction. The wafers are 0.5–1.5 mm thick and up to 40 mm long. (a) Symmetric version with a pair of samples $S_1$ and $S_2$ for for $b_c$ measurement. $S_1$ and $S_2$ are synchronously rotated. With the geometry shown, about ± 0.5 cm change of effective pathlength in the sample material can be accomplished between extreme sample orientations, which corresponds to 100 interference orders for a typical $t$ of 0.1 mm. Instead of rotating a pair of samples, one could use a single sample which is transferred from one beam to the other. With a sample 2 cm thick 400 orders could be covered. (b) Skew symmetric version; a wider neutron beam is realized at the expense of thinner samples. The sense of rotation has to be opposite for the samples. A sample about 3.5 cm thick could be transferred from one beam into the other resulting in a typical phase shift of about 700 interference orders.

given in Fig. 3. As samples a pair of equal wedges could be employed which would be scanned in the manner indicated in Fig. 3, or a pair of parallel-sided samples could be shifted between beams.

From the foregoing it follows that a wavelength range between 0.1 Å and 6 Å or an energy range between 8.2 eV and 2.3

TABLE 1

*Value of the Debye–Waller factor at T = 293 K for some reflections in silicon (Debye temperature 538 K)*

| *hkl* | $e^{-M}$ | *hkl* | $e^{-M}$ |
|---|---|---|---|
| 111 | 0.9886 | 12 0 0 | 0.5761 |
| 220 | 0.9698 | 8 8 8 | 0.4794 |
| 400 | 0.9406 | 10 10 0 | 0.4649 |
| 422 | 0.9122 | 16 0 0 | 0.3752 |
| 440 | 0.8847 | 12 12 0 | 0.3319 |
| 444 | 0.8321 | 14 14 0 | 0.2229 |
| 800 | 0.7826 | 20 0 0 | 0.2161 |
| 822/660 | 0.7590 | 12 12 12 | 0.1912 |
| 844 | 0.6924 | 16 16 0 | 0.1407 |
| 880 | 0.6125 | 18 18 0 | 0.0836 |

meV is in principle accessible with a silicon LLL interferometer. Approximately the same ranges could be covered with germanium. At present it seems unlikely that suitable crystals of other materials will become available.

## 2.2. *Intensity considerations*

Apart from the geometrical restrictions implied by Bragg's law there is also the previously mentioned limitation that the angular acceptance $\Delta\theta$ in the horizontal plane of the interferometer crystal becomes too small, which combined with the falling specific flux $\delta^2\phi/(\delta\lambda/\lambda)\delta\Omega$ (neutrons $cm^{-2}$ $s^{-1}$ $st^{-1}$) (i.e. normalized for $\lambda$ resolution $\Delta\lambda/\lambda$) of reactor sources, may result in intensities at short wavelengths which are too low.

In the following we give a rough estimate of the maximum neutron flux $\phi_{ac}$ (neutrons $cm^{-2}$ $s^{-1}$) which an actual interferometer would accept from a neutron beam from the Grenoble high-flux reactor (HFR) as a function of $\lambda$ and for two given $\lambda$ resolutions, namely $5 \times 10^{-4}$ ($\equiv \phi_{ac1}$) and $10^{-2}$ ($\equiv \phi_{ac2}$) respectively. The results are presented in Table 2.

For the estimate we assume that for all $\lambda$ the acceptance $\Delta\theta$ is essentially determined by the crystal and does not depend on whether or not a neutron guide is interposed between the source and the interferometer. This implies that the guide's reflection range is generally equal to or larger than $\Delta\theta$ which

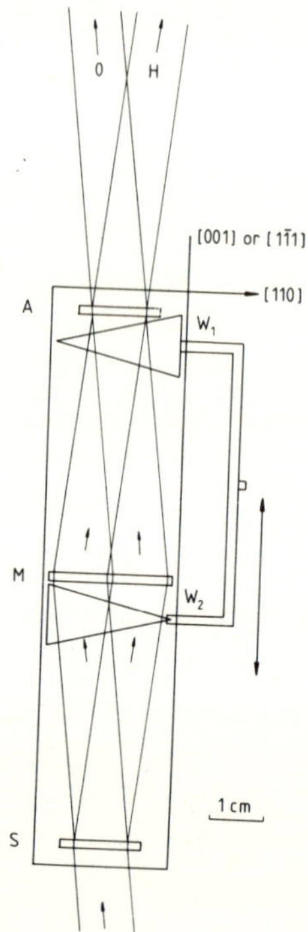


FIG. 3. Possible neutron interferometer for $E$ = 8.2 eV ($\lambda \approx 0.1$ Å), $\theta = 6°$ and 880 reflection cut from a silicon crystal 5 cm in diameter, 11 cm long and with growth direction ⟨001⟩ or ⟨111⟩. The wafers are 1.5–2 mm thick and up to 20 mm high. They should not be thinner than 1.5 mm because at 0.1 Å $\Delta_e \approx 1.3$ mm for silicon. (See §3.3 for optimized values of $t_S$, $t_M$, and $t_A$ at short wavelengths.) The sample consists of a pair of wedges $W_1$ and $W_2$ which are moved in the manner indicated. The maximum effective change of pathlength in the material is about ± 0.5 cm. Note that with wedges complete separation of the interfering beams is not necessary. At the expense of 10% beam width near crystal M two samples of about 1 cm thickness each could be transferred from one beam to the other resulting into an effective change of pathlength of ± 2 cm. 400 reflection orders could then be measured for a typical $t_\lambda$ of 0.1 mm.

is true in most cases. Furthermore, if $\Delta\Omega = \Delta\xi\Delta\theta$, where $\Delta\xi$ is the vertical divergence, we approximate $\Delta\xi \approx 10^{-2}$ independently of $\lambda$. Under these assumptions we use the $\delta^2\phi/\delta\lambda\delta\Omega$ (neutrons $cm^{-2}\ s^{-1}$ Å$^{-1}$ $sr^{-1}$) flux data calculated by Ageron (1972) for the hot, thermal, and cold source beam holes of the Grenoble reactor (bottom row of Table 2). For a particular $\lambda$ the estimate is made with the source giving the highest flux and only favourable combinations of reflection orders $hkl$ and Bragg angles $\theta$

are considered (third and fourth row respectively). The horizontal divergence $\Delta\theta$ (fifth row) was calculated for silicon with $b_c = 4.15 \times 10^{-13}$ cm, the lattice parameter $a_o = 5.4305$ Å, and the Debye–Waller factors given in Table 1.

Table 2 shows that over the $\lambda$ range 6–0.1 Å, $\Delta\theta$ decreases by more than a factor of 1500. Nevertheless from the magnitude of the estimated accepted fluxes $\phi_{ac1}$ and $\phi_{ac2}$ (rows 6 and 7 respectively), we conclude that down to 0.3 Å (0.91 eV) interferometry appears to be feasible even at a resolution as high as $\Delta\lambda/\lambda = 5 \times 10^{-4}$. Below 0.3 Å $\Delta\lambda/\lambda \approx 10^{-2}$ appears to be attainable.

The present measurements at Grenoble (Bauspiess, Bonse, and Rauch, 1978) have been performed at a thermal neutron guide with radius of curvature 19 km and $\lambda \approx 1.8$ Å. At $\Delta\lambda/\lambda \approx 10^{-2}$ the flux accepted by a 220 silicon fore crystal was measured to be about $5.6 \times 10^{3}$ neutrons cm$^{-2}$ s$^{-1}$ which is roughly one-quarter of the calculated value given in Table 2 for $\lambda = 2$ Å. A reduction compared with the expected flux values of this order of magnitude seems to be quite typical for neutron guide tubes. However, even with a reduction of a factor of 4 the main arguments used to obtain the estimate presented here remain valid.

In the epithermal region spallation sources are expected to give a higher flux than even a high-flux reactor. Hence, below 0.3 Å a spallation source is probably better suited for interferometry than the HFR.

### 2.3. *Importance of a wide wavelength range for neutron interferometry*

The energy range covered by the crystal neutron interferometer contains many scattering and absorption resonances. Therefore a main application for energy scanned interferometry appears at present to be to measure the drastic change in the scattering occurring at resonances within a small energy band. The scattering is determined by a potential term and a resonance term, and the relative phase of these terms can be measured. From the measured phase shift the resonance parameters and the superposition of various resonances can be determined. Although in the thermal region the ratio of coherent to

TABLE 2

*Neutron fluxes after perfect crystal reflection with the Grenoble HFR for various wavelengths λ and for resolutions $5\times10^{-4}$ and $10^{-2}$*

| $\lambda$ (Å) | 6 | 5 | 4 | 3 | 2 | 1 | 0.5 | 0.3 | 0.2 | 0.1 |
|---|---|---|---|---|---|---|---|---|---|---|
| $E$ (meV) | 2.3 | 3.3 | 5.1 | 9.1 | 20 | 82 | 327 | 909 | 2045 | 8180 |
| $hkl$ | 111 | 111 | 111 | 111 | 220 | 220 | 440 | 440 | 440 | 880 |
| $\theta$ (deg) | 73.1 | 52.9 | 39.6 | 28.6 | 31.4 | 15.1 | 15.1 | 9.0 | 6.0 | 6.0 |
| $\Delta\theta \times 10^{6}$ (rad) | 60 | 24 | 15 | 9.9 | 5.9 | 2.55 | 0.58 | 0.34 | 0.23 | 0.039 |
| $\phi_{ac1}$ (neutrons $cm^{-2}$ $s^{-1}$) $\Delta\lambda/\lambda = 5\times10^{-4}$ | 7500 | 3900 | 2760 | 1490 | 1300 | 900 | 90 | 5 | 1.5 | 0.25 |
| $\phi_{ac2}$ (neutrons $cm^{-2}$ $s^{-1}$) $\Delta\lambda/\lambda = 10^{-2}$ | $1.5\ 10^{5}$ | $7.8\ 10^{4}$ | $5.5\ 10^{4}$ | $3.0\ 10^{4}$ | $2.6\ 10^{4}$ | $1.8\ 10^{4}$ | 1800 | 100 | 30 | 5 |
| $\delta^{2}\phi\ \delta\lambda\delta\Omega \times 10^{-12}$ (neutrons $cm^{-2}$ $s^{-1}$ $Å^{-1}$ $sr^{-1}$) | 4.3 | 6.5 | 9.2 | 10 | 22 | 70 | 60 | 10 | 6.5 | 13 |
| | Cold source | | | | Thermal source | | Hot source | | Thermal source | |

incoherent scattering is almost constant, in the resonance region a marked energy dependence is expected. Suitable scattering resonances in bismuth, for example, are at 0.8 and 2.31 eV. Absorption resonances with a significant scattering contribution occur at 1.457 eV in indium and at 1.058 eV in hafnium.

## 3. Bragg diffraction optics of LLL neutron interferometers

For the coherent splitting of beams and their subsequent superposition in crystal interferometers it is essential to use combined dynamical diffraction by several perfect crystals (beam splitter S, mirror M, and analyzer A). The dynamical theory of diffraction of X-rays by a *single* perfect crystal was worked out thoroughly many years ago (Ewald 1916, 1917, von Laue 1931). A recent presentation of the theory which also includes some experimental results is given by Pinsker (1978).

The combined diffraction by several perfect crystals was treated much later, mostly for the direct interpretation of specific experimental results already at hand (Bonse and Hart 1965b, 1969, Authier, Milne, and Sauvage 1968, Bonse and te Kaat 1971, Bauspiess, Bonse, and Graeff 1976, Petrascheck 1976, Petrascheck and Folk 1976, Bonse and Graeff 1977. With diffraction by just one crystal plate the only essential geometric parameter entering the results is the plate thickness $t$. With more than one crystal not only more than one thickness $t$ but also relative positions, relative orientations, and possibly differences in the lattice parameters of the various crystals can influence the result in a more or less complicated manner.

For simplicity we shall assume that all diffracting crystals have exactly the same lattice parameter and, for the LLL interferometer to be discussed here, also precisely the same orientation, although in practice these assumptions are seldom strictly justified, mainly because of the presence of strains due to residual lattice defects which frequently cause a 'built-in' phase pattern to be observed in the empty interferometer.

A special feature observed with neutrons is zero absorption

in silicon (as in most other materials), which implies that both dynamical wavefields, i.e. that with antinodes (type 1) and that with nodes (type 2) on the atomic planes, have to be taken into account. As a result, with zero absorption, conditions for good fringe contrast are usually more difficult to maintain.

### 3.1. *Plane-wave treatment of diffraction in the LLL interferometer*

Since laser neutron sources do not exist, in most experiments we are not dealing with strictly plane incident waves. Nevertheless it is worth while discussing the results of a plane-wave theory for the following reasons:

(a) the plane-wave treatment is not really complicated and yet helps very much in understanding the rather involved case of arbitrary incident waveforms;

(b) the solution for the spherical incident wave can be obtained by Fourier composition of plane-wave solutions.

We introduce the transmission coefficients (Bauspiess, Bonse, and Graeff 1976)

$$\langle m|j|n\rangle_P \equiv (D_n^{\mathrm{e}}/D_m^{\mathrm{i}})_P \tag{6}$$

for the amplitude ratio of the incoming wave $D_m^{\mathrm{i}}$ and of the outgoing wave $D_n^{\mathrm{e}}$ via wavefield $j$ ($j$ = 1,2) in Laue diffraction by a plate P (Fig. 4). $m$ and $n$ are O or H depending on whether the wave is directed in the forward or in the diffracted direction respectively. The amplitude $D_n^{\mathrm{e}}$ of the whole interferometer resulting from an incident wave $D_m^{\mathrm{i}}$ can then easily be calculated by applying equation (6) consecutively at S, M, and A and over both interfering beam paths I and II. Defining the transition factor for the complete LLL interferometer as

$$(D_n^{\mathrm{e}}/D_m^{\mathrm{i}})_{\mathrm{LLL}} \equiv \langle m\ \mathrm{LLL}\ n\rangle \tag{7}$$

we obtain ($m$, $n$ = O, H)

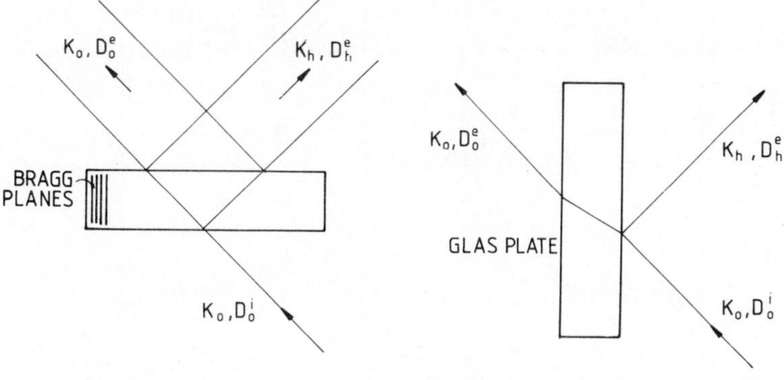


FIG. 4. Comparison of optical reflection and beam splitting (right) with the same processes for neutrons by perfect crystal diffraction (left).

$$\langle m|\mathrm{LLL}|n\rangle = \sum_{i,j,k=1,2} [\langle m|i|\mathrm{O}\rangle_{\mathrm{S}}\langle \mathrm{O}|j|\mathrm{H}\rangle_{M_{\mathrm{I}}}\langle \mathrm{H}|k|n\rangle_{\mathrm{A}} + \langle m|i|\mathrm{H}\rangle_{\mathrm{S}}\langle \mathrm{H}|j|\mathrm{O}\rangle_{M_{\mathrm{II}}}\langle \mathrm{O}|k|n\rangle_{\mathrm{A}}]. \quad (8)$$

In equation (8) the first part is the contribution via path I and the second part that via path II. We see that a total of 16 plane waves is contributing to either the outgoing beam O or H.

We shall not discuss the analytical expressions for $\langle m|j|n\rangle_{\mathrm{P}}$ or for $\langle m|\mathrm{LLL}|n\rangle$ but shall rather illustrate their complexity by comparing them with the corresponding factors occurring at the semi reflecting glass plate when it is used for amplitude splitting of light beams (Fig. 4).

In the neutron case, transmitted and reflected intensities oscillate rapidly with incident angle variations of only fractions of arcseconds (Fig. 5). Furthermore, with the incident angle fixed to the centre of the diffraction range, similar oscillations occur with a variation of the plate thickness $t$ where the extinction length $\Delta_e$ is the oscillation period (about 64 μm for $\lambda$ = 2 Å, $y$ = 0, and the silicon 220 reflection). For the symmetric Laue case

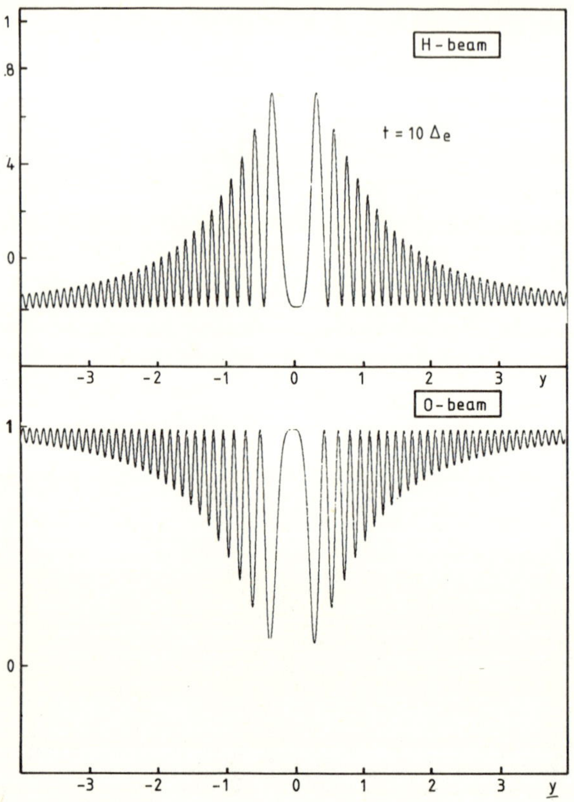


FIG. 5. Plane-wave transmission $|\langle O|1|O\rangle + \langle O|2|O\rangle|^2$ and reflection $|\langle O|1|H\rangle + \langle O|2|H\rangle|^2$ curves for a single-crystal plate; zero absorption. $y = \pm 1$ corresponds to the diffraction range, e.g. about 1.2" with $\lambda = 2$ Å neutrons and the silicon 220 reflection. Note the Pendellösung oscillations. The O and H beams are antiphase.

$$\Delta_e = \frac{\pi \cos\theta\, V_c}{D_h F_h \lambda\, b_c} (y^2+1)^{-\frac{1}{2}}$$

where $V_c$ is the volume of the unit cell. $D_h$ is the Debye–Waller factor, $F_h$ is the neutron structure factor, and $y$ is the usual parameter describing the point within the reflection range which extends from $y = -1$ to $y = +1$; in the centre $y = 0$. The oscillation is known as Pendellösung. It is antiphase in the two outgoing beams. In the case of light the outgoing intensities vary only very slowly with incident angle and do not vary at all with plate thickness.

As is expected from plane-wave single-plate diffraction, the corresponding combined diffraction by S, M, and A displays a complicated oscillatory structure. The situation for a special case, namely for the outgoing O beam and when $t_A = t_M = t_S = 10\Delta_0$ ($t_S$, $t_M$, and $t_A$ are the thicknesses of S, M, and A respectively), is shown in Fig. 6. Two other examples are shown in Fig. 7 ($t_S = t_A = 10.25\Delta_0$ and $t_M = 10.5\Delta_0$) and Fig. 8 ($t_S = t_A = 7.75\Delta_0$, $t_M = 15.5\Delta_0$). In Fig. 7 the strong peak in the centre and the fast decrease of the wings should be noted. A strong centre peak is also obtained in Fig. 8 where the special geometry $\frac{1}{2}t_M = t_S = t_A$ is realized. From the spherical wave or ray discussion given below it is found that with this geometry special beam focusing occurs.

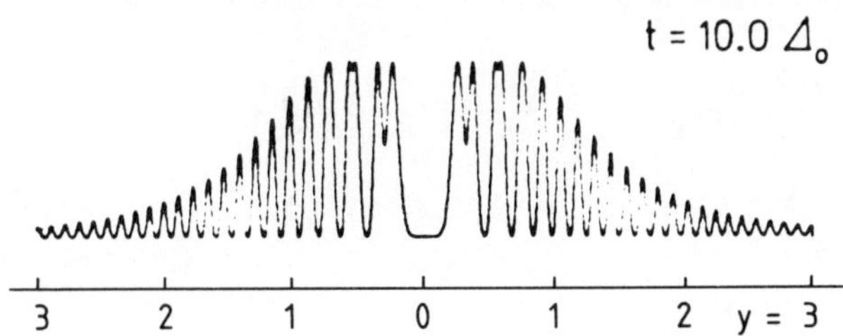


FIG. 6. Plane-wave transmission curve for an LLL combination with $t_S = t_M = t_A = 10\Delta_e$. Note the plateau of the envelope when compared with the upper curve of Fig. 5.

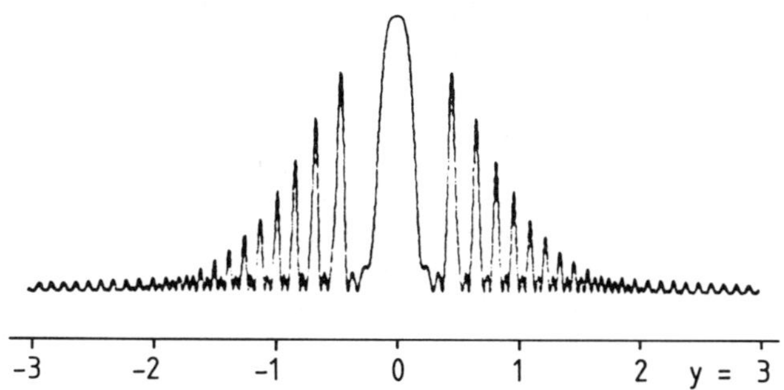


FIG. 7. As Fig. 6 but with $t_S = t_A = 10.25\Delta_0$ and $t_M = 10.5\Delta_0$. The optimized values are as shown in §3.3. Most of the intensity is in the center peak. Note the faster fall-off of the wings and also that the envelope no longer has a plateau.

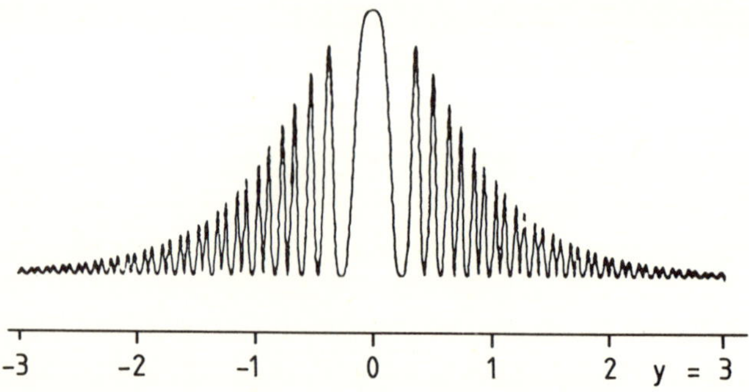


FIG. 8. As Fig. 7 but with $2t_S = 2t_A = t_M = 15.5\Delta_0$, i.e. the focusing geometry corresponding to the spherical wave curve of the inset of Fig. 12.

### 3.2. *Spherical wave treatment of diffraction in the LLL interferometer*

Contrary to the case of an incident plane wave where a single point or a pair of single points on the dispersion surface is excited, an incident wave packet generally causes the coherent excitation of a certain range of the dispersion surface. We assume incident spherical waves, i.e. completely uniform excitation of the whole dispersion surface. Following the treatment of the spherical wave diffraction by a single-crystal plate as developed by Kato (1961, 1968) we use the Fourier expansion of the incident spherical wave packet

$$\mathrm{D}_m{}^{\mathrm{i}}(\mathbf{r},t) = \pi^{-1}\int \mathrm{d}^3K\int \mathrm{d}\omega(K^2-k^2)^{-1}\delta(\omega-\omega_0)D_m{}^{\mathrm{i}}\exp(2\pi\mathrm{i}\mathbf{K}\cdot\mathbf{r}-\mathrm{i}\omega t). \quad (9)$$

From equation (9) we obtain the diffracted wave packet $\mathrm{D}_n{}^{\mathrm{e}}(\mathbf{r},t)$ behind the interferometer by multiplying the plane-wave components $D_m{}^{\mathrm{i}}$ in the integral with the corresponding interferometer transition factors $\langle m|\mathrm{LLL}|n\rangle$:

$$\mathrm{D}_n{}^{\mathrm{e}}(\mathbf{r},t) = \pi^{-1}\int \mathrm{d}^3K\int \mathrm{d}\omega(K^2-k^2)^{-1}\delta(\omega-\omega_0)\langle m|\mathrm{LLL}|n\rangle \times D_m{}^{\mathrm{i}}\exp(2\pi\mathrm{i}\mathbf{K}\cdot\mathbf{r}-\mathrm{i}\omega t). \quad (10)$$

For quantitative expressions we again refer to the literature

(Bauspiess, Bonse, and Graeff 1976, Petrascheck and Folk 1976, Bonse and Graeff 1977), and give here a qualitative discussion of the interferometer diffraction with spherical waves.

The discussion is very much simplified if we assume a point source and consider only the plane which is defined by $\mathbf{K}_0$ and $\mathbf{K}_n$. We call this plane the **v**,**w** plane and let the source also lie in that plane.

For fixed values of $\lambda$, rays which are diffracted by the interferometer illuminate a narrow line of width $w = R\Delta\theta$ normal to the **v**,**w** plane on the entrance surface of the splitter crystal. With typical values $R = 10$ m and $\Delta\theta = 10^{-6}$ (Table 2), we have $w = 10$ μm, i.e. practically zero when compared with a typical lamella thickness $t \approx 0.5$–4 mm. Hence, in comparison with its geometrical width, the incident beam may effectively be considered to be a mathematical line. However, with regard to its divergence the beam must still be thought of as being divergent enough to excite the complete dispersion surface coherently.

In the splitter S the beam is spread out over the full range $2\theta$ of the Borrmann fan, so that the width of beams 0 and

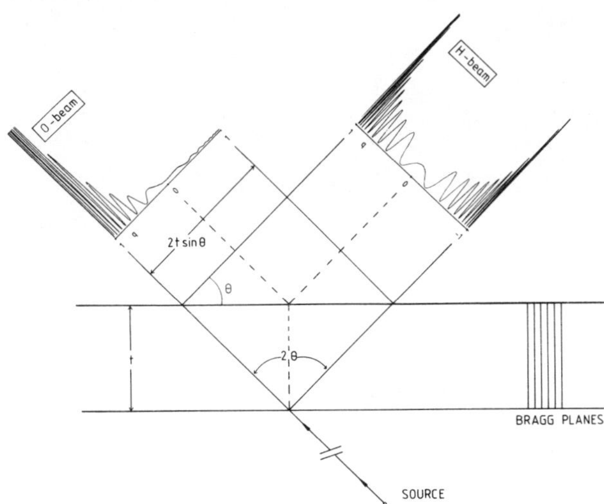


FIG. 9. Spatial intensity distributions behind a single diffracting plate with spherical waves. With a narrow incident beam the intensity is smeared over a width of $2t \sin\theta$.

H behind S becomes $2t \sin\theta$ (Fig. 9). The amplitude distributions in beams O and H are approximately proportional (Bonse and Graeff 1977) to the Fourier transforms of $\sum_{j=1,2}\langle 0|j|0\rangle$ and of $\sum_{j=1,2}\langle 0|j|H\rangle$ respectively. Consequently, as is indicated in Fig. 9, the intensity distributions in beams O and H oscillate rapidly across the beams. Note that the H-beam distribution is symmetric about the centre of the fan whereas the O-beam distribution is not. The beam spreading is continued in crystals M and A. Figure 10 illustrates the beam-widening effect through the complete LLL interferometer for the special 'ideal' geometry:

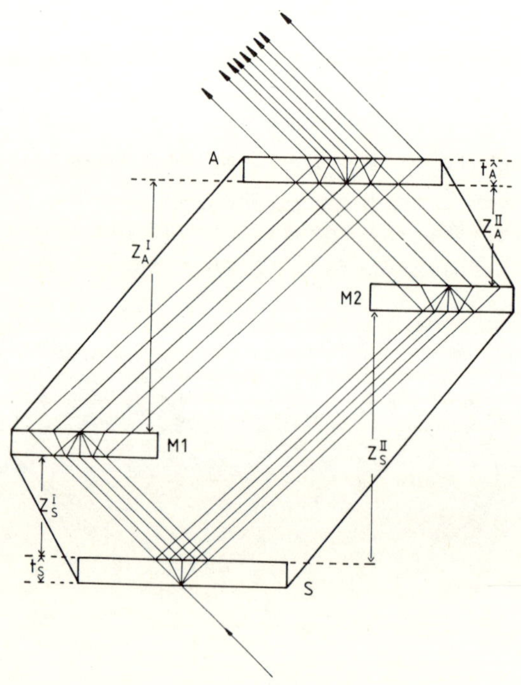


FIG. 10. Beam-widening effect in the interferometer with $t = t_S = t_{M1} = t_{M2} = t_A$.

$$t_S = t_A = t \tag{11}$$

$$t_{M1} = t_{M2} = at \tag{12}$$

$$a = 1 \tag{13}$$

$$z_A^{\,I} = z_S^{\,II}, \qquad z_A^{\,II} = z_S^{\,I} \tag{14}$$

Four additional energy flow directions have been drawn within the Borrmann fans in order to indicate very roughly the energy distributions resulting over the various beam cross-sections. However, when interpreting these distributions one should keep in mind that in principle oscillatory profiles of the kind shown in Fig. 9 are generated at every crystal. More rigorously the profiles in the outgoing beams are found as convolutions of single-plate Fourier transforms of combinations of transmission factors $\langle m|j|n\rangle$ obtained at crystals S, M, and A.

We see from Fig. 10 that the intensity originally concentrated within the narrow incident beam is finally spread out over a beam cross-section of width $6t \sin\theta$. However, most of the intensity is found in the centre third of each beam, as can also be concluded from the following simple argument.

Part of the beam incident on S is focused at the exit surface of M so that we can interpret the intensity profiles in beams O and H of Fig. 10 as the superposition of Pendellösung oscillations of crystal plates with thicknesses $t$ and $3t$. Since the number of rays contributing to the inner part $2t \sin\theta$ is three times larger than those spreading over the full width $6t \sin\theta$, the intensity ratio between the inner and outer parts is roughly 9:1.

Calculated and measured beam profiles (Bauspiess, Bonse, and Graeff 1976) confirm this conclusion, as shown in Fig. 11. The O profile calculated with $t = 10\Delta_0$ is shown at the top. Apart from the fine oscillations, the profile has two major peaks both occurring within the centre third of the cross-section. The measured profile below agrees very well with the calculations. In addition, beams I and II were measured separately, each with the other beam blocked off (bottom and top respectively). In the overlap region of I and II the intensity gain by constructive interference is clearly visible. On the other hand, the measured H beam displays very different beam profiles in I and II, thus making it less favourable for interferometry. The essential cause of this deficiency is the asymmetry of transition factor combinations between path I and path II occurring for the H beam which may be understood by setting

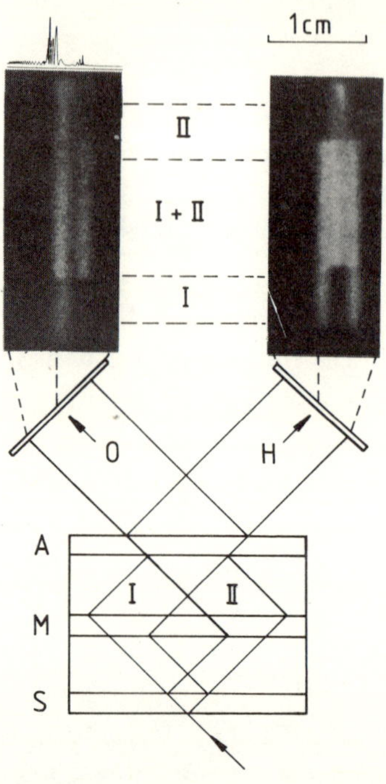


FIG. 11. Qualitative comparison of calculated ($t = 10\Delta_0$, O beam only) and measured $t = 62\Delta_0$ beam profiles; $t_S = t_M = t_A = 4$ mm, $\lambda \approx 2$ Å and silicon 220 reflection. The interferometer is not drawn to scale. Top, path II only; bottom, path I only; centre, overlap of I and II. (See text for further details.)

$m = 0$ and $n = H$ in equation (8). On the other hand, equation (8) is symmetric with respect to I and II for the outgoing 0 beam, i.e. $m = 0$ and $n = 0$, provided equations (11) and (12) hold (the validity of equation (13) is not necessary). As a consequence, for zero absorption, even with ideal geometry, the interference contrast $\gamma$ of the H beam defined as

$$\gamma = \frac{I_{max} - I_{min}}{I_{max} + I_{min}} \tag{15}$$

is never much higher than 0.5, whereas for the 0 beam it is unity (see below).

### 3.3. *Geometrical tolerances, influence of defocusing, and interference contrast*

We assume that all crystals are strictly perfect and completely parallel oriented. Also, for the sake of simplicity we assume that the net planes are normal to the plate surfaces. For optimum performance with an incident plane wave, geometrical constraints need only be applied to the thicknesses of S, M, and A: to distribute the incident wave evenly between the forward and diffracted beams, S and A should have a thickness of $(m\pm\frac{1}{4})\Delta_e$, where $\Delta_e$ is again the extinction length given above ($m$ may be different for S and A). On the other hand, M can be made totally reflecting by making $t_M = (m\pm\frac{1}{2})\Delta_e$. However, it should be remembered that $\Delta_e$ and hence the geometric conditions depend on $\lambda$, e.g. at $\lambda = 0.1$ Å $\Delta_e \approx 1.3$ mm. With $m = 1$ we obtain $t_S = t_A \approx 1.6$ mm and $t_M \approx 1.9$ mm.

No other important geometrical requirements exist in the plane-wave case. In particular the distances $Z_A^{\,I}$, $Z_A^{\,II}$, $Z_S^{\,I}$ and $Z_S^{\,II}$ (Fig. 10 and equation (14)) are not critical. In the empty interferometer there would always be a wide field of view with homogeneous phase and the fringe contrast $\gamma$ in the O beam would be unity.

With spherical waves 'focusing conditions' of the kind given by equations (11)–(14) have to be fulfilled in order to maintain maximum $\gamma$. Another ideal geometry is obtained if $a = 2$ in equation (12) (Fig. 12). In this case there are two focusing points in the centres of $M_I$ and $M_{II}$ and one in the middle of the outgoing beam.

The intensity profile can be considered as a superposition of Borrmann fans of crystal plates with thickness $4t$, $2t$, and zero.

The dependence of $\gamma$ for the O beam of an interferometer with $t = t_s = t_m = t_A = 0.6$ mm on deviations from ideal geometries is shown in Figs. 13–15. It is found that $\gamma$ depends on different kinds of geometrical aberrations from the ideal geometry in different ways. The obvious aberration is defocusing, which is illustrated in Fig. 13: rays originating from a common point on the exit surface of S no longer meet on the entrance surface of A but at a distance $\Delta z$ (defocus) in front of it. Other aberrations are $\Delta t = |t_A - t_s|$ and $\Delta t' = |t_{M1} - t_{M2}|$. Contrast

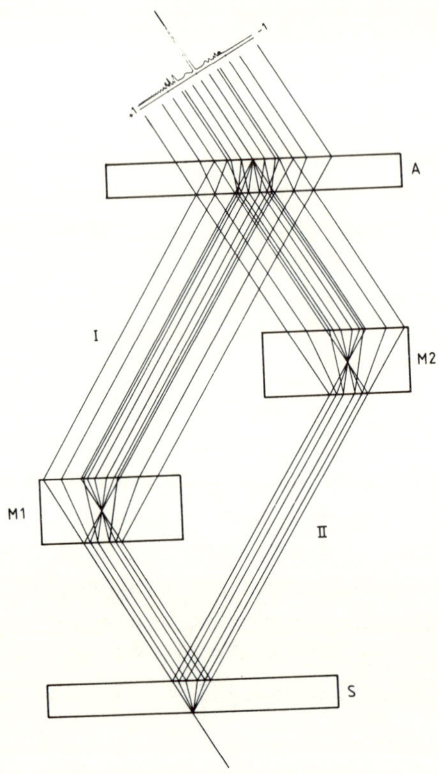


FIG. 12. *t–2t–t* interferometer. Note the focusing in the centre of M1 and M2, and in the exit beam calculated with $t = 15.1\Delta_0$.

calculations for the 0 beam assuming $t = t_S = t_M = t_A \approx 0.6$ mm are shown in Figs. 14 and 15.

Curve 1 in Fig. 14 depicts $\gamma(\Delta z)$ if $\Delta t = 0$ is maintained: $\gamma$ decreases monotonously and reaches $\gamma \approx 0.1$ at $\Delta z = 25$ μm. If, however, $\Delta z = |\Delta t|$ is maintained, e.g. if $t_A \neq t_S$ because of incorrect positioning of the *inner* surfaces of *S* and/or *A*, then $\gamma = 0$ at $\Delta z \approx 22$ μm but increases again for larger $\Delta z$ (curve 2 in Fig. 14). Obviously a partial compensation of pure defocusing type 1) occurs.

Curve 3 in Fig. 15 shows $\gamma(\Delta t)$ with $\Delta z$ kept zero. Type 3 shows the steepest decrease in $\gamma$ for low $\Delta t$ values but increases again for larger values. The last type of deviation, which is shown in curve 4 of Fig. 15, assumes that there is a differ-

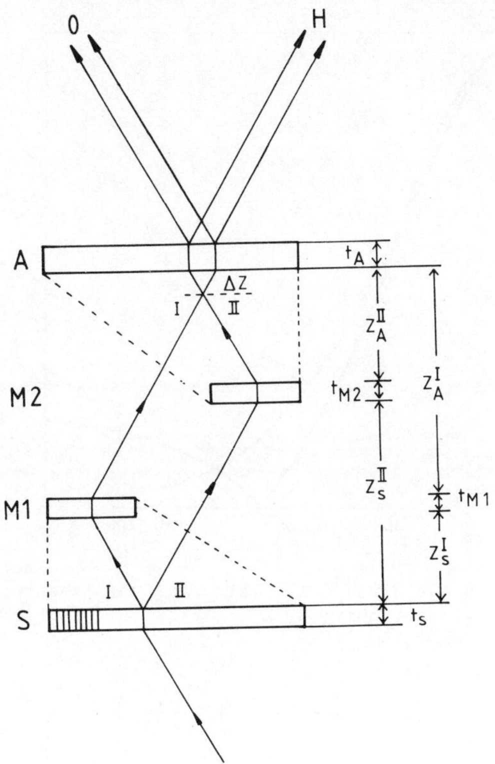


FIG. 13. Definition of the defocusing $\Delta Z$. (See text for further details.)

ence between $M_I$ and $M_{II}$ which causes a defocus $\Delta z = \frac{1}{2}\Delta t = |t_{M1} - t_{M2}|/2$. As expected, because of the factor ½, the influence on $\gamma$ is less compared with other types.

It is worth mentioning that with increasing absorption the loss of contrast is generally less (Bonse and Graeff 1977). Roughly speaking the effect of absorption is to make a spherical wave more planar because the wave components propagating parallel to the Bragg planes are absorbed less than the rest. This is one of the reasons why it is generally easier to secure good contrast of interference fringes with X-rays.

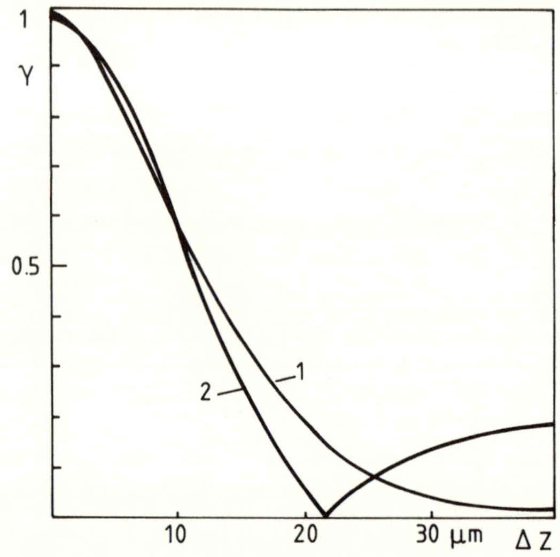


FIG. 14. Contrast γ in the 0 beam calculated for zero absorption $t = 9.3\Delta_e$: curve 1, $\Delta Z \neq 0$ but $t_A = t_S$; curve 2, $\Delta Z = |t_A - t_S|$. (See text for further details.)

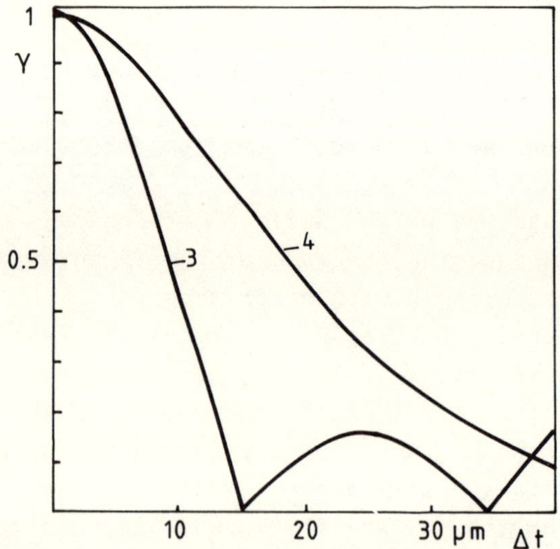


FIG. 15. As Fig. 14 but with γ as a function of $\Delta t \equiv |t_{M1} - t_{M2}|$: curve 3, $\Delta Z = 0$; curve 4, $\Delta Z = \Delta t/2$. (See text for further details.)

## 4. Measurement of coherent scattering lengths $b_c$

The LLL interferometer is well suited to the measurement of $b_c$ for neutrons. A number of different sample arrangements are shown in Figs. 2 and 3 which use parallel-sided specimens rotated about axes normal to the plane of rays or a pair of equal wedges which traverse the ray plane (Bonse and Hellkötter 1969). Similarly, parallel-sided samples can be transferred from beam I into beam II or moved in and out of one of the beams. The technique used depends on the shape of the available samples as well as on the $\lambda$ range within which $b_c$ is to be measured. Parallel-sided vanadium vessels are suitable for fluid samples since $b_c$ of vanadium is very small. For gas measurements (Kaiser *et al.*, 1977) parallel-sided aluminium vessels have been used. With gases it is more convenient to vary the pressure than to rotate the vessel or to move it in and out.

So far interferometric measurements have been performed on gaseous hydrogen, nitrogen, oxygen, helium-3, helium-4, neon, argon, krypton and xenon. $b_c$ has been measured for the following solids (Bauspiess, Bonse, and Rauch 1978): aluminium, magnesium, tin, vanadium, zinc, bismuth and niobium. The results are presented in papers I.5 and II.1.

The precision of measurement attained so far is about $10^{-4}$ for solid samples and about $10^{-3}$ for gases. The precision can be improved by at least an order of magnitude with a more stable interferometer. It appears that the sample definition (composition, homogeneity) is frequently the limiting factor rather than the measuring technique itself.

$b_c$ can also be measured for deviated scattering if a diffracting sample (perfect crystal) is used (Graeff *et al.* 1978).

## 5. Neutron phase topography

In phase topography an image is formed of the locally varying phase shift by an inhomogenous sample. With neutrons variations of not only scattering length $b_c$ but also magnetization can be of interest for the investigation of structures in solids. Because of the generally low absorption of thermal neutrons the method is not, as in the X-ray case, limited to thin samples.

Investigations of phase transformations, precipitates, or magnetic domain structure with neutron phase topography appear to be worth attempting.

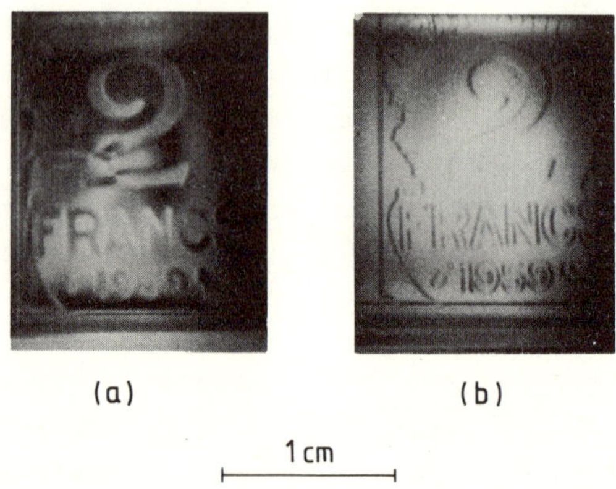


FIG. 16. (a) Neutron phase topograph of an aluminium coin taken with the interferometer of Fig. 17. The coin has thickness variations of the order of $\frac{1}{2}t_\lambda$. Gadolinium foil technique; exposure time about 40 h; $\lambda \equiv 1.8$ Å; the contrast is mainly 'area' contrast (see text). (b) As (a) but the other interfering beam is blocked off with a cadmium foil. The contrast is 'contour' contrast occurring only at the edges of the script characters (see text).

As a first example Fig. 16(a) shows the neutron phase topograph of an aluminium coin with thickness structures of the order $\frac{1}{2}t_\lambda$ where $t_\lambda$ is the $\lambda$ thickness (about 165 μm for aluminium) (Bauspiess *et al.* 1978). The interferometer was especially made to meet the requirements essential in phase topography. To accommodate larger samples easily without entering both interfering beams simultaneously a skew-symmetric LLL version was chosen with dimensions and crystallographic orientation as shown in Fig. 17. The hatched region in the centre is cut out about 5 mm lower in order to allow free manipulation of the sample without danger of touching the base of the interferometer or the mirror wafers $M_I$ and $M_{II}$. The sample and film positions within the respective beams are indicated in Fig. 17.

As can be seen, the method yields very good contrast. Regarding the nature of the contrast, there appear to be two different kinds. One kind, which we call 'area contrast', is

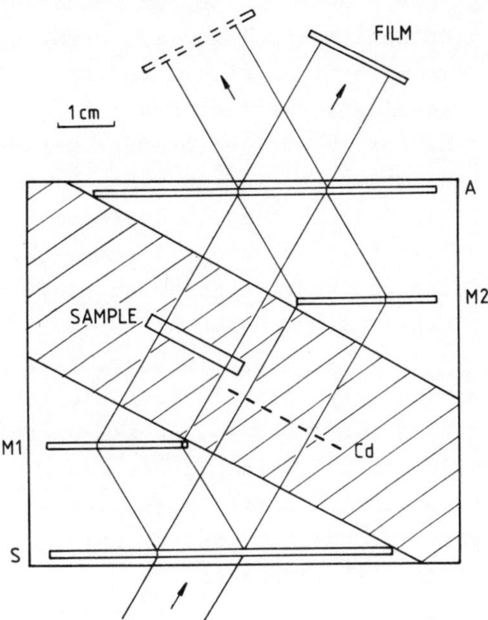


FIG. 17. Interferometer for neutron phase topography: $\lambda \approx 2$ Å; Silicon 220 reflection; wafer thickness, about 0.6 mm; wafer height, 20 mm; base thickness, about 20 mm. The hatched area is cut out 4 mm lower to allow free manipulation of the sample; the sample and film position for the topograph in Fig. 16(a) are shown; the broken lines refer to the condition under which topograph 16(b) was taken. (See text for further details.)

directly caused by the thickness difference between the areas in the topograph belonging to the script characters and those belonging to the background. The other kind, which we call 'contour contrast', occurs typically at the edges of the characters, i.e. where there are steps on the surface. The contour contrast is particularly distinct when the interfering beam not containing the sample is blocked off, i.e. when the area contrast is absent as shown in Fig. 16(b). The contour contrast is assumed to arise from a relative phase shift introduced between different portions of the coherent Borrmann fan between the crystal wafers $M_I$ and A when a surface step lies within the range of this fan. Crystals $M_I$ and A represent a two-wafer LL interferometer similar to that mentioned by Kikuta *et al.* (1975).

It is obvious that in phase topography the image quality

obtained is strongly dependent on the beam-widening ($2t \sin \theta$) and side-inverting mechanisms necessarily taking place in the Borrmann fans at zero absorption within the crystals. One way of keeping this image blurring effect small is to manufacture the wafers as thin as possible. The interferometer used in taking the topographs shown in Fig. 16 had a wafer thickness of about 600 μm. It is felt that a thickness of 300 μm could possibly be attained with wafers about 20 mm high. With lower wafers the limit which can be reached by conventional techniques appears to be near 150–200 μm. At $\theta = 30°$, 14.5°, and 5.7° the blurring is $t$, $0.5t$, and $0.1t$ respectively. Hence unless $\theta$ values considerably below 30° are used the blurring will be of the order of the wafer thickness. On the other hand, the spatial resolution of photographic neutron recording is that of fast or medium-fast X-ray film, e.g. 50–10 μm, which gives about the same limit as the geometrical image blurring with a very thin wafer interferometer.

Image diffusion is also caused if a wavelength range $\Delta\lambda$ is used, as illustrated in Fig. 18. However, there are geometries for which $\lambda$ focusing can be accomplished. For instance, the image on film 1 in Fig. 18(a) is $\lambda$ focused whereas that on film 2 is $\lambda$ blurred. To obtain the best focusing the combined effects of Borrmann fan widening and of $\lambda$ defocusing have to be considered.

Finally a strain-induced built-in pattern of the empty interferometer can deteriorate the phase topograph. Bauspiess *et al.* (1978) demonstrated the possibility of compensating most of the built-in pattern with a suitably moulded phase plate. Such a phase plate was successfully employed in obtaining Fig. 17(a).

## 6. Phase disturbances

As with X-rays crystal imperfections are a major cause of irregular phase relationships and thus incoherence. However, the crystal perfection required for neutrons is higher since larger crystals are usually necessary and the reflection range is roughly half as wide as that for X-rays.

Another source of incoherence (which is absent with X-rays) is the phase sensitivity to *rigid-body* displacements of the

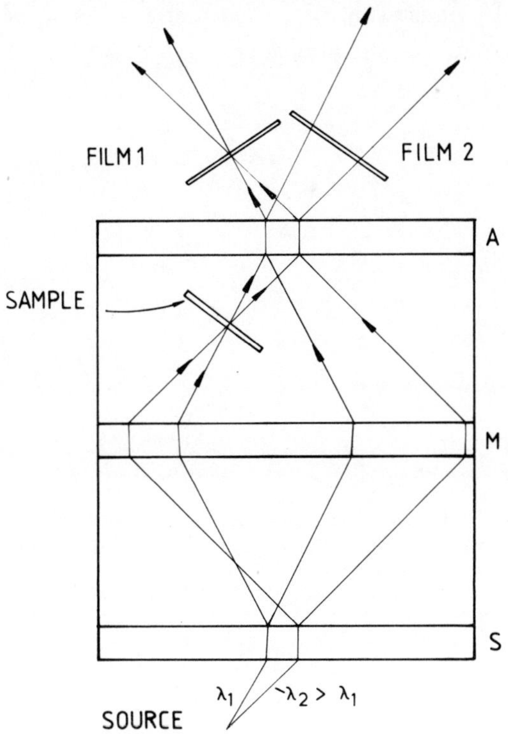


FIG. 18. $\lambda$ focusing for obtaining sharp images with a range $\Delta\lambda$ of wavelength. (See text for further details.)

interferometer as a whole. The reason is that while X-rays travel with light velocity $c$ thermal neutrons are fairly slow, e.g. $2 \times 10^3$ m s$^{-1}$ at wavelength 2 Å, so that during their flight time through the interferometer of a few microseconds they 'experience' a shift in the Bragg planes of S, M, and A even at moderate speeds of the interferometer. The phase shift introduced with every reflection by a crystal displacement $\mathbf{u}$ can be shown to be $2\pi\mathbf{u}\cdot\mathbf{g}$, where $\mathbf{g}$ is the diffraction vector. With the silicon 220 reflection a phase shift of $2\pi$ is caused by a displacement of less than 2 Å. Consequently a displacement velocity of 2 Å/5 μs = 40 μm s$^{-1}$ can reduce the interference contrast. Therefore, in neutron interferometry fairly good vibration protection is necessary even with monolithic interferometers. A somewhat more rigorous treatment (Bauspiess

1977) shows that in addition to displacement (static) phase shifts Doppler (dynamic) phase shifts also have to be accounted for.

## Acknowledgment

Support for this work was given by the Bundesminster für Forschung und Technologie, 03-45 A 09 I and 03-41 A 09 P and is gratefully acknowledged.

## References

AGERON, P. (1972). *Bull. Inf. sci. tech. Commis. Energ. At. (Fr.)* **166**, 1.
AUTHIER, A., MILNE, A.D., and SAUVAGE, M. (1968). *Phys. Status Solidi* **26**, 469.
BAUSPIESS, W. (1977). Ph.D. Thesis, University of Dortmund.
BAUSPIESS, W., BONSE, U., and GRAEFF, W. (1976). *J. Appl. Crystallogr.* **9**, 68.
BAUSPIESS, W., BONSE, U., GRAEFF, W., SCHLENKER, M., and RAUCH, H. (1978). *Acta Crystallogr.*, **A34**, S239.
BAUSPIESS, W., BONSE, U., and RAUCH, H. (1978). *Nucl. Instrum. Meth.*, **157**, 495.
BAUSPIESS, W., BONSE, U., RAUCH, H., and TREIMER, W. (1974). *Z. Phys.* **271**, 177.
BONSE, U. (1969). Present state of X-ray interferometry. In *Proc. 5th Int. Congr. on X-Ray Optics and Microanalysis* (eds. G. Moellenstedt and K.H. Gaukler), p. 1. Springer, Berlin, Heidelberg, New York.
—— (1975). *Neutron and X-Ray Interferometry: Int. Summer School on X-Ray Dynamical Theory and Topography, Limoges* (ed. F. Balibar). Laboratoire de Minéralogie et Cristallographie, Paris.
BONSE, U. and GRAEFF, W. (1977). X-ray and neutron interferometry. In *Topics in applied physics, Vol. 22, X-ray optics* (ed. H.-J. Queisser), p. 93. Springer, Berlin, Heidelberg, New York.
BONSE, U., GRAEFF, W., and MATERLIK, G. (1976). *Rev. Phys. Appl.* **11**, 83.
BONSE, U. and HART, M. (1965a). *Appl. Phys. Lett.* **6**, 155.
—— (1965b). *Z. Phys.* **188**, 154.
—— (1966). *Z. Phys.* **194**, 1.
—— (1968). *Acta Crystallogr.* **A24**, 240.
—— (1969). *Phys. Status Solidi* **33**, 351.
BONSE, U. and HELLKÖTTER, H. (1969). *Z. Phys.* **223**, 345.
BONSE, U. and te KAAT, E. (1971). *Z. Phys.* **243**, 14.
EWALD, P.P. (1916). *Ann. Phys. (Lpz.)* **49**, 1, 117.
—— (1917). *Ann. Phys. (Lpz.)* **54**, 519.
GABOR, D. (1948). *Nature (Lond.)* **161**, 777.
GRAEFF, W., BAUSPIESS, W., BONSE, U., and RAUCH, H. (1978). *Acta Crystallogr.*, A34, S238.
HART, M. (1971). *Rep. Prog. Phys.* **34**, 435.
—— (1975). *Proc. R. Soc. A* **346**, 1.
KAISER, H., RAUCH, H., BAUSPIESS, W., and BONSE, U. (1977). *Phys. Lett. B* **71**, 321.
KATO, N. (1961). *Acta Crystallogr.* **14**, 525, 627.

KATO, N. (1968). *J. Appl. Phys.* **39**, 2225, 2231.
KIKUTA, S., ISHIKAWA, I., KOHRA, K., and HISHINO, S. (1975). *J. Phys. Soc. Japan* **39**, 471.
von LAUE, M. (1931). *Ergeb. exakten Naturwiss* **10**, 133.
LANDKAMMER, F.J. (1966). *Z. Phys.* **189**, 113.
LEITH, E.N. and UPATNIEKS, J. (1962). *J. opt. Soc. Am.* **52**, 1123.
MAIER-LEIBNITZ, H. and SPRINGER, T. (1962). *Z. Phys.* **167**, 386.
MAIMAN, T.H. (1960). *Phys. Rev. Lett.* **4**, 564.
PETRASCHECK, D. (1976). *Acta Phys. Austriaca* **42**, 217.
PETRASCHECK, D. and FOLK, R. (1976). *Phys. Status Solidi* (a) **36**, 147.
PINSKER, Z.G. (1978). *Dynamical scattering of x-rays in crystals*. Springer, Berlin, Heidelberg, New York.
RAUCH, H., TREIMER, W., and BONSE, U. (1974). *Phys. Lett. A* **47**, 369.

# 2. VARIATIONS IN INTERFEROMETER TYPES

W. GRAEFF

*University of Dortmund, Federal Republic of Germany;*
*Institut Laue–Langevin, Grenoble, France*

## 1. Introduction

There are two reasons for the search for interferometer configurations other than the LLL type that has been proved to work satisfactorily: first some disadvantages of the LLL interferometer such as the smearing of the beams over the Borrmann fan or the loss of half of the intensity in the Laue case mirror may be overcome by changing the geometry; secondly, problems of neutron optics such as details of wave propagation in perfect and nearly perfect crystals, coherence, etc. can be investigated by varying the properties of the interferometer components.

We restrict the discussion to those interferometer types where Bragg diffracting single crystals are used for beam handling. Since the first operation of an LLL interferometer with X-rays (Bonse and Hart 1965) many variations have been proposed and some have been attempted experimentally. This subject has been extensively reviewed (Bonse 1969, 1975, Hart 1971, 1975, Bonse and Graeff, 1977) and so the purpose of this paper is to discuss the salient features of these variations with respect to neutron interferometry.

## 2. General considerations

Figure 1 shows the two different geometries of Bragg diffraction using a plane parallel crystal slab of thickness $t$, i.e. the Laue case and the Bragg case, together with the qualitative shape of the intensity profiles of the beams emerging from the crystal when a spherical monochromatic wave is incident. The absorption is assumed to be practically zero as is the case when thermal neutrons are diffracted by a silicon crystal.

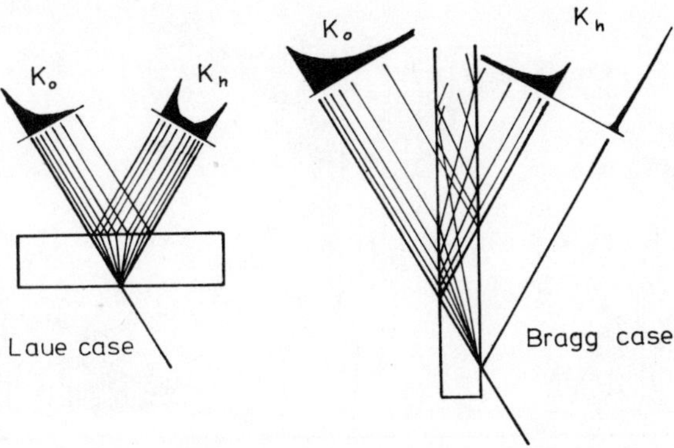


FIG. 1. Coherent beam splitting with Bragg diffracting single crystals for symmetrical Laue and Bragg geometry. The intensity profiles of the outgoing beams are indicated schematically. Pendellösung oscillations are neglected.

It is well known from ordinary light interferometry that interference contrast is a maximum if the two interfering beams are of equal intensity. Obviously the profiles given in Fig. 1 show quite different intensity distributions in the two partial beams for both geometries. This is also shown by the single-crystal reflection curves (Fig. 2), which are essentially the Fourier transform of the spatial amplitude distributions in the emerging beams. Here the intensities $T(y)$ and $R(y)$ of the transmitted and reflected plane waves are plotted against the normalized deviation $y$ of the direction of the incident plane wave from the kinematic Bragg angle. The fine structure of these curves, the so-called Pendellösung oscillations, is strongly governed by the thickness of the crystal slabs expressed in units of the extinction length $\Delta_e$:

$$\Delta_e = \frac{\pi V_c \alpha}{2D_h F_h b_c d_{hkl}} \tag{1}$$

$$\alpha = \begin{cases} \cot \theta_B & \text{(symmetrical Laue case)} \\ 1 & \text{(symmetrical Bragg case)} \end{cases}$$

where $V_c$ is the volume of the unit cell, $D_h$ is the Debye-Waller factor, $F_h$ is the structure amplitude, $b_c$ is the coherent scattering length and $d_{hkl}$ is the Bragg spacing of the

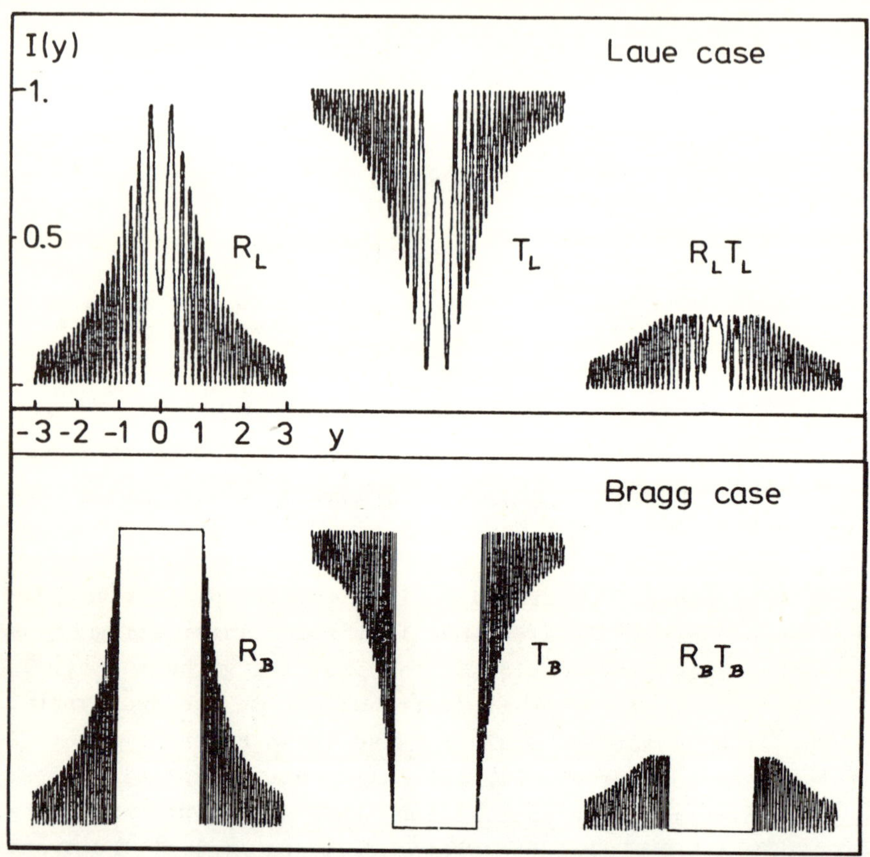


FIG. 2. Single reflection (intrinsic) curves of the components shown in Fig. 1 assuming a crystal thickness $t = 10.2\Delta_e$. R is the reflected H beam and T is the transmitted O beam. The *RT* product of R and T is valid for consecutive reflection and transmission in two crystals of equal thickness.

reflecting net planes. Typical values with Si 220 reflection and a neutron wavelength of 2 Å are 64 μm (Laue case) and 39 μm (Bragg case).

Owing to the rapid oscillations and the fact that $R(y) + T(y) = 1$, the condition $R(y) \approx T(y)$ is fulfilled at very distinct $y$ values only. This is a completely different situation from that with X-rays and absorbing crystals where the Borrmann transmission ensures $R(y) \approx T(y)$ for all $y$.

In order to obtain maximum interference contrast it is therefore evident that the analyser crystal must act upon the amplitudes in an exactly complementary manner to that of the beam splitter, i.e. the beam transmitted through the beam splitter must be reflected by the analyser and *vice versa* where beam splitter and analyser have equal thickness and orientation. In addition the reflectivity of the mirrors must be the same. Consequently the efficiency of Bragg and Laue case beam splitters and analysers can be judged by comparing the products $RT$. Owing to the range of total reflection the Bragg case is obviously less suited than the Laue case. With a crystal thickness $t \approx 10\Delta_e$ the integral $\int R(y)T(y)\mathrm{d}y$ is smaller by a factor of 1.3 for the Bragg case. The situation changes drastically when the Pendellösung oscillations are suppressed by using extremely thin crystal lamellae with thickness $t = \Delta_e/\pi$ (Fig. 3), i.e. 10–20 μm. However, the manufacture of such crystal plates requires some further technological effort.

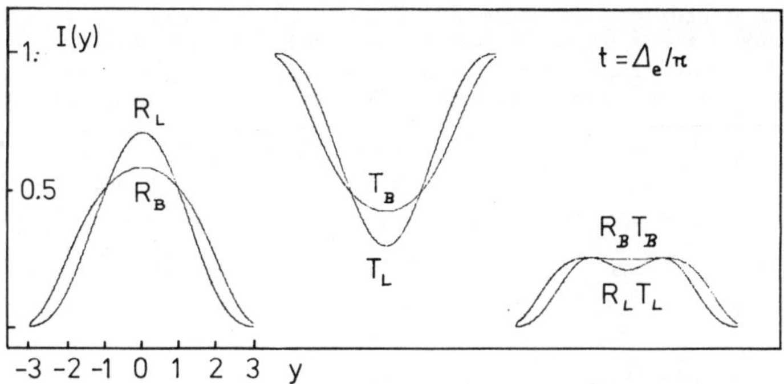


FIG. 3. As for Fig. 2 but with $t = \Delta_e/\pi$. Note the flat top of $R_B T_B$.

The above considerations yield another general result for non-absorbing interferometers. Only one of the two beams leaving the analyser has 100% interference contrast because the other one is composed of partial beams with $RR$ and $TT$ products, respectively.

The design of interferometers is also influenced by the available spectral distribution of the neutron source. Figure 4 shows the neutron flux obtainable with different silicon reflections plotted against the Bragg angle $\theta_B$. The curves are calculated from values given by Ageron (1972) for the H25

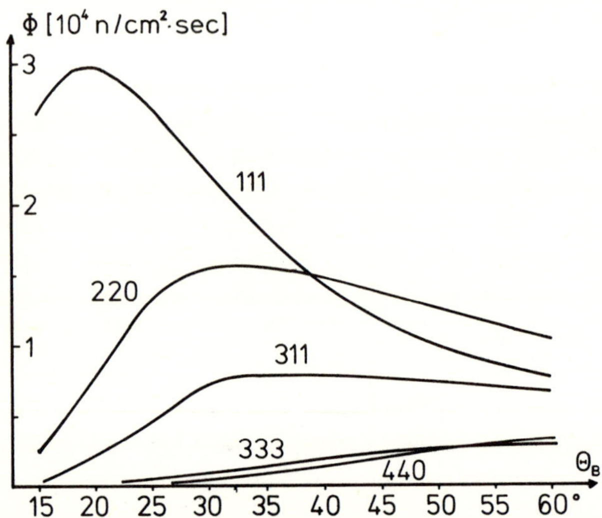


FIG. 4. Neutron flux at the neutron guide H25 (ILL, Grenoble) attainable with different reflections of a perfect crystal silicon monochromator plotted against the Bragg angle (theoretical values). Note the low contamination with higher harmonics (111/333 and 220/440) for angles up to 35°.

thermal neutron guide. The angular range corresponds to the available θ range of the D18 instrument. Reasonable fluxes can be achieved with 111, 220, and 311 reflections. At present the LLL interferometer used at the ILL is operated at the maximum of the 220 curve.

To compare the fluxes attainable with different interferometer types we calculate the normalized integral reflectivity of each type on the $y$ scale taking the value for the LLL interferometer as unity. The additional weighting factors for different reflections and Bragg angles can be taken from Fig. 4. If, moreover, the spectral width accepted by the interferometer is restricted, as is the case with some chromatic interferometer types, this has of course to be taken into account.

## 3. Interferometers with two components

In two-component interferometers the different energy flow directions of the two modes excited inside the crystal is used to separate the interfering beams. In the Laue case the H beam is symmetric about its centre, and with sufficiently thick crystals the separation is large enough to introduce a phase object into one-half of the beam cross-section. The beam is focused on the exit surface of another crystal slab of equal thickness. This arrangement was theoretically investigated by Indenbom, Slobodetskii, and Truni (1975) and experimentally demonstrated by Suvorov and Polovinkina (1975). This so-called LL interferometer was proposed for neutrons by Kikuta *et al.* (1975). Experiments using the LL interferometer have been reported elsewhere. The use of an entrance slit confining the cross-section of the transmitted beam of an asymmetric fore crystal which cuts the centre of the Laue reflection curve produces two well-separated beams (Fig. 5). The integral reflectivity normalized to the LLL reflectivity varies between 0.5 and 0.8 depending on the crystal thickness.

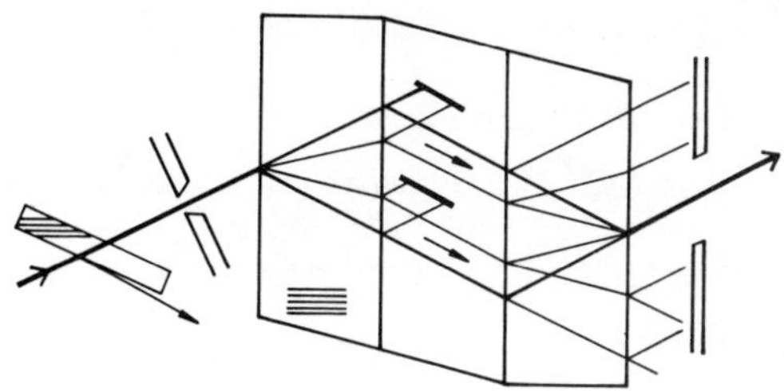

FIG. 5. LL interferometer using the effect of diffraction focusing. The asymmetric fore crystal deflects that part of the incoming beam that would generate wave fields in the middle of the Borrmann fan.

The corresponding Bragg case BB interferometer (Hart 1971, Kikuta *et al.* 1975) is sketched in Fig. 6. The optical analogue is the Jamin interferometer. The second propagation mode is generated by internal reflection at the rear surface of the crystal plate. Although the reflectivity factor is 0.22, the BB interferometer has the remarkable feature of producing a surface-reflected beam the cross-section of which is not broadened by a Borrmann fan and which can thus be reduced at will. By scanning a sample put into that beam path, for instance, a one-dimensional diffusion profile could be determined. Both types of interferometer are achromatic.

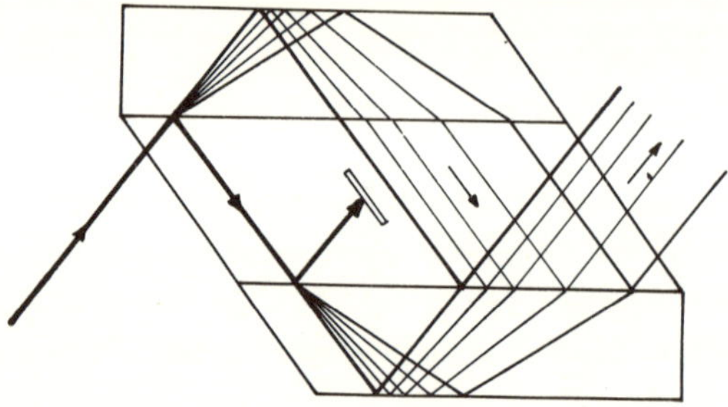

FIG. 6. BB interferometer. Note the narrow cross-section on one beam path.

## 4. The three-module Bragg case interferometer

The BBB interferometer (Fig. 7) was experimentally tested (Bonse and Hart 1966) soon after the first operation of the LLL interferometer. The geometrical tolerances were found to be almost equal to those obtained for the Laue case interferometers.

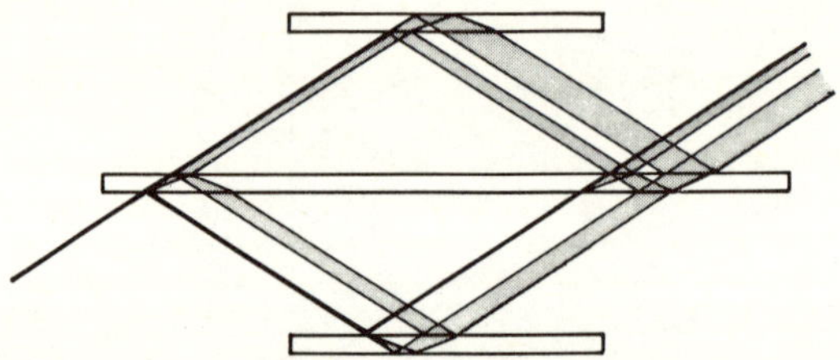

FIG. 7. BBB interferometer with three lamellae of equal thickness. Owing to internal reflections several BB interferometers are involved and the consequence is a multi-peaked intensity profile of the outgoing beam. (The two strongest contributions are displayed.)

Although Bragg case mirrors are used the total reflection range cannot contribute to the interfering beams owing to the splitting process which can only take place outside that range. Therefore the reflectivity factor is 0.65 when three lamellae of equal thickness $t \approx 10\Delta_e$ are used. If one uses rather thick mirror plates so that the surface reflected beams reach the detector solely, the reflectivity factor drops to 0.33. On the contrary, use of an extremely thin lamella for beam splitting and analysing, as discussed above, together with fairly thick mirror plates increases the reflectivity to 2.7. If mirrors with $t \approx 10\Delta_e$ are used the reflectivity factor is as high as 3.0.

## 5. The four-module Laue case interferometer

If a fourth lamella is added to the LLL interferometer and the centre gap width is chosen to be twice as wide as the outer ones the ray geometry shown in Fig. 8 results. This geometry,

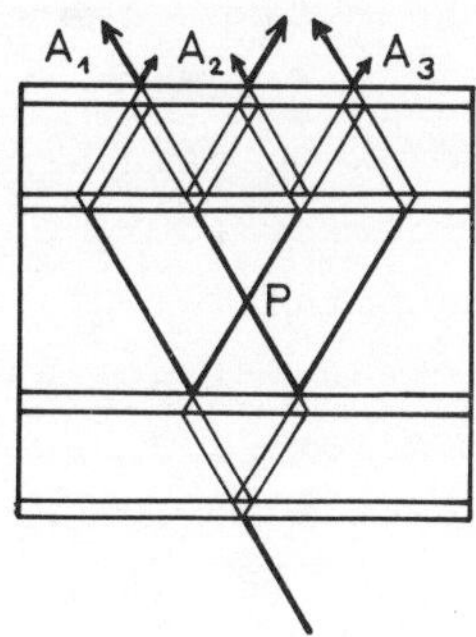


FIG. 8. LLL interferometer. Only strong focused rays are displayed. Phase shifts introduced around P show up at $A_1$, $A_2$, and $A_3$, where $A_2$ is the difference between $A_1$ and $A_3$.

which was proposed by Besirganyan, Eiramdshyan, and Truni (1973), contains three interferometers superimposing beams at $A_1$, $A_2$, and $A_3$.

A closer examination yields the result that the contrast has a maximum in the O beams emerging from $A_1$ and $A_3$, and in the H beam from $A_2$. This geometry has two possible applications: the crossing of interfering beams at P can be used to reduce the necessary sample size considerably if the sample plate is to be rotated within both beams for precise $b_c$-determination; with sufficiently thin crystal lamellae, all of equal thickness $t$ (a few tenths of a millimetre), favoured by diffraction focusing samples placed in the centre gap should be imaged quite satisfactorily. The three phase topographs which are obtained at the same time contain different information about the sample placed at the crossing P. The topographs from $A_1$ and $A_3$ can be interpreted as a stereo pair and give some three-dimensional information, whereas the one from $A_2$ images the difference of $A_1$ and $A_3$ thus eliminating sample layer where coherent overlap of the partial beams occurs. If the structure is not too complicated, it may be possible to reconstruct the structure within that layer from all three topographs, i.e. to carry out phase contrast tomography.

## 6. Mixed Bragg–Laue case interferometers

With the use of Bragg and Laue case components (Bonse and Hart 1968) in one interferometer, refraction comes into play. As long as all crystal surfaces were parallel a plane wave propagating through the interferometer could be described by the same value of the incidence parameter $y$ in all components.

We shall briefly discuss the refraction effects for the symmetrical Bragg and Laue cases. The boundary conditions for the neutron waves at the crystal surface require the continuity of tangential components of the wave vectors $\mathbf{K}_o$ and $\mathbf{K}_h$ respectively. For the symmetrical Bragg case these tangential components are identical for the $\mathbf{K}_o$ and $\mathbf{K}_h$ vectors. If the angle $\theta_o$ between the incident wave vector $\mathbf{K}_o$ and the net planes increases, the angle $\theta_h$ of the exit wave vector does the same. This is not the case for the symmetrical Laue case where the difference between the tangential components equals the reciprocal lattice vector according to Bragg's law and hence an increase in $\theta_o$ causes a decrease in $\theta_h$.

For even reflections in silicon the centre of the Laue reflection curve is shifted to one edge of the total reflection range of the Bragg curve leading to the relation

$$y_B = y_L + 1 \tag{2}$$

if the wave between the two components is travelling in the O direction and

$$y_B = 1 - y_L \tag{3}$$

if the wave is travelling in the h direction. As an example we calculate the intensities of the partial O beams for the LBL geometry:

$$I_1(y) = T_L(y)R_B(y+1)R_L(-y) \tag{4}$$

$$I_2(y) = R_L(y)R_B(1-y)T_L(-y). \tag{5}$$

If $R_B(y) = 1$ for $|y| \leqslant 1$ we see immediately that $I_1(y)$ has appreciable values between $-2 \leqslant y \leqslant 0$ whereas $I_2(y)$ is mainly confined to $0 \leqslant y \leqslant 2$. Although a plane wave passing any mixed Laue–Bragg geometry leaves in the same direction via both possible paths, a spherical wave is split into two divergent beams.

It is clear that interferometers composed of Bragg and Laue modules show poor, if any, interference contrast unless we succeed in correcting the refractive shift. If the Bragg reflection curve is centred about the Laue reflection curve the sign of $y$ still changes under certain conditions according to relation (3) but does not influence the result because $R(y)$ and $T(y)$ are symmetric about $y$ provided absorption can be neglected.

In the following we consider as the most promising mixed interferometer type the LBBL version shown in Fig. 9 which has been operated with X-rays in both an untuned (Bonse and Hart 1968) and a tuned (Spieker 1977) version. Two possible ways of tuning have been suggested: heating of the Bragg parts (or cooling of the Laue parts) or rotating the Bragg mirrors in pairs with respect to the Laue beam splitter and analyser. Both methods suffer from severe stability problems.

With neutrons, however, we can find a more elegant method of correcting for refraction without touching the crystal. Insertion of a rectangular prism made from a material with the same refractive index as silicon, e.g. aluminium whose refractive index differs by 0.25% from that of silicon or polycrystalline silicon, into the transitional regions between the Laue and Bragg modules as indicated in Fig. 9 will deviate the beams by the appropriate amount.

Figure 10 illustrates the effect. The single reflection curve of the h beam is plotted in Fig. 10(b) for both

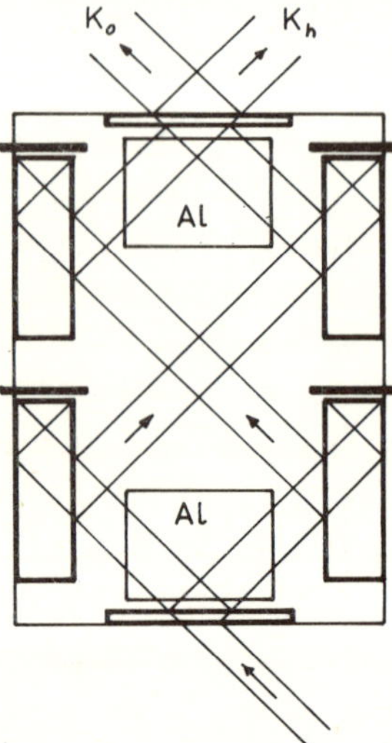


FIG. 9. LBBL interferometer. Two aluminium blocks are inserted into the beam paths to compensate for the refraction between the Laue and the Bragg parts. Note the beam stops to absorb internally reflected waves which cannot be focused at the analyser.

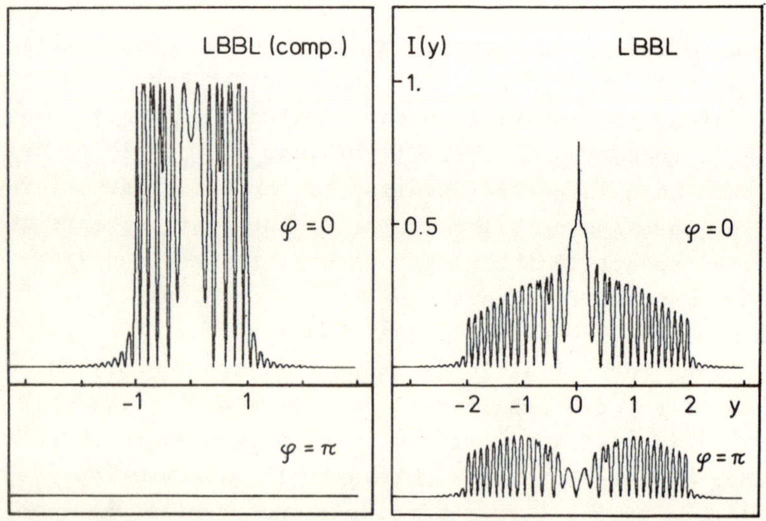


FIG. 10. Intrinsic curves for the LBBL geometry illustrating the effect of tuning: (a) the tuned case with 100% interference contrast; (b) the untuned case where a phase shift π reveals the incoherent background.

constructive and destructive interference of the untuned version. A phase shift of $\pi$ mainly affects the centre part leaving a strong incoherent background. It is this part which is favoured by Borrmann transmission and so the untuned version could be operated with X-rays. Figure 10(a) shows the intrinsic curve if refraction is fully compensated. The partial waves have equal amplitudes for all values of $y$ and can interfere with 100% contrast.

When designing an LBBL interferometer one has to take care that the Bragg parts are thick enough to separate the internally reflected waves from the surface-reflected waves. As they are not focused at the analyser they must be absorbed by beam stops as indicated in Fig. 9. For a maximum cross-section $S$ the minimum crystal thickness $t$ of the Bragg parts is simply

$$t = S/2 \cos \theta_B . \tag{6}$$

Furthermore the focusing of the surface-reflected beams is possible for one wavelength $\lambda$ only which is fixed by the length to width ratio $L/W$ of the interferometer. A change $\delta\lambda/\lambda$ in wavelength moves the focus away from the entrance surface of the analyser by a distance $\Delta z$ related to the geometry by

$$\Delta z = - \frac{4W}{\sin 2\theta_B} \frac{\delta\lambda}{\lambda} \tag{7}$$

From coherence calculations for the LLL geometry we know that $\Delta z$ should not exceed $\Delta_e/6$. Using equations (1) and (7) we calculate the allowed spectral window

$$\frac{\Delta\lambda}{\lambda} = 2\frac{\delta\lambda}{\lambda} = \frac{\pi}{12} \frac{V_c \cos^2 \theta_B}{W D_h F_h b_c d_{hkl}} \tag{8}$$

If $\theta = 27.5°$ for silicon 220 reflection we find a value of $3.6 \times 10^{-4}$ for an interferometer width $W = 15$ mm. This must be compared with $\Delta\lambda/\lambda = 5 \times 10^{-4}$ for the LLL interferometer when using the bicrystal arrangement.

The reflectivity of the LBBL is 1.6 times higher than the LLL reflectivity, leaving a net factor of 1.15 in favour of the LBBL interferometer. This makes it a valuable instrument for precision measurements of $b_c$ for small samples which can be placed in the beam crossing. If we compare the reflectivity with that of the LLLL interferometer which also has a coherent overlap accessible by samples, the intensity gain is even higher.

## 7. Three-beam interferometer

Beam splitting in a more symmetric way can also be achieved by simultaneous reflection at two different sets of net planes. After two consecutive coplanar reflections the partial beams are focused at the analyser and leave the interferometer in the 0 direction. Figure 11 shows the sketch of the silicon 440, 404 version which has been operated successfully with synchrotron radiation (Graeff and Bonse 1977).

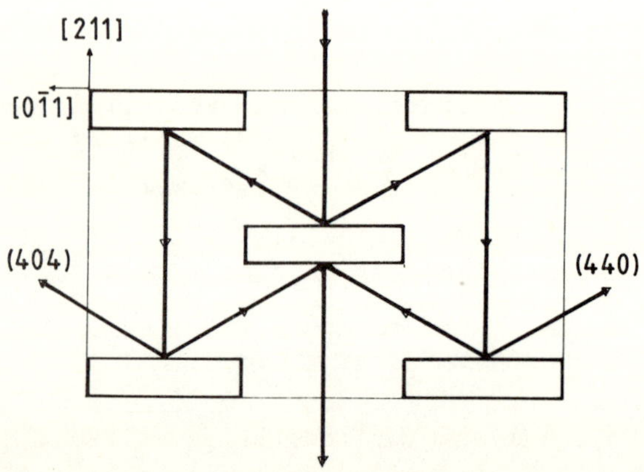


FIG. 11. Three-beam interferometer with coplanar beams to be used with silicon 440, 404 and λ = 1.6627 Å or silicon 220, 202 and λ = 3.3255 Å.

The unique feature of this arrangement is the complete absence of any beam widening by Borrmann triangles, which makes it suitable for high resolution phase contrast topography with a typical resolution of several microns. Unfortunately the built-in spectral window of the coplanar three-beam interferometer, which is of the order of $10^{-5}$, requires the use of an intense source such as a synchrotron. Even with a three-dimensional ray pattern when all wavelengths can be focused the small acceptance range confined to a solid angle of $10^{-8}$ st or less is a severe limit to the luminosity.

Similar arguments hold for interferometers of the Michelson type proposed for Ångström range interferometry and resonators (Deslattes 1968, Hart 1975). They all use multiple-beam cases for beam handling.

Most of the interferometer types discussed so far have never been used with neutrons. Some of them should not be but some appear to complete the neutron interferometry instrumentation quite well.

## References

AGERON, P. (1972). *Bull. Inf. sci. tech. Commis. Energ. At. (Fr.)* **166**, 1.
BESIRGANYAN, P.A., EIRAMDSHYAN, F.O., and TRUNI, K.G. (1973). *Phys. Status solidi A* **20**, 611.
BONSE, U. (1969). *Proc. 5th Int. Congr. on X-ray Optics and Microanalysis*, p. 1. Springer, Berlin, Heidelberg, New York.
—— (1975). Int. Summer School on X-ray Dynamical Theory and Topography, Limoges, ed. by F. Balibar (Laboratoire de Minéralogie et Cristallographie, Paris 1975).
BONSE, U. and GRAEFF, W. (1977). *Topics in Applied Physics*, Vol. 22, p. 93. Springer, Berlin, Heidelberg, New York.
BONSE, U. and HART, M. (1965). *Appl. Phys. Lett.* **6**, 155.
— (1966). *Z. Phys.* **194**, 1.
— (1968). *Acta Crystallogr.* **A24**, 240.
DESLATTES, R.D. (1968). *Appl. Phys. Lett.* **12**, 133.
GRAEFF, W. and BONSE, U. (1977). *Z. Phys.* **B27**, 19.
HART, M. (1971). *Rep. Prog. Phys.* **34**, 435.
—— (1975). *Proc. R. Soc.* **A346**, 1.
INDENBOM, V.L., SLOBODETSKII, I.Sh., and TRUNI, K.G. (1975). *Sov. Phys.—JETP* **39**, 542.
KIKUTA, S., ISHIKAWA, I., KOHRA, K., and HISHINO, S. (1975). *J. Phys. Soc. Japan* **39**, 471.
SPIEKER, P. (1977). Doctorate thesis, University of Dortmund.
SUVOROV, E.V. and POLOVINKINA, V.I. (1975). *JETP Lett.* **20**, 145.

# 3. TWO-CRYSTAL NEUTRON INTERFEROMETRY

A. ZEILINGER[†], C.G. SHULL, M.A. HORNE[‡], AND G.L. SQUIRES[§]

*Massachusetts Institute of Technology, Cambridge, Mass. 02139, U.S.A.*

## 1. Introduction

Perfect-crystal neutron interferometry was first demonstrated by Rauch, Treimer, and Bonse (1974) with a three-crystal system in Laue geometry. The wide separation between the coherent beams in this type of interferometer has made possible a number of significant experiments, most notably those dealing with the neutron–gravitational potential interaction and the Fermion spin rotation. Interpretation of these experiments was not critically dependent upon a detailed description of the diffraction processes occurring in the three crystal plates that form this type of interferometer. Although several authors (Rauch and Suda 1974, Bauspiess, Bonse, and Graeff 1976, Petraschek and Folk 1976) have developed theories of this interferometer type, no detailed experimental confirmation of the theory has been attempted. On the other hand, the theoretical predictions of Kato (1960, 1961) of wave propagation in a single plate of perfect crystal in Laue geometry have been confirmed in considerable detail for the neutron case by Shull (1968, 1973) and Shull and Oberteuffer (1972).

Preliminary to performing the presently reported investigations of the diffraction processes occurring in the simpler case of a two-crystal interferometer, it was observed that there was interference action within one beam of a normal three-crystal interferometer as illustrated in Fig. 1. Here the forward diffracted beam was stopped behind the first crystal plate by a cadmium absorber thus leaving only one beam

---

†Permanent address: Atominstitut der Österreichischen Universitäten, Wien, Austria.

‡Permanent address: Stonehill College, North Easton, Mass., U.S.A.

§Permanent address: Cavendish Laboratory, Cambridge, England.

path open. The interference action within this beam was studied by transverse scanning through this beam of a multiple step phase plate of aluminium and observing the intensity released from the interferometer in the two characteristic directions. The step heights in this phase plate were consecutively $\frac{1}{2}D$, $D$, $\frac{3}{2}D$,..., where $D$ is the $\lambda$ thickness, and the plateau width was wider than the beam width of 1.25 mm. In this interferometer the individual crystal plate thickness was 2.464 mm with silicon (220) reflection being used in symmetrical arrangement with a neutron wavelength of 1.46 Å.

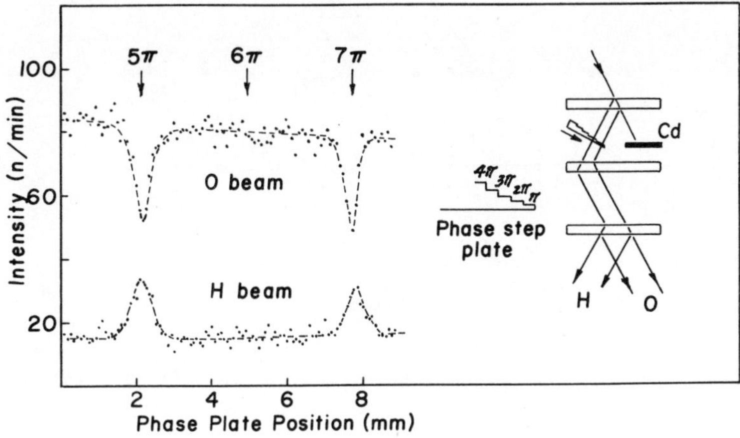


FIG. 1. Interference action within one beam of a three crystal interferometer.

It is seen from the intensity patterns of Fig. 1 that complementary effects are being sensed in the two emergent beams whenever step edges corresponding to odd-π phase difference are passed through the beam with, however, no effect at positions corresponding to even-π phase difference. Moreover the fact that the intensity perturbations in the two emergent beams are not quantitatively complementary suggests the presence of additional interference action in the other beam leaving the second crystal plate (not shown in the figure). Additional experiments sensing the intensity in this beam indeed confirmed this.

In view of this demonstration of internal interference action within one beam, it was decided to prepare a two-crystal interferometer with thicker crystal plates and larger separation distance so that the effects could be studied more carefully. For this purpose, an interferometer was cut in the usual monolithic fashion from a silicon crystal ingot which had been grown parallel to the [111] axis (see Fig. 2).† To obtain a

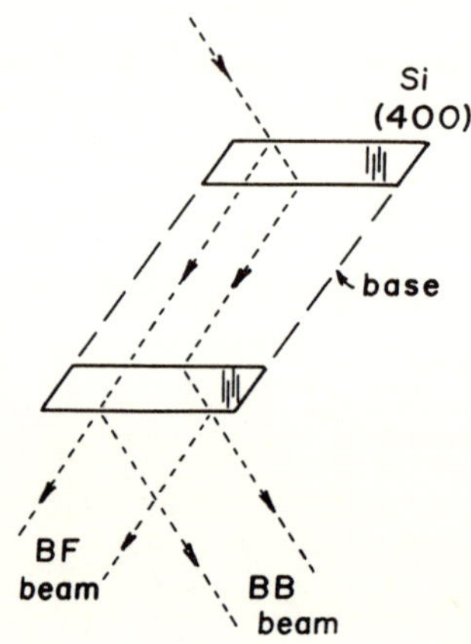


FIG. 2. Two-crystal interferometer showing the Borrmann fan limitation of radiation within the system.

long path length between crystal plates in our experiments, the (400) diffracting planes in symmetric Laue geometry were selected for use with 1.46 Å neutrons. The plate thickness was 9.210 mm and the separation distance was 37.6 mm. The surfaces were lightly etched to a depth of 6 μm with monitoring by neutron surface reflectivity to ensure the removal of the mosaic layer. With this geometry the full width of the Borrmann fan at the exit face of the first crystal was 11.75 mm and the transverse width of the beam travelling to the second crystal was 9.90 mm. For the (400) reflection and the above neutron wavelength of 1.46 Å (obtained from a pyrolytic graphite monochromator), the Pendellösung period is calculated to be 92.9 μm. All of the present experiments have been performed with an

†Here it should be mentioned that a monolithic two-crystal system was used by Kikuta *et al.* (1975) to study some dynamical diffraction phenomena.

entrance beam of width 1.07 mm as defined by tapered gadolinium edges.

As is common experience with neutron interferometer systems it was necessary to provide vibration isolation, and this was accomplished by supporting the interferometer on a lead pad of weight 25 kg which floated on an air-filled rubber tube. Experience with the system showed it to be very sensitive to temperature gradients and therefore it was necessary to enclose the system in a thermal shield box with temperature controlled surroundings. The present experimental observations have concentrated on the distribution of intensity released from the interferometer and on some features of its performance as an interferometer.

## 2. Intensity distribution from interferometer

The entrance position at the first surface is being defined by an entrance slit opening, and the intensity released from the interferometer has been studied by scanning a slit opening across the exit face of the second crystal with intensity measurements being made simultaneously in both the BF and BB directions. It is to be expected that diffraction focusing will produce a localized intensity peak at the centre of the double Borrmann fan at the exit face. This has been predicted for X-rays by the theoretical treatment of Indenbom, Slobodetskii, and Truni (1974) using a Green's function approach and verified experimentally by Suvorov and Polovinkina (1974). The effect can also be understood using a ray theory approach. Incoming plane waves slightly off the exact Bragg direction generate two wave fields in the first crystal which travel in different directions $\pm\Omega$ relative to the lattice planes and leave the first crystal at positions $\pm\Gamma$ defined as

$$\Gamma = \frac{\tan \Omega}{\tan \theta_B}.$$

These rays propagate parallel to each other to the second crystal where at the entrance face each again generates two wave fields again travelling in $\pm\Omega$ directions. Thus those rays which travelled as different wave fields in the two crystals meet at a focal point on the exit face of the second crystal providing the two crystal plates are of the same thickness. This is true over a range of incident directions and is illustrated in Fig. 3. One expects a distributed intensity in the region surrounding the focal point and within the double Borrmann fan.

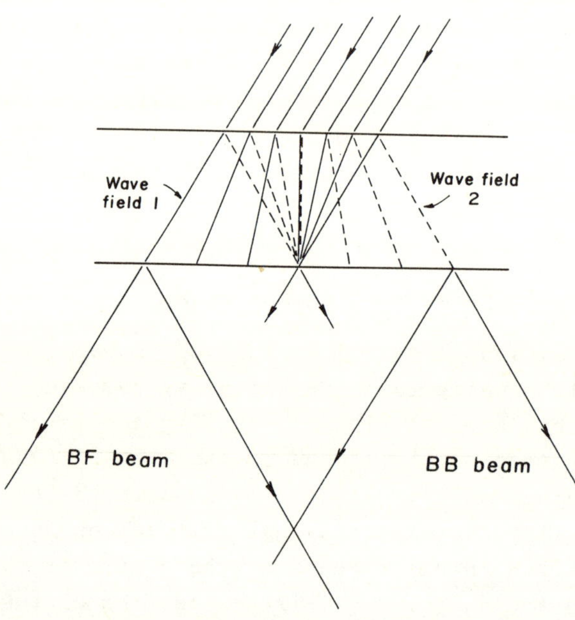


FIG. 3. Detail of the double Borrmann fan passage of rays in the second crystal plate. Focusing action occurs at the centre of the exit face and the distributed intensity is found outside the central region.

It is evident that divergence effects associated with different wavelength components will cause focusing at slightly different positions on the exit face of the second crystal. In the approach of Indenbom *et al.* (1974) the intrinsic width of the focal point for purely monochromatic radiation and a $\delta$-function entrance slit is calculated to be 12 μm for our conditions and this is much smaller than the present experimental resolution.

Figure 4 shows the measured intensity distributions for the two exiting beams in the present interferometer as a function of exit slit position. Strong central peaks (in ratio about

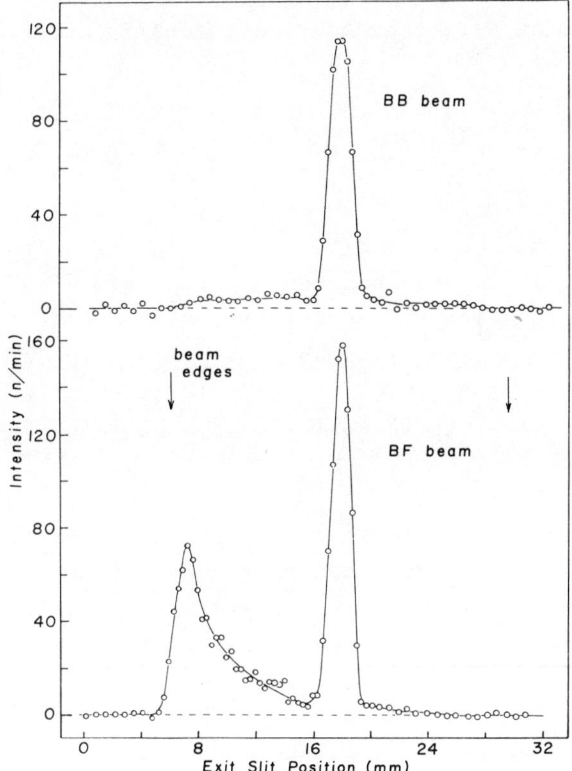


FIG. 4. Distribution of intensity (corrected for background) released from the exit face in the two directions with scanning of an exit slit across the exit face of the interferometer.

2 to 1) are seen in both beams whose widths are accounted for by the scanning resolution which is much larger than the intrinsic width. A weak symmetrically distributed intensity out to the edges is seen in the BB beam and a stronger asymmetric distribution is characteristic of the BF beam. The enhanced distributed intensity in the left side of the BF beam (the same orientation sense as in Fig. 2) is characteristic of both the Green's function approach and the plane-wave ray approach. Likewise the symmetric distributed intensity in the BB beam collapsing to zero intensity at the edges is predicted by both theories. In the plane-wave ray approach it is necessary to assume incoherence between different incident directions

but, of course, coherence between the two wave fields in order to obtain agreement with the observed intensity profiles.

## 3. Interferometer action

As a first demonstration of the interference action of this two-crystal interferometer, the intensity of the central peak in the two beams has been studied by translating a single-step phase plate across the beam between the two crystals. This plate was aluminium with a step height corresponding to a phase difference $3\pi$. The intensity is shown as a function of step position in Fig. 5 with the exit slit held fixed at the central position. The oscillatory intensity patterns, which are complementary in the two beams, demonstrate the presence of interference action.

The character of this oscillation pattern was rather unexpected. It is evident that the observed oscillations are not Pendellösung fringes, simply because the spacing of the observed oscillations is much larger than the spacing of Pendellösung fringes for our crystal thickness. Moreover, we have found difficulty in associating the effect of Pendellösung action with the observations. However, the pattern can be explained phenomenologically by assuming the existence of an intrinsic phase gradient in the interferometer across the Borrmann fan. If this gradient is taken as being constant with the phase changing linearly with $\Gamma$ and of magnitude $\phi_0$ in phase difference between the Borrmann fan edges, the intensity distribution in the BB central peak beam is expected to be

$$I_{BB}(\Gamma_0) = C\left[\int_0^{\Gamma_0}(1-\Gamma^2)^{\frac{1}{2}}\{1+\cos(\phi_0\Gamma)\}d\Gamma + \int_{\Gamma_0}^{1}(1-\Gamma^2)^{\frac{1}{2}}\{1+\cos(\phi_0\Gamma+\pi)\}d\Gamma\right]$$

where $\Gamma_0$ defines the position of the step edge and $C$ is a constant. In this formulation it is assumed that there exists a uniform intensity distribution for all incident directions. Figure 6 shows the theoretical predictions of the BB intensity

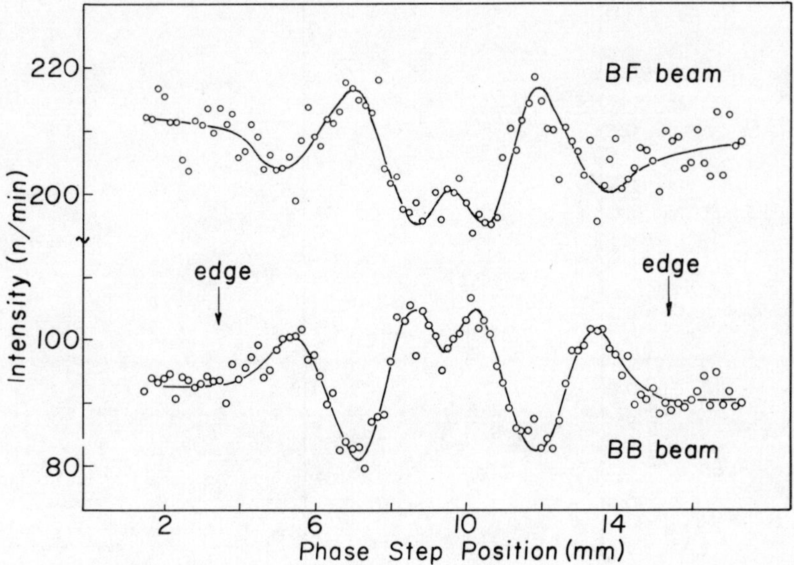


FIG. 5. Intensity in the central peak with scanning of a 3π phase-step plate across the beam travelling between the two crystal plates.

for various values of the intrinsic phase difference $\phi_0$. A similar expression for the BF beam shows that the two beams are complementary to each other. By comparing the measured intensity variation of Fig. 5 with these calculated curves it can be established that the intrinsic phase difference $\phi_0 = 3.7\pi$ was characteristic of the present interferometer. It is seen that the position of the minimum away from the centre is a very useful criterion of this phase difference.

A further test on the reality of this intrinsic phase gradient was obtained by compensating for it through insertion of an aluminium wedge in the beam between the two crystals. The wedge had an apex angle of 1.63° and the effective wedge angle could be varied by tipping around a horizontal axis. With optimum wedge orientation (which agreed with that expected for $\phi_0 = 3.7\pi$) the intensity distribution as a function of step edge position is shown in Fig. 7. This clearly exhibits the required shape from Fig. 6 for a phase-compensated interferometer. It should be mentioned that, although this intrinsic phase gradient arises from the interferometer fabrication or

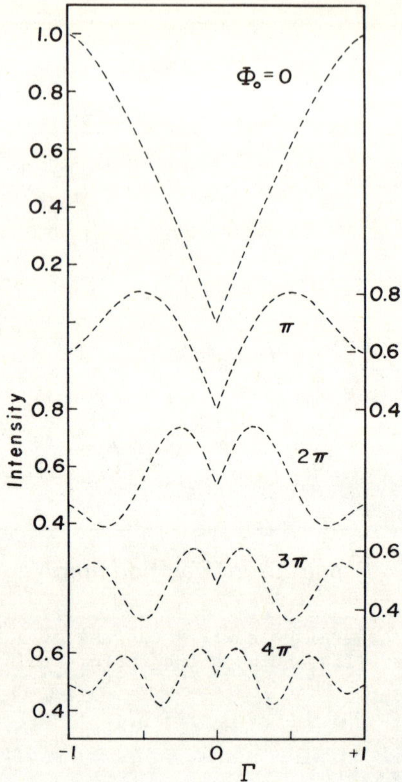


FIG. 6. Calculated variation of central peak intensity (BB beam) with $3\pi$ phase-step scanning as a function of the magnitude of the interferometer intrinsic phase gradient.

means of support, it is also very sensitive to temperature gradients of the order of mK $cm^{-1}$.

As a second demonstration of the interferometer action, the relative phase of the left and right halves of the beam between the two crystals was varied continuously while the compensating wedge was in place. This was accomplished by inserting a parallel-faced aluminium plate of thickness 1.539 mm in half of the beam and tilting it about a horizontal axis perpendicular to the direction of the beam. Figure 8 shows the continuous intensity oscillations obtained, with the exit slit again fixed at the central focus position. The damping of these oscillations arises because the tilt angle of the plate

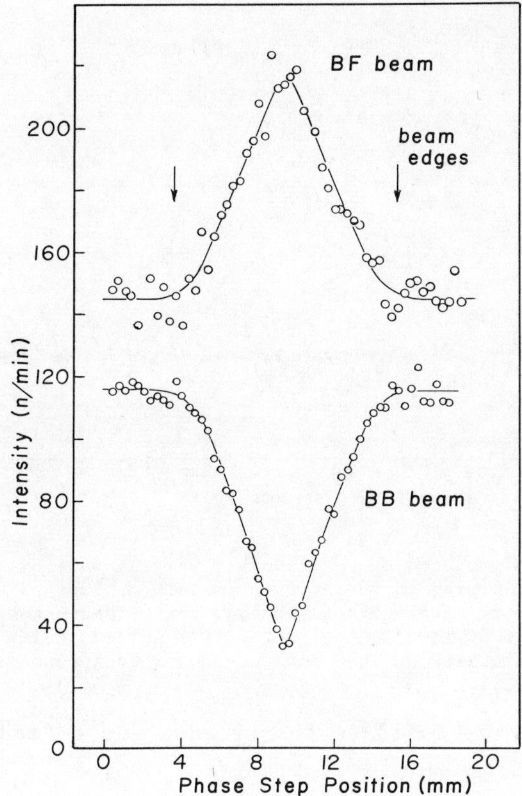


FIG. 7. Intensity in the central peak (corrected for background) with scanning of a $3\pi$ phase-step plate with compensation of the interferometer intrinsic phase gradient by aluminium wedge refraction by $5.6 \cdot 10^{-3}$ arcsec.

was increased continuously and linearly with time, and therefore the thickness and the relative phase did not increase linearly with time. Consequently, data points at increasing aluminium thickness correspond to averages over increasing increments in relative phase, thereby reducing the oscillation visibility. Calculation of the expected damping results in the solid curves shown in Fig. 8.

The approach used to interpret both the intensity distribution from the exit face and the interference profiles with phase-step scanning has been plane-wave ray theory assuming no coherence between rays approaching the interferometer from different directions. Future experiments with smaller entrance

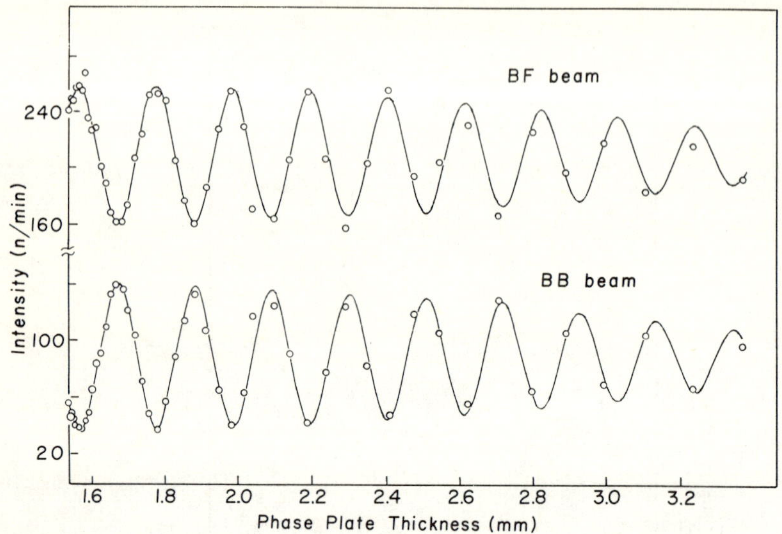


FIG. 8. Fringe pattern produced by continuously varying the relative phase of the two halves of the beam in the interferometer. A plane-parallel plate of aluminium was inserted into half of the beam in the phase-compensated interferometer and tilted around a horizontal axis perpendicular to the beam direction. The intensity has been corrected for background.

slits are expected to elucidate the effects due to coherence between waves of different incident directions.

## Acknowledgements

We wish to acknowledge our indebtedness to J. Callerame who was instrumental in preparing the three-crystal interferometer referred to above, to R. Deslattes and A. Henins for assistance in its fabrication, to D. Atwood for help in the experimentation, and to A. D'Addario for assistance in fabricating both the two-crystal interferometer and the associated apparatus. The research programme reported here has been supported by grants from the National Science Foundation and the Department of Energy.

## References

BAUSPIESS, W., BONSE, U., and GRAEFF, W. (1976). *J. appl. Crystallogr.* **9**, 68.

INDENBOM, V.L., SLOBODETSKII, I., and TRUNI, K.G. (1974). *Sov. Phys.—JETP* **39**, 542.

KATO, N. (1960). *Acta Crystallogr.* **13**, 349.

—— (1961). *Acta Crystallogr.* **14**, 526, 627.

KIKUTA, S., ISHIKAWA, I., KOHRA, K., and HOSHINO, S. (1975). *J. Phys. Soc. Japan* **39**, 471.

PETRASCHEK, D. and FOLK, R. (1976). *Phys. Status Solidi A* **36**, 147.

RAUCH, H. and SUDA, M. (1974). *Phys. Status Solidi* **25**, 495.

RAUCH, H., TREIMER, W., and BONSE, U. (1974). *Phys. Lett.* **47A**, 369.

SHULL, C.G. (1968). *Phys. Rev. Lett.* **21**, 1585.

—— (1973). *J. appl. Crystallogr.* **6**, 257.

SHULL, C.G. and OBERTEUFFER, J.A. (1972). *Phys. Rev. Lett.* **29**, 871.

SUVOROV, E.V. and POLOVINKINA, V.I. (1974). *JETP Lett.* **20**, 145.

# 4 STUDIES ON DYNAMICAL DIFFRACTION PHENOMENA OF NEUTRONS USING WAVE FAN AND TOTAL REFLECTION PROPERTIES

SEISHI KIKUTA

*Institute of Industrial Science, University of Tokyo, Tokyo, Japan*

## 1. Introduction

The diffraction phenomena of neutrons in nearly perfect crystals are in many respects the same as those of X-rays. It is, however, noteworthy that thermal neutrons are only weakly absorbed in most materials while X-rays are very strongly absorbed. Accordingly, the dynamical theory of diffraction for the case of zero absorption can be applied to most cases in neutron diffraction. In Bragg case diffraction the intrinsic diffraction curve has a profile of the so-called silk hat type and a reflectivity $R = 1$ in the angular range of total reflection $-1 \leqslant W \leqslant 1$, where $W$ is a parameter indicating the angular deviation from the diffraction condition. Even if absorption is taken into account, the reflectivity is very close to unity. For instance, in the case of a 111 reflection of silicon with $\theta_B = 45°$, $R = 0.9998$. Further, the partial reflection from the back surface of the crystal can be observed easily. In Laue case diffraction the energy flows of two wave fields inside the crystal have appreciable intensities over the whole angular range of the wave fan (Borrmann fan) even for a thick crystal. When a beam with an angular deviation of $W$ is incident on a crystal with a symmetric Laue case geometry, the paths of the wave fields split into two directions making angles of $\pm\, \Theta$ to the diffracting planes which satisfy the following relation:

$$\tan \Theta = \frac{W}{(1+W^2)^{\frac{1}{2}}} \tan \theta_B$$

It is noticeable that a minute angular change of the incident beam causes an extremely large angular change in the paths of the energy flow in the wave fan. The angular amplification ratio at the exact Bragg condition is

$$G = \frac{2\Delta_0}{\lambda} \sin \theta_B \tan \theta_B$$

where $\Delta_0$ is the extinction distance. $G$ is of the order of $10^5$. In the present case of 111 reflection of silicon with $\lambda$ = 2.44 Å, $G = 1.1 \times 10^5$.

Hitherto experimental studies on the dynamical diffraction phenomena of neutrons have been made by many investigators, such as measurements of rocking curves in various arrangements (e.g. Knowles 1956, Sippel, Kleinstück, and Shulze 1964), observations of energy flows and Pendellösung fringe structures (e.g. Shull 1968) and construction of interferometers (Rauch, Treimer, and Bonse 1974).

In the present paper four studies are described in which the properties of wave fan and total reflection are positively used. First, the Bragg case rocking curves from the front and back surfaces of a plane-parallel plate are observed for plane-wave incidence, the conditions for which are provided by selecting a part of the wave fan. Secondly, a method of confining neutrons for a finite time in a closed orbit by making total reflections consecutively in several crystals is proposed. Thirdly, a method of measuring a minute deviated angle of the beam by using the action of angular amplification of the wave fan is attempted. Finally, a two-component interferometer is constructed in which the spreading of wave fields in the wave fan is utilized for making two coherent beam paths.

## 2. Measurement of the intrinsic rocking curve in Bragg case diffraction

The rocking curve is usually measured by using a double-crystal diffractometer in a parallel setting. To obtain a rocking curve close to the intrinsic curve, the angular spread of the incident beam should be much narrower than the angular range of diffraction from the specimen. This can be achieved by using an asymmetric Bragg case diffraction or by selecting the central part of the wave fan at the Laue case setting. Here the latter method is used (Kikuta *et al.* 1975, 1978).

When a ribbon-like beam impinges on the surface of a crystal with symmetric Laue case geometry and two slits are placed on the entrance and exit surfaces of the crystal opposite to each other, beams which nearly satisfy the exact Bragg condition are selected. If the crystal thickness $t$ is much larger than the slit width $s$, i.e. $s/t \ll 1$, the angular spread of the diffracted beam is given as

$$\Delta W = \frac{s}{t \tan \theta_B}$$

In the present case, 111 reflection of silicon with $\lambda$ = 2.44 Å is used. Under conditions of $t$ = 13 mm and $s$ = 1 mm, $\Delta W$ is estimated at 0.18. Since the angular width corresponding to $W = 1$ is 0.790″, the angular spread of the exploring beam is 0.14″ if the angular spread due to the geometrical diffraction by the slit, i.e. 0.045″, is neglected. In the case of zero absorption, the intrinsic Bragg case diffraction curve has a half-width of $3/\sqrt{2}$ in the $W$ scale. In the present case it is 1.68″ which is a factor of 12 greater than the angular spread of the exploring beam obtained above.

The rocking curve was measured by using a monolithic double-crystal system, which is shown schematically in Fig. 1. The first and second crystals, C1 and C2, are arranged in a parallel setting in which the symmetric Laue case and symmetric Bragg case diffractions take place successively. Two crystals are connected to each other with a thin bridge b so that the minute rotation of the second crystal is carried out by utilizing elastic bending of the bridge. A monolithic system effectively ensures mechanical stability. A plate spring s attached to the edge of the second crystal is pushed at the centre by a micrometer m. The intensities $I_h$, $I_h'$ and $I_d$ of the beam reflected from the front surface, the beam reflected from the back surface and then transmitted through the front surface, and the beam transmitted through the back surface respectively were measured. The rocking curve of $I_h$ has a half-width of 2.0″ and a reflectivity of about unity (Fig. 2). The calculated half-width of the intrinsic curve is 1.68″. The agreement between experiment and theory is fairly good. The rocking curve of $I_h'$ has small peaks at the angles near $W = \pm 1$. When the

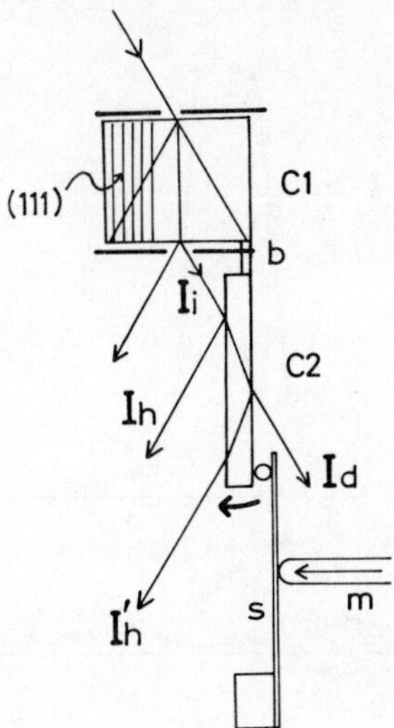


FIG. 1. Monolithic double-crystal system for measuring the Bragg case rocking curve.

incident beam makes angles corresponding to the flanks just outside the range of total reflection, the beam can penetrate inside the crystal, be partially reflected at the back surface, and then be partially transmitted through the front surface. As a result, small peaks are observed. It should be noted that in this case the crystal is 3 mm thick. The rocking curve of $I_d$ is close in profile to a mirror reflection of that of $I_h$.

## 3. Proposal for the construction of a crystal system using multiple total reflection with a Bragg angle of $\pi/2$

In the X-ray case it has been proposed that closed beam paths which form polygons can be obtained by an appropriate choice of crystals, net planes, and X-ray wavelengths (e.g. Bond, Duguay, and Rentzepis 1967). Such crystal systems seem to be more effective in the neutron case, since the reflectivity in the total reflection range is very close to unity and neutron beams of any wavelength can be used. Various arrangements can be considered, e.g. two crystals, three crystals, etc. (Kikuta

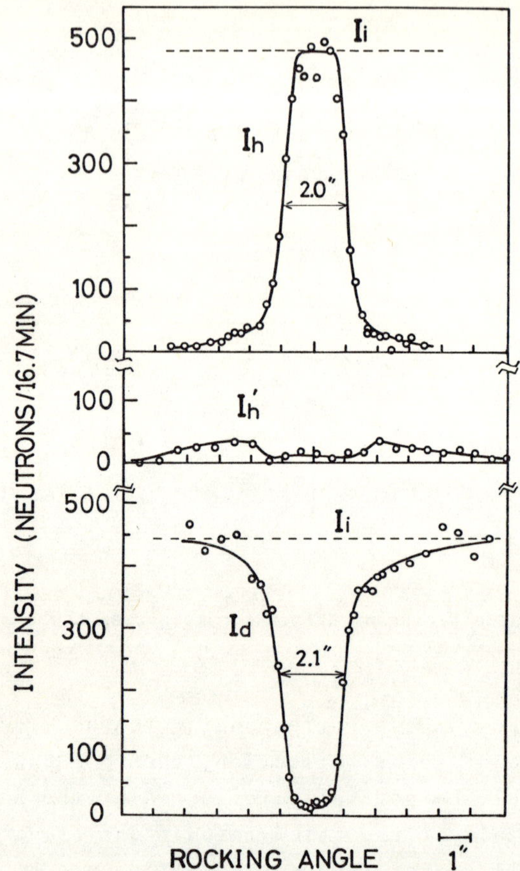


FIG. 2. Rocking curves of the reflected beam from the front surface ($I_h$) and the back surface ($I_h'$), and a rocking curve of the transmitted beam ($I_d$).

1976). In this paper a two-crystal arrangement is discussed because it is the most simple case and has interesting features arising from diffraction of the $\pi/2$ Bragg angle.

When the Bragg angle is $\pi/2$, the angular range of total reflection is given by

$$\Delta\theta = 2(|\psi_h|)^{\frac{1}{2}}$$

while for the usual condition of $\theta_B \ll \pi/2$

$$\Delta\theta = \frac{2|\psi_h|}{\sin 2\theta_B}$$

assuming that the diffracting planes are parallel to the front surface (Kohra and Matsushita 1972). Here $\psi_h$ denotes the Fourier coefficients of the dielectric susceptibility of the crystal for X-rays and corresponds to $\lambda^2 N_c F_h/\pi$ for neutrons. $N_c$ is the number of unit cells per unit volume and $F_h$ the crystal structure factor for neutrons. Figure 3 shows DuMond diagrams for these cases. It should be noted that the angular range of total reflection in the case $\theta_B \approx \pi/2$ is $10^2$–$10^3$ times as broad as the one in the case where $\theta_B \ll \pi/2$. For

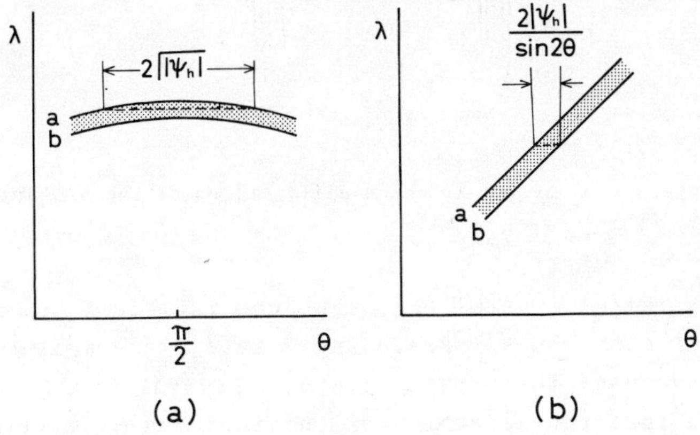


FIG. 3. DuMond diagrams for (a) $\theta_B \approx \pi/2$ and (b) $\theta_B \ll \pi/2$. The region between the two curves a and b represented by $\lambda = 2d \sin\theta \pm d|\psi_h|/\sin\theta$ satisfies the Bragg condition.

instance, in the case of 422 reflection of silicon with $\theta_B = \pi/2$, where $\lambda = 2.217$ Å is used, $\Delta\theta = 710''$. On the other hand, the wavelength spread of the diffracted beam is given by $\Delta\lambda = 2d|\psi_h|$ in the case of $\theta_B = \pi/2$, which is the same order of magnitude as in the case of the usual conditions. In this case $\Delta\lambda = 6.57 \times 10^{-6}$ Å.

The property of the wide angular range of total reflection is useful for the construction of a crystal system using multiple reflection. Figure 4 shows such a system, which involves

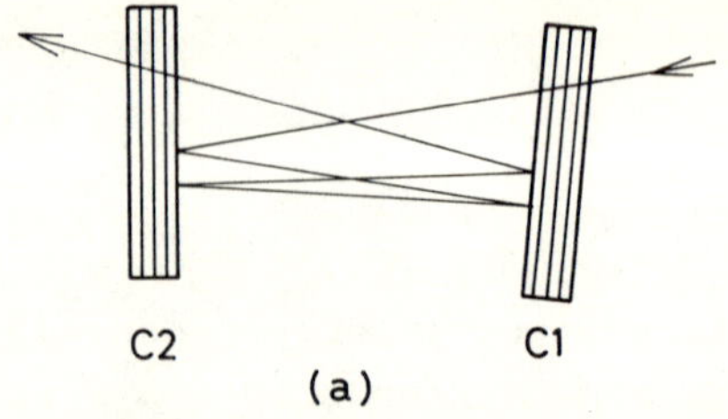


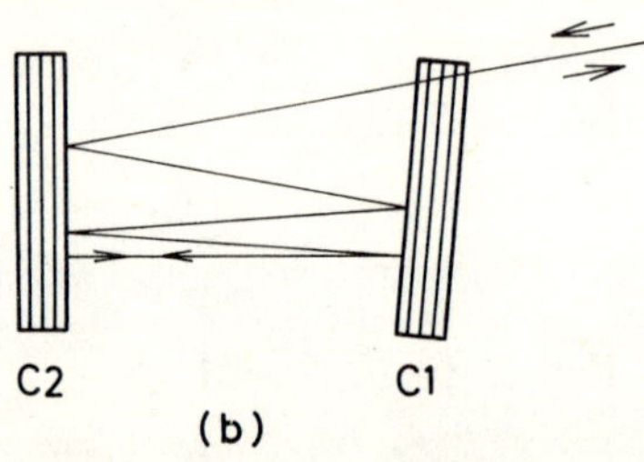


FIG. 4. Double-crystal system in which diffractions of the $\pi/2$ Bragg angle take place consecutively.

two plane-parallel crystal plates placed face to face, and diffractions near the $\pi/2$ Bragg angle take place consecutively on the surfaces of the crystal plates. Crystal C1 is arranged so that the direction of the incoming beam lies just outside the total reflection range. As a result, the beam can be transmitted through crystal C1. Crystal C2 is arranged such that it is slightly inclined with respect to the setting where the surfaces of two crystals are parallel. Then the beam reflects many times between the crystals C1 and C2 as long as the beam falling on each crystal remains within the total reflection range. Finally, when the beam is deviated from the total reflection range of crystal C1 or C2, it can be transmitted through crystal C1 or C2 and leaves from the crystal system. This behaviour is represented by means of a DuMond diagram, as shown in Figure 5. The incident beam is shown as a band parallel to the ordinate for simplicity. The incident beam makes an angle $A$ with the surface normal. If the angle $A$ is outside the range of total reflection, the incident beam can be transmitted

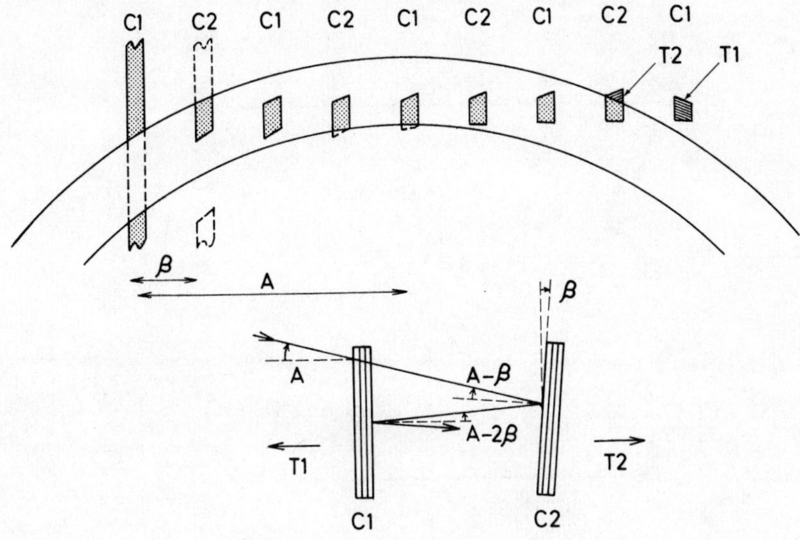


FIG. 5. DuMond diagram illustrating the behaviour of the beam in the double-crystal system expressed in Fig. 4.

through crystal C1. The transmitted beam corresponds to a band outside two curves. On crystal C2, which is rotated by an angle β from the parallel setting, part of the band lying inside two curves can reflect, and in turn we can proceed to another edge of the range of total reflection. Part of the band, T2, leaves C2 and part of the band, T1, leaves C1. Reflections of the neutron beam between two crystals can possibly be repeated at least several tens of times, estimating from the value of the angular range of total reflection.

The proposed experimental arrangements are sketched in Fig. 6. In the first arrangement (Fig. 6(a)) the crystal system is placed between the double-crystal diffractometer. The double-crystal diffractometer is frequently used for the measurement of the deviated angle of the beam produced by a certain effect. In this arrangement the deviated angle of the beam is expected to increase in proportion to the number of times that the beam is reflected between the two crystals and a very minute deviated angle can be measured with a high angular sensitivity. This method will possibly be applied to the search for a neutron charge (Shull, Billman, and Wedgwood 1967).

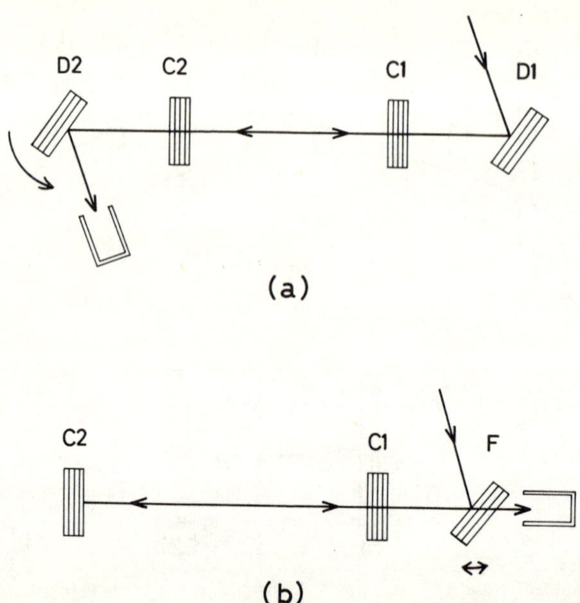


FIG. 6. Proposed experimental arrangements using the double-crystal system expressed in Fig. 4 (a) to improve the sensitivity of detecting minute angular changes of the beam and (b) to make the flight time of neutrons longer.

In the second arrangement (Fig. 6(b)) the fore crystal F is placed in front of the crystal system. If the fore crystal is time modulated by oscillating, it reflects and transmits the beam alternately. Hence the time of flight of neutrons within the fore crystal can be measured. The neutron wavelength is known from the lattice spacing of the crystal. Thus the ratio of Planck's constant to the neutron mass can be determined accurately, since the time of flight is increased by multiple reflection. Similar methods for the same purpose have been proposed previously (Stedman 1968, Weirauch 1975).

In practice it will be a severe problem that the neutron flux available is very weak, since the experimental condition requires a beam with very small wavelength and angle spreads. It should be pointed out that when the beam satisfies the exact diffraction condition of the $\pi/2$ Bragg angle, several simultaneous reflections always occur. For instance, for normal incidence on the 422 plane of silicon, 400 and 220 reflections also take place. This can, however, be obviated by tilting the crystal slightly as in the present case. It is also necessary to tilt the two crystals upwards through several arcseconds to prevent neutrons from falling owing to the gravitational force.

## 4. A method of measuring a minute angular deviation of a beam

The double-crystal diffractometer with parallel setting and Bragg case geometry is used to measure the ray deviations caused by, for instance, the refraction effect due to a prism and to obtain neutron refractive indices. Asymmetric Bragg case diffraction can be used as well as symmetric diffraction in order to improve the angular sensitivity of the instrument. In the present study symmetric Laue case diffraction is used for both crystals in the double-crystal arrangement and the action of the angular amplification of the wave fan is utilized (Kikuta *et al.* 1975). In X-ray diffraction a similar experiment has been made previously (Malgrange, Velu, and Authier, 1968)

The double-crystal system used is shown in Fig. 7. The first and second crystals, C1 and C2, are connected to each other with a groove. The first crystal acts as a collimator and the second as an analyser in angle. An almost parallel beam is obtained by setting slits on the entrance and exit surfaces of the first crystal in the same way as in the experiment described in §2. The 111 reflection of silicon with $\lambda$ = 2.44 Å was used. The thickness of the first crystal was 19 mm and the width of both slits was 0.4 mm so that the angular spread of the diffracted beam is estimated at 0.15 in the $W$ scale or 0.039″. When this beam is incident on the second crystal, which is parallel to the first crystal, the paths of the wave fields are along the diffracting planes because the beam satisfies the exact Bragg condition. If a wedge-shaped specimen is inserted between the two crystals, the beam will be refracted and the paths of the wave fields inside the second crystal split into two directions making different angles to the diffracting planes. The deviated angle of the beam can be obtained from the separation of the paths on the exit surface. In the present case a wedge-shaped crystal of germanium with an apex angle of 20° is used as a specimen, the deviation angle is

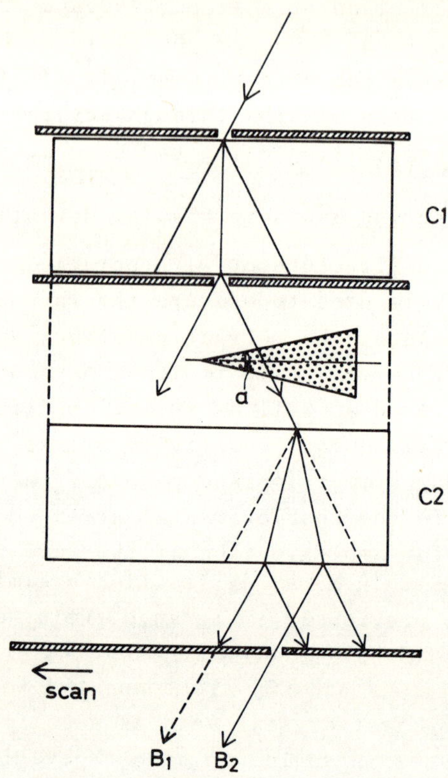


FIG. 7. Monolithic double-crystal system for measuring the deviated angle of a beam by means of the action of the angular amplification in the wave fan.

calculated to be 0.0324″ or 0.125 in the $W$ scale. Accordingly, the paths in the second crystal make an angle $\Theta = \pm 59'$ to the diffracting planes. The separation of the paths on the exit surface is $2t \tan \Theta$ for crystal thickness $t$. In the present case the separation is calculated to be 0.65 mm, since $t$ = 19 mm. This distance was measured by translating stepwise a slit 0.4 mm wide placed behind and parallel to the exit surface. The experimental result is reproduced in Fig. 8. The intensity distribution of the diffracted beam from the crystal system has one peak A at the centre in the case without a specimen. This peak splits into two peaks $B_1$ and $B_2$ on both sides of the centre when the specimen is inserted. The peak heights of $B_1$ and $B_2$ are about half of that of A, as expected.

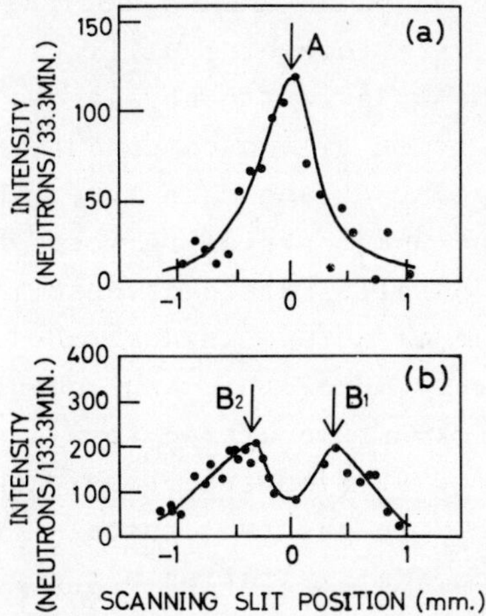


FIG. 8. Intensity distributions of the diffracted beam (a) without the specimen and (b) with the specimen.

The agreement between calculation and experiment is fairly good. This preliminary result suggests that the angular resolution of the system will be improved further if the separation of the paths on the exit surface is increased by using a thicker crystal, a high-order reflection, or a neutron beam of longer wavelength.

## 5. Two-component neutron interferometer

An X-ray interferometer LLL type was developed initially (Bonse and Hart 1965) and various modifications have subsequently been made. A neutron interferometer of the LLL type has recently been constructed (Rauch *et al.* 1974) and the mechanism of interference has been analysed (e.g. Bauspiess, Bonse, and Graeff 1976).

All types of interferometers constructed to date have three or more crystal components. In the present study the construction of a neutron interferometer composed of two crystal components is attempted (Kikuta *et al.* 1970, 1975, 1978, 1979; Hart 1971). In terms of the beam paths in the interferometer the two-component interferometer corresponds to the optical Jamin interferometer while the usual three-component interferometer corresponds to the Mach–Zehnder optical interferometer.

The two-component interferometer utilizes the spreading of the wave fields in the wave fan to obtain two coherent beam paths. There are two types of two-component interferometer, the Bragg case type and the Laue case type. The interferometer of the Bragg case type uses partial reflection from the back surface of the plane-parallel plate as described in §2. In the present paper the Laue case type is described.

The interferometer is monolithic and composed of two thick crystals of equal thickness separated some distance from each other, as shown in Fig. 9. Symmetric Laue case diffractions take place successively in the two crystals. Thick crystals are necessary to produce a separation between the two beam paths which is wide enough to allow the insertion of phase-shifting materials. In the first crystal two paths I and II

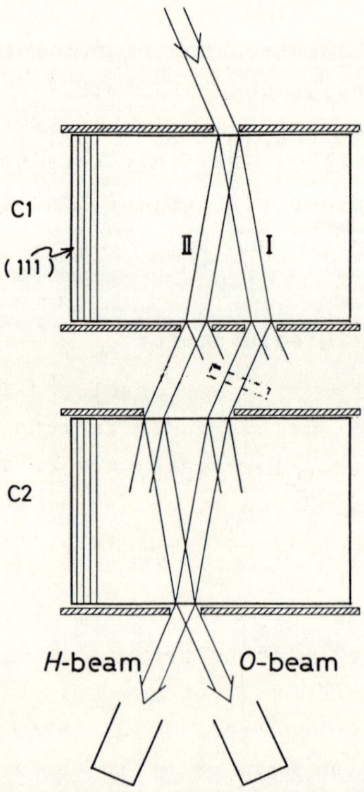


FIG. 9. Two-component interferometer of the Laue case type.

making equal angles to the diffracting planes are selected by slits placed on the exit surface of the crystal. Either the reflected beams or the transmitted beams from the first crystal can be used for interference; the former are used in the present study. Two incident beams impinge on the second crystal with some angles deviated from the exact Bragg angle so that paths I and II in the second crystal propagate in directions making the same angles as in the first crystal. Since the first and second crystals have the same thickness, two of the four paths created in the second crystal overlap on the exit surface and interference takes place. On the exit surface of the second crystal the wave fields split into the forward O beam and the deviated H beam. It is shown that the interference fringe pattern formed in the H beam has a slightly better contrast than that in the O beam and both fringe patterns change complementarily. These situations are the same as in the usual three-component interferometer.

The interferometer was produced from a crystal block of silicon, the thickness of both the first and second crystals was 11.7 mm and the gap distance was 14.2 mm. Symmetric 220 Laue case diffraction was used with $\lambda = 2.44$ Å. Beams having an angular deviation $W \approx \pm 0.5$ were selected by means of the slit system, the slit width being 2 mm. Interference fringes were obtained by inserting aluminium parallel plate $4.980 \pm 0.002$ mm thick in path I between the first and second crystals and rotating it in a step of 0.216°. As shown in Fig. 10, the intensity oscillation of the O beam and H beam exhibit high contrasts and complemental changes. The solid curve in Fig. 10 is determined by fitting the observed points to a sinusoidal curve. As a result the specimen thickness producing a phase shift of $2\pi$ was determined as $0.125 \pm 0.001$ mm and the coherent scattering length of aluminium as $3.45 \pm 0.02 \times 10^{-13}$ cm.

The two-component interferometer has a simple design and can be constructed easily since the requirement is that two crystals have an equal thickness. As the incident beam has a narrow angular spread, the available neutron flux is rather weak.

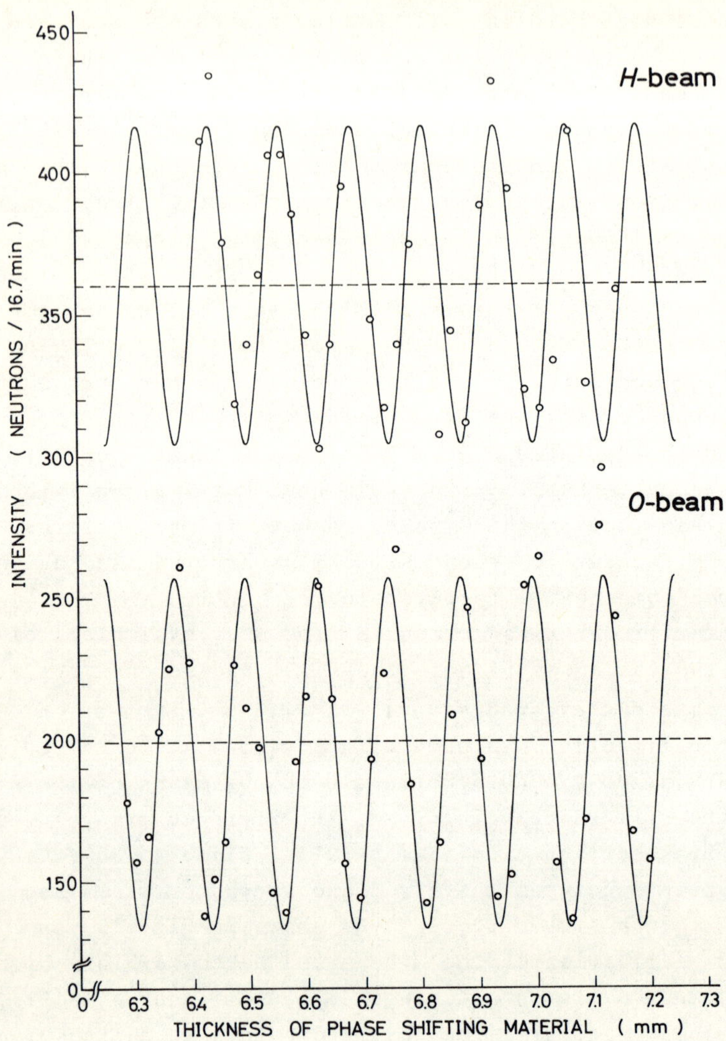


FIG. 10. Intensity modulations of the deviated (H) beam and the forward (O) beam with changing optical paths.

## References

BAUSPIESS, W., BONSE, U., and GRAEFF, W. (1976). *J. appl. Crystallogr.* **9**, 68–80.

BOND, W.L., DUGUAY, M.A., and RENTZEPIS, P.M. (1967). *Appl. Phys. Lett.* **10**, 216–18.

BONSE, U. and HART, M. (1965). *Appl. Phys. Lett.* **6**, 155–6.

HART, M. (1971). *Rep. Prog. Phys.* **34**, 453–90.

KIKUTA, S. (1976). Presented at 31st Annual Meeting of Phys. Soc. Japan, Nagoya.

KIKUTA, S., ISHIKAWA, I., KOHRA, K., and HOSHINO, S. (1975). *J. Phys. Soc. Japan* **39**, 471–8.

KIKUTA, S. and KOHRA, K. (1970). Presented at the 25th Annual Meeting of Phys. Soc. Japan, Tokyo.

KIKUTA, S., TAKAHASHI, T., NAKAYAMA, K., FUJII, Y., and HOSHINO, S. (1978). *J. Phys. Soc. Japan* **45**, 715–6.

KIKUTA, S., TAKAHASHI, T., NAKAYAMA, K., FUJII, Y., and HOSHINO, S. (1978). *J. Phys. Soc. Japan* **45**, 1065–6.

KIKUTA, S., TAKAHASHI, T., NAKAYAMA, K., FUJII, Y., and HOSHINO, S. (1979). *J. Phys. Soc. Japan* **45**, 1024–5.

KNOWLES, J.W. (1956). *Acta Crystallogr.* 9, 61–9.

KOHRA, K. and MATSUSHITA, T. (1972). *Z. Naturforsch.* **27a**, 484–7.

MALGRANGE, C., VELU, E., and AUTHIER, A. (1968). *J. appl. Crystallogr.* **1**, 181–4.

RAUCH, H., TREIMER, W., and BONSE, U. (1974). *Phys. Lett. A* **47**, 369–71.

SHULL, C.G., BILLMAN, K.W., and WEDGWOOD, F.A. (1967). *Phys. Rev.* **153**, 1415–22.

SHULL, C.G. (1968). *Phys. Rev. Lett.* 2, 1585–9.

SIPPEL, D., KLEINSTÜCK, K., and SCHULZE, C.E.R. (1965). *Phys. Lett.* **14**, 174.

STEDMAN, R. (1968). *J. sci. Instrum.* (*J. Phys. E.*) **1**, 1168–70.

WEIRAUCH, W. (1975). *Nucl. Instrum. Methods* **131**, 111–17.

# 5. THE D18 DIFFRACTOMETER FOR NEUTRON INTERFEROMETRY AT THE ILL

W. BAUSPIESS

*University of Dortmund, Federal Republic of Germany;*
*Institut Laue-Langevin, Grenoble, France*

## 1. Introduction

Three things are needed for neutron interferometry experiments: an interferometer (a crystal in the case of Bragg diffraction interferometry), a neutron source, and a device to select and handle the neutrons that will be used. In this paper this last essentially technical aspect of neutron interferometry will be discussed, using as an example the new diffractometer for neutron interferometry that is being built at the ILL. Experience gathered during operation of a prototype version for several years (Bauspiess, Bonse, and Rauch 1976, Rauch *et al.* 1976 Badurek *et al.* 1976, Bauspiess 1977) has been used in its design. The new diffractometer, called D18, will be operational in the near future, and will then be available to 'external' users who wish to carry out neutron interferometry but do not have a high flux reactor.

We cannot yet present any results of performance tests but have to describe the characteristics of the interferometer as inferred from the design. Experimental results presented in this paper refer to experiments with the prototype machine or are extrapolations from those experiments.

## 2. Instrument position and beam path through the instrument

The diffractometer will be placed at the first position of a neutron guide. Of the total height of the guide (20 cm), the lower third (more accurately 5.5 cm) is available to the instrument. The guide (ILL designation H25) has a horizontal curvature of 19 km, the width of the guide is 3 cm, and its characteristic (cut-off) wavelength is calculated to be $\lambda$ = 1.5 Å. The neutron flux at the end of this guide has been calculated to be (Ageron 1972)

$$\phi = 8 \times 10^8 \text{ neutrons cm}^{-2} \text{ s Å} \qquad \text{at} \quad \lambda = 2 \text{ Å}$$

Experiments indicate a somewhat lower value (about one-third of the calculated value). As the upper part of the guide is passed on to other instruments, the interferometer cannot be placed in direct line with the guide. Instead, a fore crystal is used to deflect part of the beam towards the interferometer position (Fig. 1). This fore crystal will in most cases

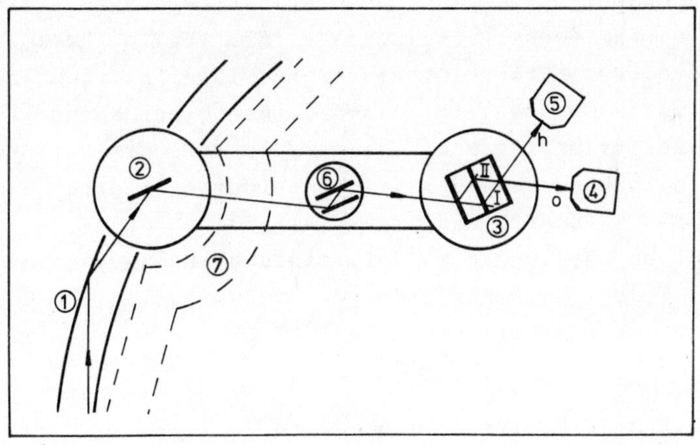


FIG. 1. Beam path within the diffractometer: 1, neutron guide; 2, fore crystal; 3, interferometer; 4, 5, neutron detectors for O and H beams; 6, intermediate (germanium-silicon bicrystal) monochromator; 7, protection shield.

be a silicon crystal with lattice planes oriented parallel to those of the interferometer in order to use most efficiently the neutron intensity available in the small volume in $k$ space that can be accepted by the interferometer.

In the interferometer the beam is coherently split up into two partial beams I and II and recombined by successive Bragg reflections. The intensity behind the interferometer, which contains the information of the rleative phase of the beams I and II, is recorded in two neutron detectors. The sum intensity counted by both detectors is expected to be of the order of

$1.5 \times 10^3$ neutrons $cm^{-2}$ $s^{-1}$ for a neutron wavelength of 2 Å.

The divergence of the neutron beam within the guide (about 10′ for 2 Å neutrons), together with the chosen fore crystal reflection (*hkl*) determines the particular wavelength band that is passed on to the interferometer. A narrowing of this band is often necessary for exact wavelength definition, partly because the spectrum has been found to be very inhomogeneous and partly because the measurements at high orders of interference (Rauch *et al.* 1976) or the use of dispersive samples (Graeff *et al.* 1978) require a certain degree of monochromaticity. To obtain such a narrowing of the wavelength spectrum a groove-shaped perfect crystal with a lattice constant different from that of the fore crystal can be inserted in the beam path between fore crystal and interferometer (as 'intermediate' monochromator). One can, for example, use a germanium crystal with the same orientation as the silicon fore crystal, reducing the wavelength band to widths of the order of $1 \times 10^{-3}$ in $\Delta\lambda/\lambda$ (Bauspiess, Bonse, and Graeff, 1977).

The neutron wavelength is determined by using the interferometer crystal as a spectrometer and comparing the positions of different Bragg reflections.

## 3. Principal diffractometer components

The use of several perfect crystals in the instrument requires high stability of the crystal orientation with respect to each other. In particular the rotation about the vertical ($\theta$) axis has to be controlled, as the width of the rocking curves is of the order of arc seconds. Each crystal has also to be aligned about the horizontal axis parallel to the reflecting lattice planes, otherwise the interferometer will not accept all of the beam height or, in the case of the intermediate monochromator, the centre wavelength will change over the height of the beam. To achieve the stability required, the goniometer units carrying the crystals are mounted onto a common steel beam – a 'neutron' optical bench (Fig. 2).

The distance between fore crystal and interferometer is 2 m, to leave sufficient space for collimators, beam polarizers, etc., as well as for 30 cm of protection shield around the

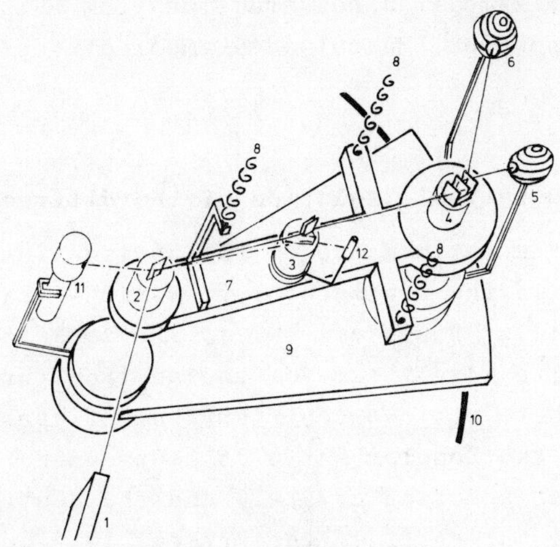


FIG. 2. Schematic drawing of D18 diffractometer: 1, neutron guide; 2, fore crystal goniometer; 3, goniometer for bicrystal monochromator; 4, interferometer goniometer; 5, 6, neutron detectors for O and H beams; 7, diffractometer bench; 8, vibration isolation springs; 9, platform; 10, rail; 11, X-ray tube; 12, X-ray detector. Sample table, diffractometer support, and fore crystal shielding are not shown.

fore crystal.

Each goniometer unit allows alignment of the crystal mounted onto it about all three axes, with a resolution of 1/400°, given by the increment of the stepping motors. Additionally, the interferometer goniometer and that for the intermediate monochromator contain a high-resolution θ drive with step resolution of $1.5 \times 10^{-3}{}''$ and $2 \times 10^{-2}{}''$ respectively. The fore crystal and the interferometer can also be translated in the horizontal plane with a displacement increment of 1 μm.

The diffractometer is suspended from an iron construction which, in turn, rests on a platform. This platform can be rotated about the fore crystal axis to allow the variation of the scattering angle of the fore crystal (i.e. variation of neutron wavelength or reflection index *hkl*). The heavy neutron detectors are not connected directly to the diffractometer, but also rest on the platform where they can be rotated about an axis concentric to the interferometer axis. A horizontal goniometer is provided to carry the sample holder and other sample-related equipment. This goniometer is fixed (20 cm above the beam level) to the same iron construction from which the diffractometer is suspended. Manual adjustment of the rotation axis is possible.

## 4. Vibrational and thermal isolation of the diffractometer

It is known that the performance of the neutron interferometer falls off rapidly in the presence of vibrations (Bauspiess, Bonse, and Rauch 1976, Bauspiess 1977). The cause of this effect is the finite flight time of the neutron through the interferometer. There are two contributions of approximately equal importance: the Doppler shift of neutron energy due to Bragg reflection at a moving crystal, and the neutron wave phase shift due to the shift of the crystal lattice. One can show that the overall apparent phase shift is given by

$$\Delta\phi = (\mathbf{a}\cdot\mathbf{g})t_0^{\,2}$$

where $\mathbf{a}$ is the acceleration, $\mathbf{g}$ the reciprocal lattice vector ($|\mathbf{g}| = 2\pi/d$), and $t_0$ is the flight time of a neutron between two interferometer plates. This estimate is valid for low frequencies (less than 1 kHz) and in that region independent of frequency. For neutrons with a wavelength of 2 Å and an interferometer with plate distances of 30 mm, the time of flight is 16.8 μs, so that the acceleration due to vibration has to be kept below $|\mathbf{a}| = 1$ cm s$^{-2}$ (approximately $10^{-3}$ g) if it is not to influence the interference contrast.

At the beam position of the instrument high-level vibration amplitudes with a main component of 25 Hz exist on the ground floor, presumably from the compressors of the nearby tritium extraction plant. Isolation from these vibrations is necessary. As a possible solution, we chose to suspend the diffractometer by means of three long springs, yielding a proper frequency of 0.7 Hz, together with a damping system based on the immersion of rigid plates in a viscous medium.

It should also be mentioned that the vibrations generated on the diffractometer bench itself, such as excitations by stepping motors or flexibly mounted large masses that can act as vibration amplifiers, also have to be controlled. For this reason a special current-controlled motor drive unit has been developed (Falaise 1977). Under certain conditions even acoustic vibration coupling has been observed.

Owing to the suspension by springs, the position of the diffractometer is strongly dependent on the load distribution on the bench. Therefore the upper suspension points of the springs can be moved up or down, controlled in a feedback loop by linear displacement sensors, monitoring the horizontality of the diffractometer bench.

As well as vibrational isolation good thermal stability is another 'environmental' requirement for successful operation of the instrument. Even with accurate mechanical design and manufacture, the 1/100″ stability of the different components can only be maintained if thermal gradients on the diffractometer bench are carefully avoided. In addition, quick changes in ambient temperature around the interferometer (as they are caused by air currents) may cause distortion of the interferometer crystal, which can induce a time-dependent fringe pattern in the interfering beams leading to a decrease in interference contrast even if the position within the rocking curve is not changed.

As an active temperature control is difficult to provide in the presence of large masses of concrete shielding and is also liable to create permanent small local temperature fluctuations, we decided to use a thermal isolation only. The whole experimental site is enclosed with walls covered with glass fibre plates, so that the temperature inside this 'cabin' follows the outside temperature slowly but quick changes are suppressed, particularly the 24 h temperature cycle. Preliminary experiments with the prototype diffractometer yielded drift values of 1″ per day, and we think that this figure can be improved further.

## 5. Neutron detectors and monitors

The two main detectors in the path of the interfering beams that leave the interferometer are cylindrical helium-3 detectors with a window at one end of the cylinder. The detectors which are 10 cm long and 5 cm in diameter are filled with helium-3 at a pressure of 3 bar.

Conventional neutron monitors such as fission chambers cannot be used to monitor the neutron beam because of the low intensity reflected from the (perfect) fore crystal. One must also be very careful about placing any scatterer in the beam path between fore crystal and interferometer because even neutrons scattered at very low angles will no longer be accepted by the interferometer.

Several monitoring methods are planned for the new instrument. Each of them has certain advantages.

(1) In the standard monitoring mode 20 per cent of the beam will be shielded from the interferometer. This region will instead be used by a small (1 cm diameter, 8 bar pressure) helium-3 detector monitoring the intensity reflected from the fore crystal.

(2) Another method of monitoring will use one of the 'parasite' beams of the interferometer, i.e. one of the beams that has undergone only one Bragg reflection. The intensity for the monitor will be lower, but the full beam height can be used for the interfering beams. Moreover, a possible decrease in intensity due to a drift off the centre of the rocking curve is accounted for by the monitor.

(3) Finally, one can use the sum intensity of the two main detectors as a monitor and later on evaluate the difference count rate only. This, of course, is only possible when both detectors 'see' exactly the same parts of the beam cross-section.

Alternatively, methods (1) and (2) can also be used to check the crystal alignment by detecting any drift off the peak of the rocking curve.

## 6. X-ray equipment

An X-ray tube (molybdenum anode) is mounted near the fore crystal and can be rotated around it. The X-ray beam will mainly be used for the stabilization of the silicon–germanium bicrystal monochromator (Bauspiess, Bonse, and Graeff 1977) by monitoring the parallelism of the silicon base crystal with respect to the fore crystal, so that the angle between the germanium groove and the fore crystal is constant and only defined by the angle at which the groove is glude to the base crystal.

Because of the narrow spectral width of the X-ray line, it can also be used as a wavelength reference either for calibration of the instrument or when operating the interferometer in a dispersive setting with respect to the fore crystal. The high voltage for the X-ray tube is supplied by a voltage- and current-controlled X-ray generator with a maximum power of 2.4 kW.

## 7. Instrument control and data acquisition

In view of the sensitivity of the diffractometer towards any disturbances from outside, a control system has been designed to allow a maximum of remotely controllable functions while maintaining high flexibility in instrument configuration (Fig. 3).

Each of the 25 motors has its own control unit, and, apart from the high-resolution drives, the position of each goniometer element can be read by an absolute linear or angular encoder. There are four independent scaler systems (one master and two slave scalers each) and the input sources to these scalers can be program selected. Two of the scaler systems will be used for continuous alignment check of the interferometer and (optionally) of the intermediate monochromator in a feedback loop with the corresponding high-resolution $\theta$-drive motors.

All these control elements are linked together by a Motorola 6800 microprocessor. This microprocessor contains in its memory the necessary driver routines as well as status tables and software limits for each goniometer element. The commands for the mircoprocessor can be entered at different levels in order to allow full use of the instrument flexibility.

(1) Via the front panel of the control cabinet: each motor and encoder is addressable by pushbuttons, the step frequency can be varied manually, and a single-step mode is possible. The reading of the encoders is displayed on an 80-character alphanumeric display each time the corresponding button is pressed.

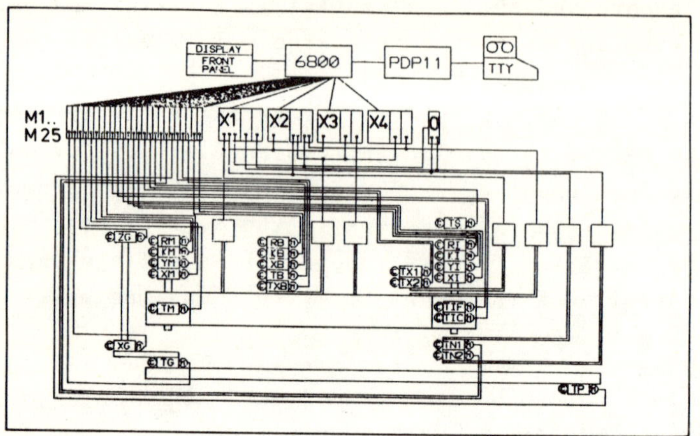


FIG. 3. Instrument control elements: M1—M25, motor drive units; C1—C25 (not drawn), absolute encoders; X1—X4, scaler systems with program selectable inputs; O, OR gate for sum intensity of neutron detectors; TP, rotation of platform; TN1, TN2, rotation of neutron detectors; TG, rotation; XG, ZG, horizontal and vertical translation of X-ray tube; TM, rotation about vertical axis; XM, YM, translation; KM, RM, rotation about horizontal axes of fore crystal; TB, rotation about vertical axis (high resolution); XB, translation; KB, RB, rotation about horizontal axes of bicrystal monochromator; TXB, rotation of X-ray detector; TIC, TIF, coarse and high-resolution rotation about vertical axis; XI, YI, translation; KI, RI, rotation about horizontal axes of interferometer; TX1, TX2, auxiliary goniometers concentric to the interferometer; TS, rotation of sample.

(2) Via a Silent 300 terminal: a simple control language has been developed to address single units, to select the configuration of a scaler system, to initialize a stabilization loop, etc. An example is given in Fig. 4. The Silent terminal is equipped with a dual cassette tape drive; one of the cassettes can contain a sequence of commands; while data can be recorded on the other.

(3) As an 'intelligent interface' to a micro 1 computer (built around a LSI 11-microcomputer card): this computer will facilitate user-programmed instrument control as it allows Fortran level programming and text editing in a foreground/background operating system. Data output will still be onto the cassette of the Silent terminal.

The cassette tape written by the Silent terminal can then be read by a similar machine connected to the ILL's central PDP10

| | positioning a motor: |
|---|---|
| MSP M1 20 | select velocity of motor M1 |
| MIN M1—1200 | preset step count for motor M1 |
| MOP M1 S | start motor M1 |
| CRD C1 | read encoder C1 (corresponds to motor M1), contents will be displayed on the terminal |
| | starting a scaler system: |
| XIN X1 3 2 2 | select inputs for scaler system X1: channel 3 of preset unit, channel 2 of slave scaler 1, and channel 2 of slave scaler 2 |
| XPR X1 100000 | set preset count of master scaler of scaler system X1 |
| XOP X1 S | start scaler system X1 |
| XRD X1 | output contents of slave scalers of scaler system X1 on the terminal |
| | further commands for: cassette — I/O, definition of the configuration and start of a scan or a stabilization loop. |

FIG. 4. Sample command sequence for direct control of the microprocessor.

computer, for data format conversion, reduction, analysis, or transfer to IBM-format magnetic tapes.

The diffractometer has already been constructed by the Institut für Physik, Universität Dortmund; the platform and diffractometer suspension and the electronic control system, both developed by the ILL services, are currently under test. Installation is planned for the reactor shutdown periods of June and August 1978. We expect that by the end of 1978 the D18 instrument will be operative.

## Acknowledgments

The help of U. Dretzler and G. Ernst, Dortmund, and of G. Gobert, Grenoble, in constructing the instrument is gratefully acknowledged. This work was supported by the Bundesminister für Forschung und Technologie, 03-45A 09 I and 03-41A 09 P.

## References

AGERON, P. (1972). *Bull. Inf. sci. tech. Commis. Energ. At. (Fr.)* **166**, 5.

BADUREK, G., RAUCH, H., ZEILINGER, A., BAUSPIESS, W., and BONSE, U. (1976). *Phys. Rev. D* **14**, 1177.

BAUSPIESS, W. (1977). Doctorate Thesis, Dortmund.

BAUSPIESS, W., BONSE, U., and GRAEFF, W. (1977). *J. Appl. Crystallogr.* **10**, 141.

BAUSPIESS, W., BONSE, U., and RAUCH, H. (1976). The perfect crystal neutron interferometer: A tool for novel and precise measurement. *In Proc. Conf. on Neutron Scattering, Gatlinburg, Tenn.* (ed. R.M. Moon), Vol. 2, p.1094. National Technical Information Service, Springfield, Virginia.

FALAISE, J.C. (1977). ILL Tech. Rep. 77148T, Instiut Laue—Langevin, Grenoble.

GRAEFF, W., BAUSPIESS, W., BONSE, U. and RAUCH, H. (1978). *Acta Crystallogr.* A **34**, S238.

RAUCH, H., BADUREK, G., BAUSPIESS, W., BONSE, U., and ZEILINGER, A. (1976). Determination of scattering lengths and magnetic spin rotations by neutron interferometry. *In Proc. Int. Conf. on the Interactions of Neutrons with Nuclei, Lowell, Mass.* (ed. E. Sheldon), Vol. II, p. 1027. University of Lowell, Lowell, Mass.

# 6. EXPERIENCE WITH THE PRODUCTION OF SINGLE-CRYSTAL INTERFEROMETERS

W. TREIMER

*Fritz-Haber-Institut der Max-Planck-Gesellschaft Berlin, Faradayweg 4-6, D-1000 Berlin 33, Federal Republic of Germany*

The first successful operation of a neutron interferometer (Rauch, Treimer, and Bonse 1974, Bauspiess *et al.* 1974, Treimer 1975) and the series of measurements whichwere made (Rauch *et al.* 1976, Treimer 1978) resulted in production of further perfect interferometers.

Neutron interferometry requires not only high quality crystal material, but also accurate, correct, and proper handling of the construction and production of working interferometers. Because of the lower intensity of neutron fluxes compared with fluxes obtained from X-ray sources neutron interferometers must usually be larger than X-ray interferometers. This means that there is a greater probability of lattice deformations such as dislocations and strain fields being present. (Treimer, 1978.)

An LLL single-crystal interferometer consists of three plates which are cut from a large dislocation-free perfect single crystal (diameter approximately 8 cm). The first crystal plate (splitter) splits the incoming neutron beam into two widely separated coherent partial beams which are diffracted by the second crystal plate (mirror). Thus four beams are produced, of which two converge and form a standing interference pattern. This pattern arises from coherent superposition of these two beams which have a fringe spacing of ångström range. This is not detectable but by moiré mapping with the third crystal plate large interference fringes are produced. This moiré mapping can be observed using photographic plates or detectors.

The beam cross-section (the area which the detector 'sees' behind the interferometer) should be as phase homogeneous as possible, particularly for measurments with detectors. Because of averaging over the whole pattern (which in the ideal case consists of only one large fringe) interference oscillations due to phase shifts are obtained if and only if the whole beam cross-section has the same phase. Otherwise the resulting signal is smeared out and interference contrast vanishes. Therefore neutron interferometers are first tested with X-rays. Testing with X-rays means different conditions for interference contrast. Because of the anomalous absorption of X-rays in thick crystals only one wavefield propagates through the crystal with rather low absorption; the other wavefield is absorbed. The transmission path through all three crystal plates is the path of lowest absorption which is parallel to the net planes. Unlike X-rays, neutrons are not usually absorbed by perfect crystals such as silicon or germanium. Because of this fact both wavefields of the dispersion surface are present in equivalent quantities in the crystal plates and both contribute to interference. The intensity is spread out over the whole Borrmann fan owing to the incoming divergence of the neutron beam which means that the sensitivity of the interferometer to point defects or small strain fields is greatly increased. The volume of the crystal penetrated by neutrons is larger than in the case of X-rays.

Therefore before buying crystal material two slices (approximately 10 mm thick) were cut from each end of the crystal cylinder and examined for lattice defects by Lang and double-crystal topographies and then by interference topography. Lang and double-crystal topography show that the crystal material is free of dislocations and coarser defects. These topographs should not show any contrast on the film plate. Mo $K_{\alpha 1}$ radiation is usually used. To ensure that the crystal is free of any disturbing inhomogeneities a small interferometer is cut from the test slices (Fig. 1). Each crystal material (including perfect crystals) deviates by $\Delta d/d$ from ideal lattice geometry as a result of striations, fluctuations of dopant concentrations, etc. Usually these defects appear all over the crystal (cylinder) but they become larger near the edge zone, i.e. near the lateral area. The small test interferometer should therefore be cut from a region where these $\Delta d/d$ variations are supposed to be large compared with regions of 'perfect' net planes (Fig. 2).

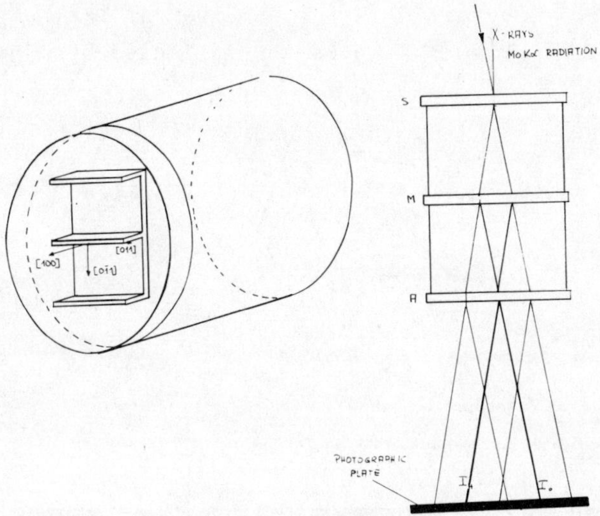


FIG. 1. Crystal cylinder showing the position of the small interferometer; left, beam paths in the interferometer.

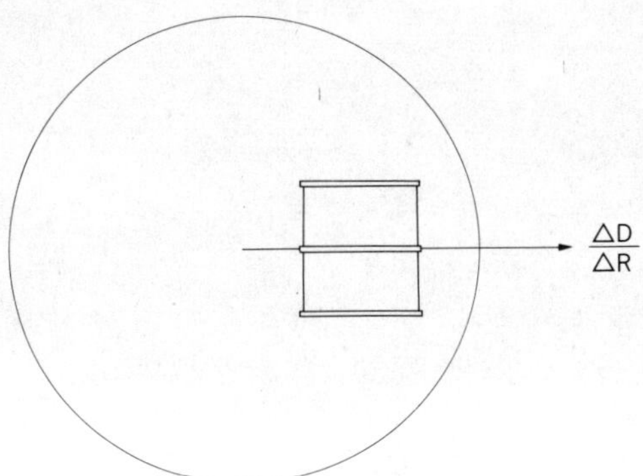


FIG. 2. Region from which the test interferometer should be cut.

The test interferometer is the same size as the test crystal plate, which is not necessarily because these defects exist symmetrically around the growing axis. The test interferometer (enlarged in Fig. 1) had the following dimensions: length, 30 mm; width, 26 mm; crystal plates height, 6 mm. After etching the interferometer was put in a conventional Lang apparatus and fixed on a suitable appliance. Then the crystals and the film plate were moved in the same way as for Lang topography (Fig. 3). Figure 4 shows the reflected beam behind

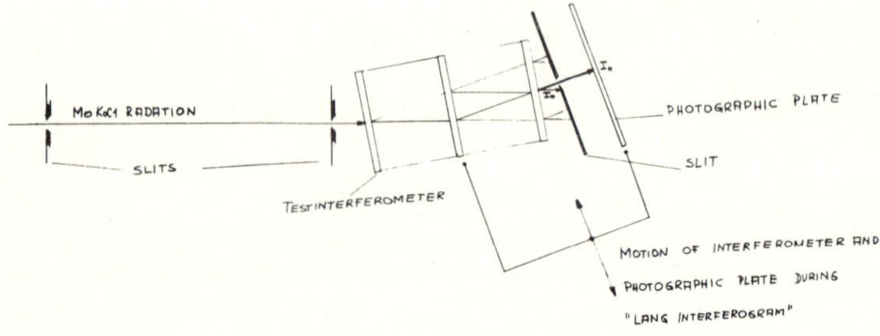


FIG. 3 Arrangement for the 'Lang interferogram'.

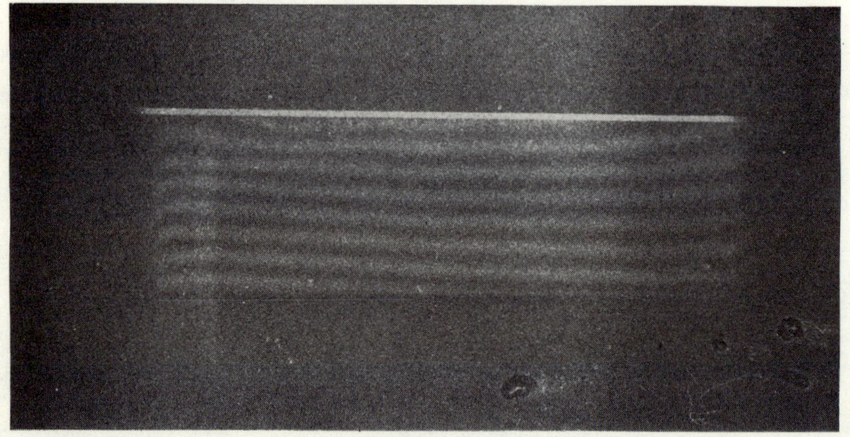


FIG. 4. Lang interferogram; Mo $K_{\alpha 1}$ radiation; rotational moiré; height, 6 mm; length, 21 mm.

the interferometer.

The crystal was elastically deformed so nevertheless rotational moiré patterns appeared. Apart from this the interferogram is free of other phase contrasts which means that this crystal material is suitable for a neutron interferometer. It should be mentioned here what kind of image pattern one can expect. Two possibilities can be distinguished: (1) the crystal material is perfect and (2) the crystal material has inherent point defects, strain fields or variations of the lattice constant, and similar defects. If the test slices show difficult, inherent, and inhomogeneous interference patterns (Fig. 5) the material is not suitable for neutron interferometry. However, even if the crystal is perfect it is not necessarily a working interferometer (Treimer, 1975). The interference image may possibly contain fringes which, however, can be balanced. Interference fringes (of perfect crystal material) can arise owing to rotational moiré and defocusing effects (different optical path

FIG. 5. Interferogram of a test interferometer showing patterns produced by strain fields and $\Delta d/d$ fluctuations.

lengths in the interferometer). Rotational moiré fringes will be observed if the standing interference pattern is rotated with respect to the parallel position of the Bragg planes of the third crystal plate. Rotational moiré mappings are shown in Figs. 6 and 7. Note that a rotation of less than 0.001″ changes the maximum of the interference beams of Fig. 6 into the minimum of Fig. 7. Because of the overlapping of the two reflected

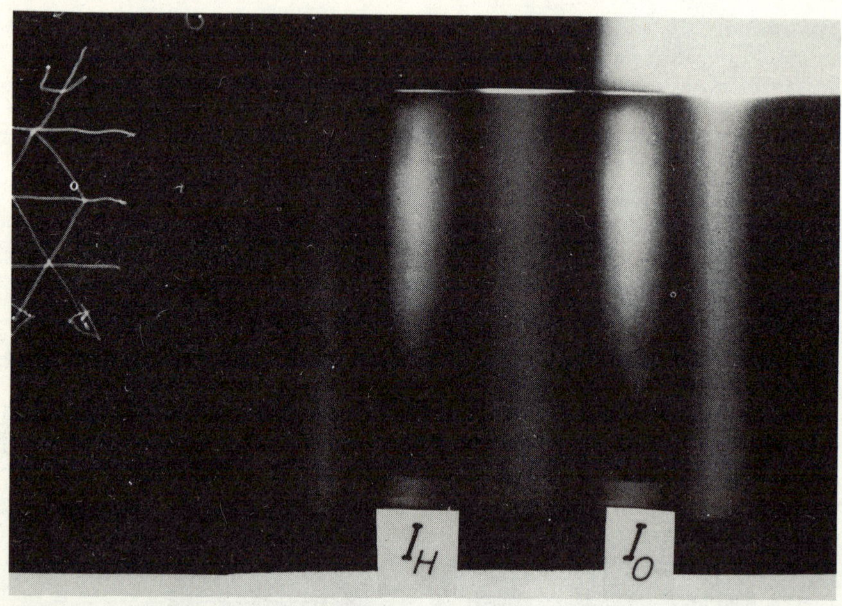


FIG. 6. X-ray interferogram with five reflexes (two are overlapping): radiation, Mo $K_{\alpha 1}$; height, 44 mm; width of one reflex, 10 mm. Note the increased intensity of the interference reflexes.

beams only five reflexes appear on the photographic plate. Figure 8 shows a perfect interferogram in which the interference reflex does not contain any phase structure.

To prove the interference ability an aluminium wedge was inserted into one of the beam paths in the interferometer. Therefore a continuous phase shift was created which produced a system of fringes of constant wedge thickness (Fig. 9). The slope of the fringes is due to inexact alignment of the wedge.

It is a necessary (but not sufficient) condition that a

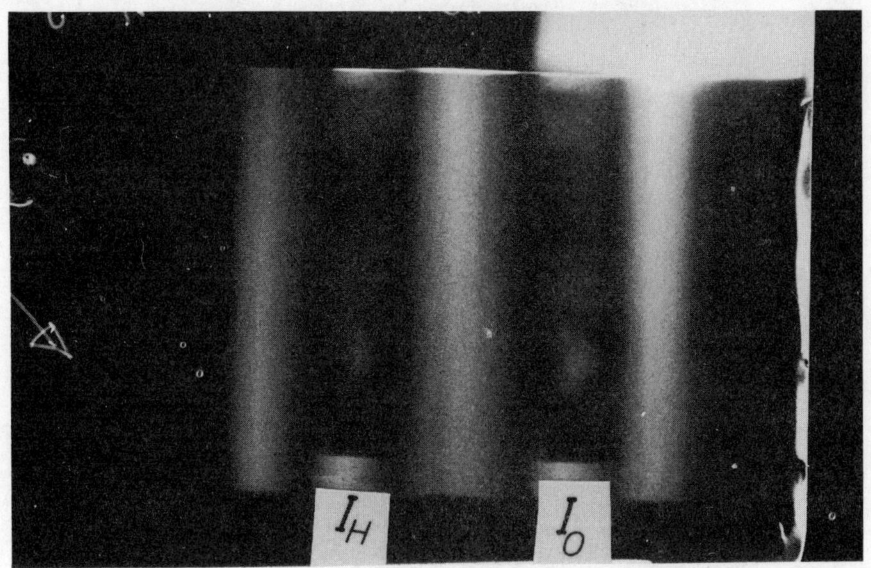


FIG. 7. As in Fig. 6 but with the analyser crystal rotated about 0.001″. Note that the interference reflexes are very weak because they are at a minimum.

neutron interferometer first has to work as an X-ray interferometer. For that reason these tests are important before starting neutron interferometry because a larger definite phase homogeneous area is needed. To estimate the two effects described above two examples are given.

The defocusing effect is rather small and can easily be managed. If $\Delta t$ is the difference between the optical path length in the interferometer the vertical pattern has a spacing (Petraschek and Rauch 1976)

$$L_{\mathrm{D}} = \frac{3D\Delta_0 tg\theta_{\mathrm{B}}}{\Delta t}$$

where $\Delta_0$ is the Pendellösung period, $\theta_{\mathrm{B}}$ is the Bragg angle and $D$ is the thickness of the plates. If $D \approx 3.5$ mm, $\Delta_0$ (silicon, $\lambda_0 = 2.0$ Å) $= 0.641 \times 10^{-2}$ cm and $\Delta t \approx 30$ μm, $L_{\mathrm{D}}$ is 1.7 cm. The accuracy of the optical path lengths after cutting the interferometer from the silicon cylinder is approximately 40 μm, after polishing it is better than 2 μm and after etching it is

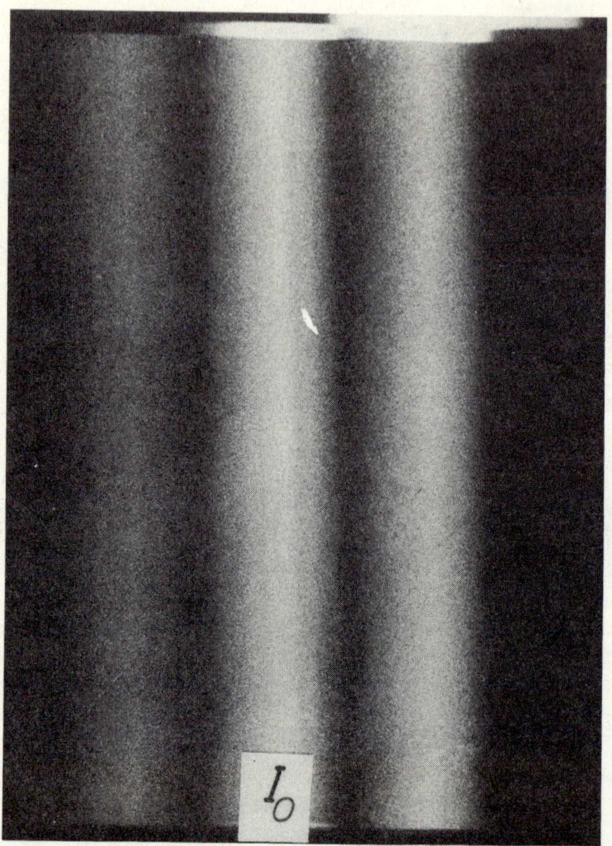


FIG. 8. A perfect homogeneous interference pattern.

of the order of 10–20 μm. The plates are usually trapezoidal which means that there is a further defocusing effect but this is smaller than the λ thickness of the crystal or the Pendellösung period $\Delta_0$.

The crystal must be supported carefully when the interferometer is cut and polished, which was done by a surface grinding machine. The crystal can be slightly elastically deformed if it is not properly fixed on its support. Therefore the 'free' interferometer is often defocused which again means that interference patterns occur.

Rotational moiré patterns easily arise if the (perfect) interferometer is badly mounted. The spacing due to this kind

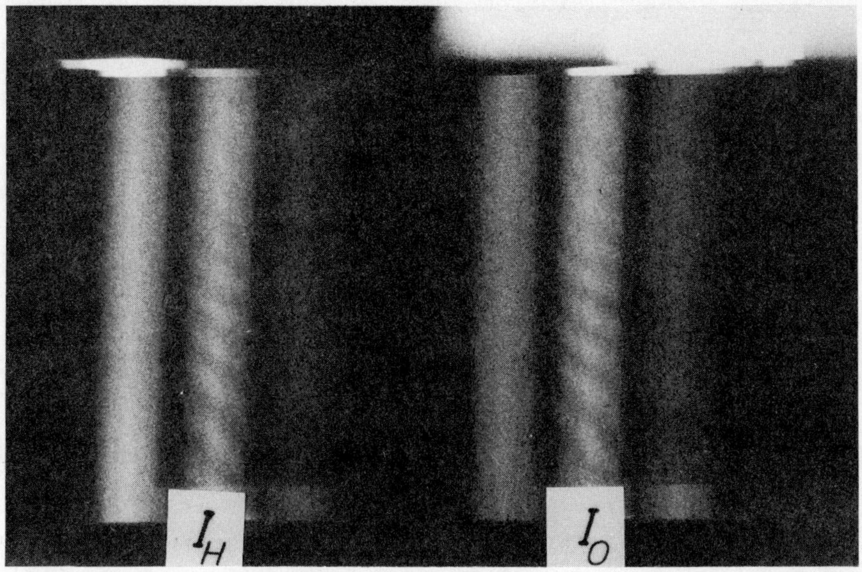


FIG. 9. Fringe patterns created by an aluminium wedge in one of the beam paths.

of moiré pattern can be calculated from $L_R = d/\alpha$, where $\alpha$ is the angle between the lattice planes of the analyser crystal of the interferometer and the standing interference pattern. A pure rotational moiré pattern consists of horizontal fringes which disperse to a coarser pattern if $\alpha$ becomes smaller. In Fig. 7 the change from maximum (Fig. 6) to minimum can be explained by further interference effects such as $\Delta d/d$ (i.e. lattice variation) interference contributions which overlap the rotational moiré pattern. A rotational angle $\alpha$ of approximately 0.002″ induces fringe patterns with a spacing of about 20 mm. This means that two fringe systems are produced in a neutron interferometer in which the plates are 40 mm high.

The final tests were carried out using neutrons. Aluminium plates were used to prove the interference ability. The neutron interferometer crystals were also tested by Rauch (1978) to obtain (a) the best mounting, (b) beam profiles at different positions, and (c) intensity ratios for the O and H beams. Although X-ray interferograms show almost homogeneous phase areas with a fringe visibility

$$F_V = \frac{I_{max} - I_{min}}{I_{max} + I_{min}}$$

of more than 90 per cent the usual fringe visibility for neutron interference is approximately 40 per cent. It was shown that vibrations and thermal fluctuations in particular decrease the modulation when the phase is shifted. By damping the support and using slits of cross-section 1 mm × 10 mm the visibility $F_V$ was increased to 79 per cent. It should be noted that the same interferometer works differently depending on which part is being used, i.e. it depends strongly on which part of the crystal volume the beams in the interferometer penetrate. The production of neutron interferometers which work well and in which such a dependence is rather small is the basis for further high resolution experiments. The quantity and quality of experimental information and results obtained decrease with decreasing interference modulation. Therefore good crystal materials are necessary but not sufficient for good neutron interferometry.

## Acknowledgments

This work is supported by the Bundesministerium für Forschung und Technologie Förderungsvorhaben 03-41A 06 P.

## References

BAUSPIESS, W., BONSE, U., RAUCH, H., and TREIMER, W. (1974). *Z. Phys.* **271**, 177–182.

PETRASCHEK, D. and RAUCH, H. (1976). AIAU 76401, Atominstitut Vienna. Rep.

RAUCH, H., TREIMER, W., and BONSE, U. (1974). *Phys. Lett. A*, **47**.

RAUCH, H., BADUREK, G., BAUSPIESS, W., BONSE, U., and ZEILINGER, A. (1976). Proc. Int. Conf. on the Interaction of Neutrons with Nuclei, Lowell (Mass. USA) CONF-760715-P2 (Vol. II) 1094.

RAUCH, H. (1978) Strahlprofile im Neutroneninterferometer. Seminar. Atominstitut Vienna.

TREIMER, W. (1975). Thesis, Vienna.

—— (1975). Rep. AIAU 75405, Atominstitut Vienna.

—— (1978). *Cryst. Research and Techn.*, **13**, 9, 1105.

# 7. APPLICATIONS OF THE FRESNEL DIFFRACTION OF NEUTRONS

A. G. KLEIN and G. I. OPAT
*School of Physics, University of Melbourne, Parkville, Victoria, Australia 3052*

## 1. Introduction

In classical optics two principal types of interference experiment are distinguished: interference by amplitude division and interference by division of wavefront. The first of these, typified by the Michelson and Mach–Zehnder interferometers and by various thin-film interference situations, relies on partial reflection, i.e. beam splitters. They are realized in the case of neutron optics in the highly successful single-crystal interferometers (Bonse and Hart 1965, Rauch, Treimer, and Bonse 1974) in which the dynamical Bragg diffraction of neutrons provides the beam splitting. Another situation of interference by amplitude division has been manifested in thin-film experiments with neutrons by Hayter, Penfold, and Williams (1976).

Interference by division of the wavefront has as its paradigm Young's two-slit experiment in which spatially separated parts of a coherent wavefront are recombined. One of the classical variants of this scheme is the Fresnel biprism experiment which has also been demonstrated for electrons by Möllenstedt and Düker (1956) and for neutrons by Maier-Leibnitz and Springer (1962). An even simpler set-up was used by us (Klein and Opat 1975, 1976, Klein, Martin, and Opat 1977). The wavefront is split by the sharp boundary between two regions of space. Fresnel diffraction from the boundary leads to the recombination of the two halves of the wavefront, which may have suffered different phase shifts, and thus interference effects may be observed. In our experiments the diffracting boundary was a magnetic domain wall in a single crystal of Fe–3%Si, contained in a thin polycrystalline foil. The two halves of the wavefront suffered phase shifts of opposite sign due to the rotation of the neutron spin in oppositely directed magnetic fields within the neighbouring domains. The incident neutron beam, of wave-

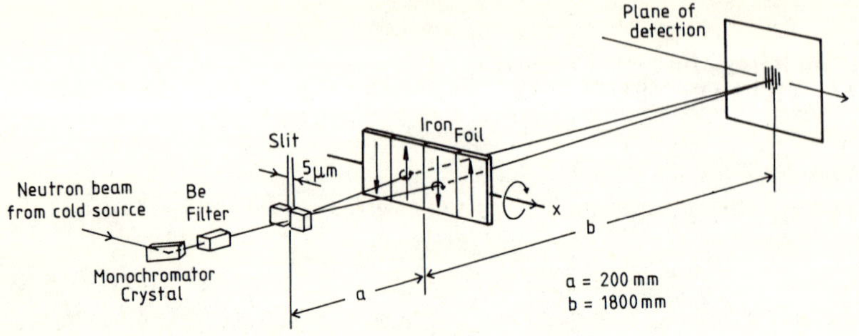


FIG. 1. Schematic layout of appratus in Fresnel diffraction experiment.

length 4.33 Å, was given lateral coherence in a 5 μm slit accurately aligned with the direction of the domains. Destructive interference, as seen in Fig. 2, was observed for the appropriate thickness of magnetic field traversed, consistent with the expected spinor behaviour of the neutrons. The experiment was simple in principle and relatively simple in execution. The question now is: are there any other applications in which the simplicity of Fresnel diffraction could be employed to advantage.

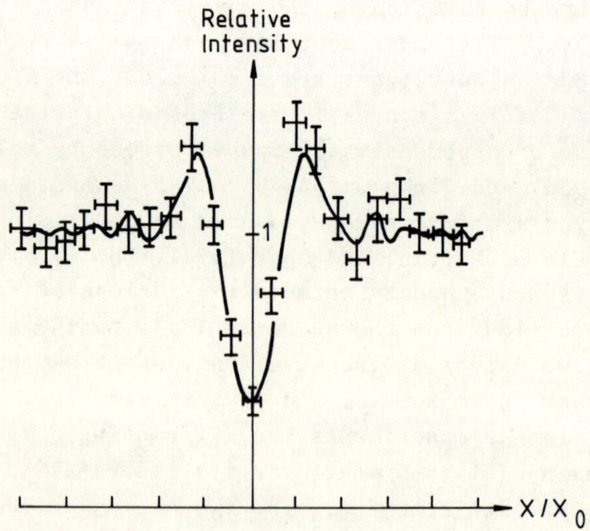


FIG. 2. Fresnel diffraction pattern showing destructive interference caused by spin precession.

## 2. Possible applications of Fresnel diffraction

The first and most obvious possibility is the visualization of magnetic domain boundaries inside ferromagnetic specimens. In a set-up similar to the one shown in Fig. 1 domain walls will cast a shadow-line (which is in reality a Fresnel diffraction pattern) provided that their plane is perpendicular to the direction of the incident neutron beam. The beam must be given spatial coherence by a small aperture (about 10 μm), rather than a slit, so that domain walls of any orientation can be observed. While this leads to a significant loss of flux, a large factor may be gained by using a full thermal beam instead of a narrowly monochromated beam of neutrons. While only some of the incident wavelengths will lead to complete destructive interference, the incoherent sum of all the patterns at different wavelengths still gives rise to an easily discernible overall pattern at a somewhat reduced contrast. Compared with neutron topography and other methods of domain visualization (Schlenker and Baruchel 1978), the proposed method may have only a single advantage, namely the magnification by projection which is apparent in Fig. 1. In our experiments a magnification factor of 10× arose naturally; this may be of advantage in examining small specimens.

The second possible application of Fresnel diffraction of neutrons concerns the measurement of nuclear scattering lengths or refractive indices of small specimens. The proposition is to observe the Fresnel diffraction patterns from the edges of small platelets or from very thin wires; the phase shifts introduced by the specimens can be extracted from the patterns. While other methods, especially the single-crystal interferometer, are obviously superior in general, the proposed approach may be advantageous in cases where only very small specimens are available. It may also be the case that a wider range of wavelengths may be explored because the scale of Fresnel diffraction patterns varies only as $\lambda^{\frac{1}{2}}$ rather than $\lambda$.

The third and possibly the most interesting application concerns the possibility of constructing Fresnel zone plates for use in monochromating and focusing neutron beams.

### 3. Zone plates for neutrons

Originally put forward by Soret (1875), the idea of zone plates is well known. Constructive interference of waves diffracted from alternate Fresnel zones occurs for particular pairs of conjugate distances whenever the optical paths to the zone boundaries differ by $\lambda/2$.

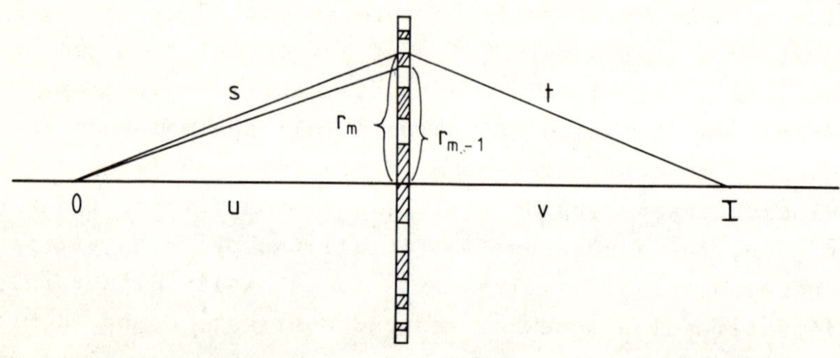


FIG. 3. Geometry of focusing by Fresnel zone plate.

Referring to Fig. 3, we have for the boundaries of the circular Fresnel zones

$$u + v = s + t - \tfrac{1}{2}m\lambda$$

i.e.

$$u + v + \tfrac{1}{2}m\lambda = (u^2+r_m^{\,2})^{\frac{1}{2}} + (v^2+r_m^{\,2})^{\frac{1}{2}} \,.$$

Expansion of this gives

$$m\lambda = r_m^{\,2}(1/u+1/v) - \tfrac{1}{4}r_m^{\,4}(1/u^3+1/v^3) + \ldots \,. \qquad (1)$$

If we neglect higher-order terms we have

$$1/u + 1/v = 1/f = m\lambda/r_m^{\,2} \qquad (2)$$

which is the equation of a thin lens of focal length $f = r_m^{\,2}/m\lambda$.

It was pointed out by Lord Rayleigh (1880) that a fourfold increase in intensity at the focal point can be obtained by giving the waves a phase reversal at alternate zones, rather than complete absorption. Experiments along these lines were reported by Wood (1898) who, by photographic means, produced phase-reversal zone plates in thin layers of bichromated gelatin and showed their effectiveness in visible light. Since that time zone plates have remained a laboratory curiosity and an undergraduate exercise, except for two recent developments. Firstly, as Gabor has shown, a zone plate is the hologram of a point object. Apart from yielding a simple explanation of the original Gabor holograms, this point has far-reaching practical consequences, namely the fact that zone plates can be manufactured holographically. They are, in fact, the interference patterns given by the superposition of a plane wave with a spherical wave. The second recent development is closely linked with this observation, in that a research group at the Göttingen Observatory (Niemann, Rudolph, and Schmahl 1974) has succeeded in producing small zone plates by holographic methods using the techniques of micro-lithography. Furthermore, Niemann, Rudolph, and Schmahl (1976) have built a microscope for soft X-rays using such holographic zone plates as imaging elements. The zone plates were produced using visible light from an $Ar^+$ or $Kr^+$ laser, and were tested with X-rays of wavelength 45 Å. A change in wavelength by a factor of about 100 affects the higher-order terms in equation (1) and gives rise to spherical aberration. However, this may be pre-corrected in the holographic process by pre-distorting the spherical waves through a plane parallel-sided slab. This allows the effective use of a large number of zones, more than 2000, and zone plates several millimetres in diameter.

Let us now examine the possible applications of this in the case of neutrons. Figure 4 shows the cross-section of a phase-reversal zone plate for use with long-wavelength neutrons. We have the following relations.

(i) Focal length:

$$f = r_m{}^2/m\lambda$$

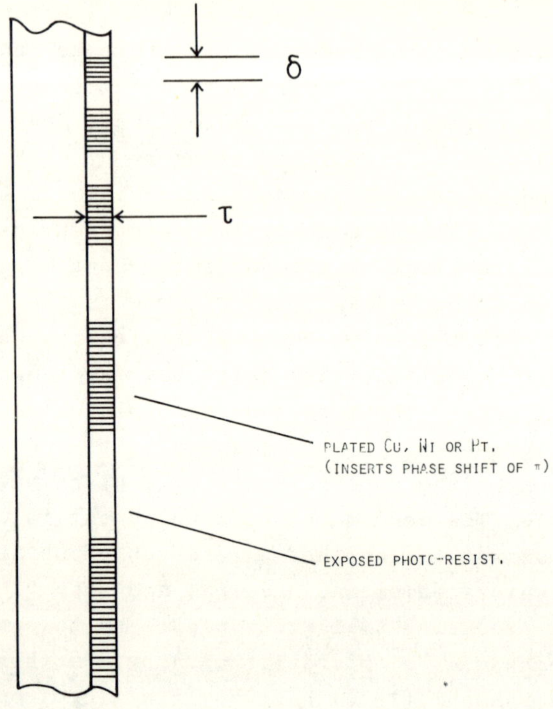


FIG. 4. Cross-section of proposed zone plate: τ, zone thickness; δ, zone width.

from equation (2), for $1 < m < M$.

(ii) Width of outer zone:

$$\delta = r_M - r_{M-1} = r_M/2M .$$

(iii) $F$-number:

$$F = f/2r_M = r_M/2M\lambda = \delta/\lambda . \qquad (3)$$

(iv) Zone thickness to give a phase shift of $\pi$:

$$\tau = \lambda/\{2(n-1)\} .$$

The refractive index, $n$, is given in terms of the particle density $N$, and the scattering length $b$ by

$$n = 1 + \lambda^2 Nb/2\pi \;;$$

thus

$$\tau = \pi/\lambda Nb \;.$$

If, in addition, we define:

(v) 'flatness' = (zone thickness)/(width of outer zone), i.e.

$$\rho = \tau/\delta$$

we can rewrite equation (3):

$$F = \tau/\rho\lambda = \pi/\rho\lambda^2 Nb \;. \qquad (4)$$

With due regard for the limitations of photolithography (as used in the manufacture of micro-electronic components and holographic gratings and zone plates), let us use the conservative estimates $\delta = 1\ \mu m$ and $\rho = 1 \rightarrow \tau = 1\ \mu m$. We then obtain from (3), using $\lambda$ in Ångstrom units,

$$F = 10\ 000/\lambda$$

as a conservative estimate. This is an extremely small aperture (cf. electron microscope lenses) unless the wavelength is rather long. For a possible practical example, consider zones made of copper ($Nb = -6.6 \times 10^{-6}$ Å$^{-2}$) and neglect the phase shift contribution from the alternate zones. (This could be of opposite sign owing to the positive scattering-length of the hydrogen in the photoresist material.) We then obtain a phase shift of $\pi$ for $\lambda = 50$ Å and thus $F = 200$. For these values of the parameters, a zone plate of 1 mm diameter will consist of $M = 250$ zones and have a focal length of 200 mm at a wavelength of 50 Å. This is quite a respectable performance for a lens, and it could even turn out to be useful provided that sufficient neutron flux were obtainable at this long wavelength.

Before commenting on the possible uses of such an object, *two* further observations are in order.

(a) Used as a lens, a zone plate is extremely dispersive, as can be seen from equation (2). This may be turned to advantage by noting that a small but finite aperture in the focal plane can serve to isolate a finite range of wavelengths from an incident beam and thus acts as a monochromator. A zone plate can therefore be used as a monochromating condensing lens to illuminate an object.

(b) We conjecture that zone plates can be 'stacked', i.e. the effective *F* number can be reduced by using several zone plates, one behind the other, spaced by a certain minimum distance. (The spacing should be such that the ragged wavefront emerging from a zone plate has enough room to 'heal', by diffraction, into a complete spherical wave. This 'healing distance' should be of order $\Delta \approx (r_2 - r_1)^2/\lambda \approx 0.17\, f$. For the design considered above, this equals 34 mm.)

Based on the above considerations, but without using the (untried) 'stacking' feature mentioned in (b), we propose the *gedanken* neutron microscope shown in Fig. 5. $Z_1$ together with

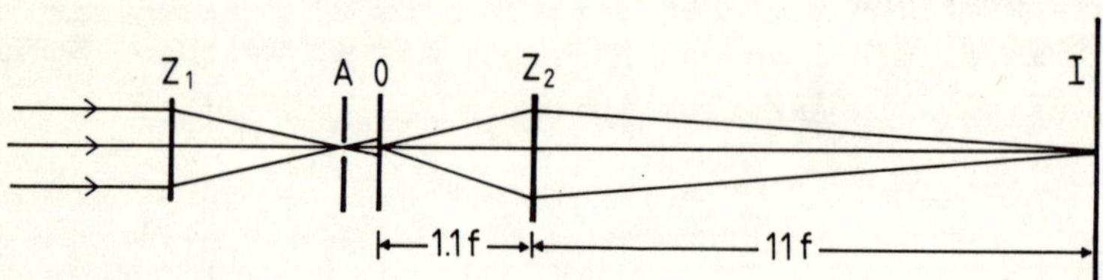


FIG. 5. Notional neutron microscope optics (see text).

the aperture stop A in its focal plane form a monochromating condenser, illuminating the object O. $Z_2$ is a projection lens giving, in the configuration shown, a magnification of 10×. With the zone plate lenses nominally designed above (1 mm diameter, 200 mm focal length) the diffraction-limited resolution should be 1.2 μm; in practice it will be limited by chromatic aberration. This could be quite an interesting system, except for one fatal snag: most of the objects whose neutron imaging may be of interest are likely to be totally transparent to the neutrons, in spite of the large depth of focus implied by a large *F* number. However there may be ways around this, for example using gadolinium or samarium staining, or, better still, by adapting some scheme of phase contrast.

The latter possibility is by no means excluded, but it would at this stage take us too far into the realm of the hypothetical.

The chromatic aberration which is likely to impair the performance of such systems can be tackled to a limited extent by the scheme inherent in the Huygens eyepiece in classical optics: a combination of two lenses, spaced by the mean focal length of the two, can be shown to have zero lateral chromatic aberration for a particular value of $\lambda$ (i.e. a separated doublet can be achromatized for a finite range of wavelengths). However, this is true only if the object distance is large compared with the focal length. Consequently, such an 'achromatic doublet' can only be used in situations requiring a de-magnified image. It could form the basis of a condensing lens which, to a first order, is achromatic over a limited range of wavelengths. Another possible application is shown in Fig. 6; it is a scanning neutron microscope which works by activation analysis of small areas on a specimen. The specimen stage S is scanned mechanically and is illuminated by the de-magnified image of the aperture stop A. Zone plates $Z_2$ and $Z_3$ form an achromatic doublet in the sense that $\partial M/\partial\lambda = 0$, where $M$ is the magnification of the doublet at the central wavelength selected by $Z_1$ and A. It is possible that resolutions superior to auto-radiography may be obtained.

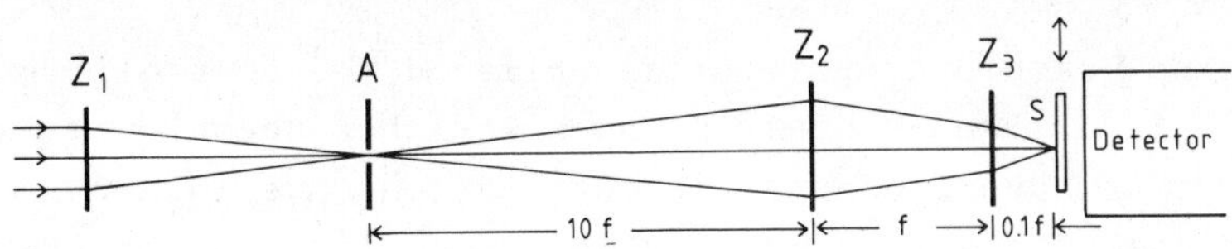


FIG. 6. Notional scanning neutron microscope (see text).

Finally, we remark that a portion of a zone plate is equivalent to a focusing grating. It is in fact a grating with variable spacing and, as before, it can be prepared holographically. It can thus be used for simply focusing a neutron beam, though in a dispersive manner as discussed in (a) above. This leads to another form of focusing monochromator, as shown in Fig. 7(a).

In accordance with our earlier observation, achromatic focusing may be obtainable by the scheme shown in Fig. 7(b).

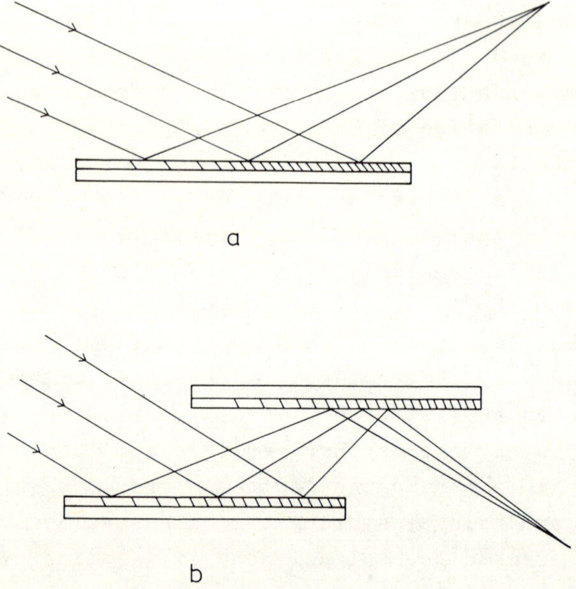


FIG. 7. Portion of zone plate as focusing grating: (a) dispersive; (b) dispersion corrected.

## 4. Conclusion

Our original simple experiment established the feasibility of Fresnel diffraction of neutrons as a workable technique. In this paper we have presented in outline some possible further applications of the technique, namely visualization of magnetic domain walls, measurement of nuclear scattering lengths with very small specimens and the use of Fresnel zone plates for focusing and monochromating long-wavelength neutron beams. Since the holographic production of suitable zone plates has been demonstrated by the Göttingen group (Niemann, Rudolph, and Schmahl 1974) and successfully used by them in X-ray microscopy, there is some hope for the feasibility of neutron microscopy along the lines that we have sketched out.

## References

BONSE, U. and HART, M. (1965). *Appl. Phys. Lett.* **6**, 155.
HAYTER, J.B., PENFOLD, J., and WILLIAMS, W.G. (1976). *Nature* (*Lond.*) **262**, 569.
KLEIN, A.G., MARTIN, L.J., and OPAT, G.I. (1977). *Am. J. Phys.* **45**, 295.
KLEIN, A.G. and OPAT, G.I. (1975). *Phys. Rev. D* **11**, 523.
—— (1976). *Phys. Rev. Lett.* **37**, 238.
MAIER-LEIBNITZ, H. and SPRINGER, T. (1962). *Z. Phys.* **167**, 386.
MÖLLENSTEDT, G. and DÜCKER, H. (1956). *Z. Phys.* **145**, 377.
NIEMANN, B., RUDOLPH, D, and SCHMAHL, G. (1974). *Opt. Commun.* **12**, 160.
—— (1976). *Appl. Opt.* **15**, 1883.
RAUCH, H., TREIMER, W., and BONSE, U. (1974). *Phys. Lett. A* **47**, 369.
RAYLEIGH, LORD, *Scientific Papers*, Vol. III, p. 112. Dover, New York, 1964.
SORET, . (1875), *Poggendorff's Annln.*
SCHLENKER, M. and BARUCHEL, J. (1978). *J. appl. Phys.* **49**, 1966.
WOOD, R.W. (1898). *Phil. Mag.* **45**, 511.

# 8. INTENSITY PROFILES AND INTERFERENCE PROPERTIES OF LLL INTERFEROMETERS

D. PETRASCHECK

*Institut für Physik, Universität Linz, A-4045 Linz, Austria*

## 1. Introduction

With the successful development of a Laue-type neutron interferometer (Rauch, Treimer, and Bonse 1974, Bauspiess *et al.* 1974), the interference properties for the 'ideal geometry' have been calculated (Rauch and Suda 1974). Later more detailed studies considering geometrical irregularities and crystal plates of different thicknesses were made (Petrascheck and Rauch 1976). Spatial intensity profiles have also been computed for the ideal geometry and the 'focusing geometry' (Bauspiess, Bonze, and Graeff 1976, Petrascheck 1976).

It has been shown that in the non-absorbing neutron interferometer the interference properties are quite different from those of an X-ray interferometer with anomalous absorption. In a theoretical treatment for the ideal geometry, the two cases have been connected (Petrascheck and Folk 1976). The aim of this paper is to include the other types of LLL interferometers, in particular the focusing geometry, in the theory.

## 2. Plane waves

The most common type of interferometer used in neutron and X-ray interferometry is the LLL interferometer. The splitter breaks up the incident beam into two coherent beams which are diffracted at the mirror plates in such a way that they recombine at the analyser (Fig. 1). For incident plane waves, the wave functions behind the interferometer are calculated by applying the dynamical theory to three successive Laue reflections. It is easy to show that the wave functions of the two paths are in phase for the forward beam, i.e.

$$\psi_0^{I} = \psi_0^{II} , \qquad (1)$$

if the following focusing conditions are fulfilled:

$$t_A^{\,I} = t_S^{\,II} = t$$

$$D_M^{\,I} = D_M^{\,II} = D_M \tag{2}$$

$$D_S = D_A = D \, .$$

For the deviated beam one obtains

$$\psi_G^{\,II} = \frac{-\nu^2 \sin^2\{A(z^2+\nu^2)^{\frac{1}{2}}\}}{\nu^2 + z^2 - \nu^2 \sin^2\{A(z^2+\nu^2)^{\frac{1}{2}}\}} \psi_G^{\,I} \, . \tag{3}$$

$z$ describes the deviation of the incident wave vector **k** from the Bragg angle $\theta_B$:

$$z = y + \mathrm{i}g \tag{4a}$$

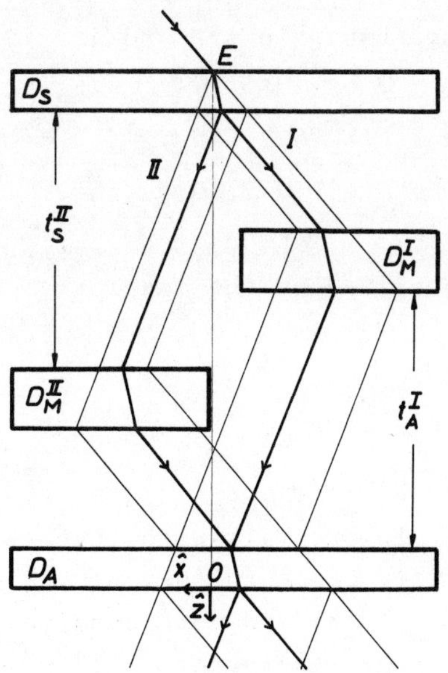


FIG. 1. Sketch of a skew and asymmetric LLL interferometer.

$$y = \frac{2b\delta\theta_B \sin 2\theta_B - (1-b)\{V(0)/E\}}{2b^{\frac{1}{2}}[\{|V(\mathbf{G})V(-\mathbf{G})|\}^{\frac{1}{2}}/E]} \tag{4b}$$

$$g = \frac{1-b}{2b^{\frac{1}{2}}} \frac{V_i(0)}{\{|V(\mathbf{G})V(-\mathbf{G})|\}^{\frac{1}{2}}} . \tag{4c}$$

$V(\mathbf{G}) = V_r(\mathbf{G}) - iV_i(\mathbf{G})$ is the Fourier transform of the interaction potential. $E$ is the energy of the incident neutrons and $\delta\theta_B = \theta_B - \theta$ is its angular deviation from the glancing angle $\theta_B$.

$$b = \frac{\cos \gamma}{\cos \gamma_G} \tag{5}$$

where $\gamma$ and $\gamma_G$ are the angles of incidence and of diffraction respectively. $A$ is a parameter for the crystal thickness $D$:

$$A = \pi D/\Delta_0 \tag{6a}$$

where $\Delta_0$ is the Pendellösung length and is given by

$$\Delta_0 = \frac{\pi\{\cos \gamma \cos \gamma_G\}^{\frac{1}{2}}}{k\{|V(\mathbf{G})V(-\mathbf{G})|\}^{\frac{1}{2}}/E} . \tag{6b}$$

The absorption is described by the imaginary part of

$$\nu^2 = \frac{V(\mathbf{G})V(-\mathbf{G})}{|V(\mathbf{G})V(-\mathbf{G})|} \approx 1 - 2i\kappa \tag{7}$$

with $\kappa = V_i(\mathbf{G})/|V(\mathbf{G})|$ in crystals with inversion symmetry. The nomenclature is very close to that of the review article by Rauch and Petrascheck (1978) and more detailed definitions of the quantities used are given there.

In the zero-absorption case (Rauch and Suda 1974) the wave functions of the deviated beam are out of phase and have different amplitudes. With increasing absorption ($\kappa A > 1$) the $\sin^2$

term dominates in equation (3) and we have $\psi_G^{II} \approx \psi_G^{I}$ as for the X-ray interferometer (Bonse and Hart 1965a, 1965b). Then, additionally, the third equation of the focusing conditions (2) $D_S = D_A = D$ becomes irrelevant.

The forward reflection wave of path I is given by the reflectivities of the transmission at the splitter and of the diffraction at the mirror and the analyser:

$$P_0^{I}(y) = \exp(-\mu_0 T_e)\left|[\cos\{A(z^2+\nu^2)^{\frac{1}{2}}\} + \frac{iz}{(z^2+\nu^2)^{\frac{1}{2}}}\sin\{A(z^2+\nu^2)^{\frac{1}{2}}\}]\right. \tag{8}$$
$$\left.\times\frac{\sin\{A_M(z^2+\nu^2)^{\frac{1}{2}}\}\sin\{A(z^2+\nu^2)^{\frac{1}{2}}\}}{z^2+\nu^2}\right|^2$$

where $\exp(-\mu_0 T_e)$ describes the normal attenuation, $T_e$ is the effective total crystal thickness, and $\mu_0$ is the linear absorption coefficient

$$\mu_0 = \frac{kV_i(0)}{E} \tag{9a}$$

$$T_e = \frac{T}{2}\left(\frac{1}{\cos\gamma} + \frac{1}{\cos\gamma_G}\right) \qquad \text{with } T = 2D + D_M . \tag{9b}$$

Similar equations are immediately obtained for each of the four paths and directions possible. If the phases between path I and path II are shifted by a (nuclear) phase shift $\alpha$, the reflection curves are

$$P_0(y,\alpha) = 4P_0^{I}(y)\cos^2\frac{\alpha}{2} \tag{10}$$

$$P_G(y,\alpha) = \left|\frac{V(\mathbf{G})}{V(-\mathbf{G})}\right|\exp(-\mu_0 T_e)\left|\frac{\sin\{A_M(z^2+\nu^2)^{\frac{1}{2}}\}}{(z^2+\nu^2)^{\frac{1}{2}}}\right. \tag{11}$$
$$\left.\left[1 - \{1+\exp(i\alpha)\}\frac{\nu^2\sin^2\{A(z^2+\nu^2)^{\frac{1}{2}}\}}{z^2+\nu^2}\right]\right|^2 .$$

For zero absorption the total intensity is independent of $\alpha$

and is given by the Laue diffraction at the mirror:

$$P(y,\alpha) = P_0(y,\alpha) + P_G(y,\alpha) = \frac{\sin^2\{A_M(1+y^2)^{\frac{1}{2}}\}}{1 + y^2} . \qquad (12)$$

In the zero absorption case in particular the reflection curves are dominated by fast oscillations, as shown in Fig. 2 for a crystal of medium thickness and ideal geometry ($D_M=D$). For sufficient wafer thickness $D, D_M \gg \Delta_0$ one can average over the Pendellösung oscillations. There one has to distinguish between three different cases.

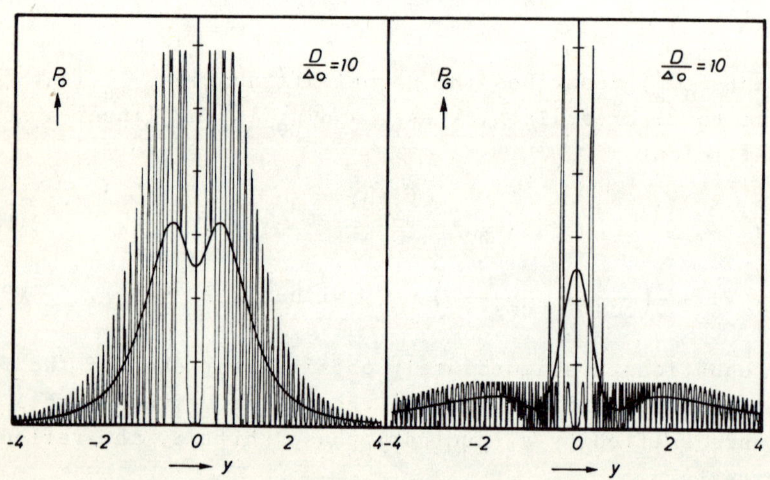


FIG. 2. Reflection curves for zero absorption, medium thickness, and ideal geometry ($D = D_M = 10\Delta_0$): (a) forward beam after Bauspiess *et al*. 1976; (b) deviated beam.

(a) The geometry where all crystal plates are equally thick, $D_M = D$, is usually called 'ideal geometry' (Rauch and Suda 1974):

$$\bar{P}_0(y) = \frac{1 + 6y^2}{4(1+y^2)^3} \qquad \bar{P}_G(y) = \frac{1 - 2y^2 + 2y^4}{4(1+y^2)^3} \; . \tag{13}$$

(b) Another kind of ideal geometry exists for $D_M = 2D$. As a diffraction focusing which is very similar to the diffraction focusing occurring after two crystal plates appears in the forward beam, we call it 'focusing geometry' (Bauspiess, Bonse, and Graeff 1976, Indenbom, Slobodetskii, and Truni 1974, Indenbom, Suvorov, and Slobodetskii 1976):

$$\bar{P}_0(y) = \frac{1.5 + 4y^2}{4(1+y^2)^3} \qquad \bar{P}_G(y) = \frac{0.5 + 2y^4}{4(1+y^2)^3} \; . \tag{14}$$

(c) The other geometrical situations can be included in a third case as long as $|D_M - D| \gg \Delta_0$ and $|D_M - 2D| \gg \Delta_0$. This case is of less interest and is only included for completeness:

$$\bar{P}_0(y) = \frac{1 + 4y^2}{4(1+y^2)^3} \qquad \bar{P}_G(y) = \frac{1 + 2y^4}{4(1+y^2)^3} \; . \tag{15}$$

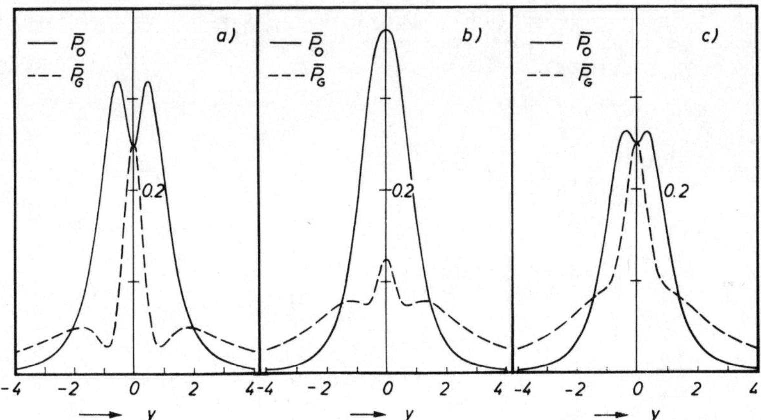


FIG. 3. Averaged reflection curves for zero absorption: (a) ideal geometry (after Rauch and Suda 1974; (b) focusing geometry; (c) other cases (after Petrascheck and Rauch 1976).

The rocking curves (13)–(15) are shown in Fig. 3. In weak absorbing crystals the average reflection curves for the symmetric case are given by

(a) $D_M = D$; ideal geometry (Petrascheck and Folk 1976)

$$\bar{P}_0(y,\alpha) = \frac{\exp(-\mu_0 T_e)}{8(1+y^2)^3}\Bigg((1+10y^2)\cosh\left\{\frac{2\kappa A}{(1+y^2)^{\frac{1}{2}}}\right\} + $$

$$+ (1+2y^2)\cosh\left\{\frac{6\kappa A}{(1+y^2)^{\frac{1}{2}}}\right\} + \tag{16a}$$

$$+ 2y(1+y^2)^{\frac{1}{2}}\left[3\sinh\left\{\frac{2\kappa A}{(1+y^2)^{\frac{1}{2}}}\right\}+\sinh\left\{\frac{6\kappa A}{(1+y^2)^{\frac{1}{2}}}\right\}\right]\Bigg)\cos^2\frac{\alpha}{2}$$

$$\bar{P}_G(y,\alpha) = \frac{\exp(-\mu_0 T_e)}{2(1+y^2)}\cosh\left\{\frac{2\kappa A}{(1+y^2)^{\frac{1}{2}}}\right\} - $$

$$- \frac{\exp(-\mu_0 T_e)}{8(1+y^2)^3}\left[3(1+4y^2)\cosh\left\{\frac{2\kappa A}{(1+y^2)^{\frac{1}{2}}}\right\} - \right. \tag{16b}$$

$$\left. - \cosh\left\{\frac{6\kappa A}{(1+y^2)^{\frac{1}{2}}}\right\}\right]\cos^2\frac{\alpha}{2}\,.$$

(b) $D_M = 2D$; focusing geometry

$$\bar{P}_0(y,\alpha) = \frac{\exp(-\mu_0 T_e)}{8(1+y^2)^3}\left[2(1+y^2)+4y^2\cosh\left\{\frac{4\kappa A}{(1+y^2)^{\frac{1}{2}}}\right\} + \right. \tag{17a}$$

$$\left. + (1+2y^2)\cosh\left\{\frac{8\kappa A}{(1+y^2)^{\frac{1}{2}}}\right\}+2y(1+y^2)^{\frac{1}{2}}\sinh\left\{\frac{8\kappa A}{(1+y^2)^{\frac{1}{2}}}\right\}\right]\cos^2\frac{\alpha}{2}$$

$$\bar{P}_G(y,\alpha) = \frac{\exp(-\mu_0 T_e)}{2(1+y^2)} \cosh\left\{\frac{4\kappa A}{(1+y^2)^{\frac{1}{2}}}\right\} - \qquad (17b)$$

$$- \frac{\exp(-\mu_0 T_e)}{8(1+y^2)^3}\left[4(1+2y^2)\cosh\left\{\frac{4\kappa A}{(1+y^2)^{\frac{1}{2}}}\right\} - \cosh\left\{\frac{8\kappa A}{(1+y^2)^{\frac{1}{2}}}\right\}\right]\cos^2\frac{\alpha}{2} .$$

The different absorption of the terms in equations (16) and (17) can be visualized by considering the wave fields as rays which travel along the directions of the energy flow (characterized by the geometrical parameter $\Gamma_{1,2}$ (Petrascheck and Folk 1976) for wave fields 1 and 2 respectively where $\Gamma_1 = -\Gamma_2$ in the symmetric case) and are differently absorbed. At each crystal plate the rays split according to the directions $\Gamma_1$, $\Gamma_2$.

For the focusing geometry any ray which travels with $\Gamma$ in the splitter and the analyser and $-\Gamma$ in the mirror contributes to the first term of equation (17a). These rays give contributions to the intensity at the focusing point in the section pattern and are absent in the direction of the deviated beam (equation (17b)) (see Fig. 6(a)). For the contributions including $4\kappa A$, the energy flows in the mirror and the splitter or analyser are parallel. Finally, rays which are parallel in all the wafers contribute to terms including $8\kappa A$ and contain the anomalous weak attenuated part of the incident wave. Interference terms between the different ray components drop out by averaging over the Pendellösung oscillations.

## 3. Integrated intensities

For monochromatic waves the integrated intensity is defined by

$$R_{0,G}(\alpha) = \int dy P_{0,G}(y,\alpha) . \qquad (18)$$

Figure 4 shows the integrated powers for the ideal geometry and zero absorption up to a thickness $D/\Delta_0 = 5$ computed by numerical integration. Approximate expressions can be found for thick crystals. For the ideal geometry one obtains[†]

[†]The corrections to equations (19) are of order $(1/A)^{3/2}$ but have quite large numerical factors.

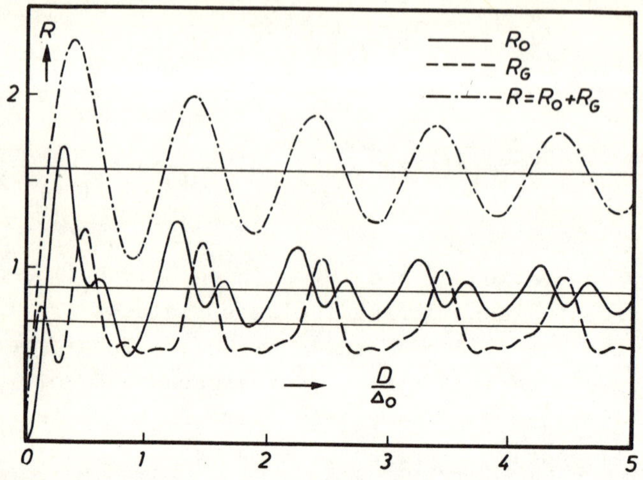


FIG. 4. Integrated intensities for $D_M = D$ and zero absorption.

$$R_0 \approx \frac{9\pi}{32} + \frac{\pi}{8}\{J_1(2A)+2J_1(4A)-J_1(6A)\} \tag{19a}$$

$$R_G \approx \frac{7\pi}{32} + \frac{\pi}{8}\{J_1(2A)-2J_1(4A)+J_1(6A)\} \quad . \tag{19b}$$

From the asymptotic behaviour of the Bessel functions we conclude that the height of the oscillations decreases in proportion to $(D/\Delta_0)^{-\frac{1}{2}}$. For $D/\Delta_0 = 10$ the intensity varies roughly within ±10 per cent.

This suggests that we should consider the average integrated powers, which we give here only for the symmetric case.

(a) Ideal geometry (Petrascheck and Folk 1976):

$$\overline{R}_0(\alpha) = \frac{\pi}{8}\exp(-\mu_0 T_e)\left\{I_0(2\kappa A)+\frac{8I_1(2\kappa A)}{2\kappa A} - 27\frac{I_2(2\kappa A)}{(2\kappa A)^2} + \right.$$
$$\left. + I_0(6\kappa A)-3\frac{I_2(6\kappa A)}{(6\kappa A)^2}\right\}\cos^2\frac{\alpha}{2} \tag{20a}$$

$$\bar{R}_G(\alpha) = \frac{\pi}{8}\exp(-\mu_0 T_e)\Big[4I_0(2\kappa A) - \Big\{3I_0(2\kappa A) + 6\frac{I_1(2\kappa A)}{2\kappa A} - 27\frac{I_2(2\kappa A)}{(2\kappa A)^2} - I_2(6\kappa A) - 3\frac{I_2(6\kappa A)}{(6\kappa A)^2}\Big\}\cos^2\frac{\alpha}{2}\Big] . \tag{20b}$$

(b) Focusing geometry

$$\bar{R}_0(\alpha) = \frac{\pi}{8}\exp(-\mu_0 T_e)\Big\{1 + \frac{4I_1(4\kappa A)}{4\kappa A} - 12\frac{I_2(4\kappa A)}{(4\kappa A)^2} + I_0(8\kappa A) - 3\frac{I_2(8\kappa A)}{(8\kappa A)^2}\Big\}\cos^2\frac{\alpha}{2} \tag{21a}$$

$$\bar{R}_G(\alpha) = \frac{\pi}{8}\exp(-\mu_0 T_e)\Big[4I_0(4\kappa A) - \Big\{4I_0(4\kappa A) - 12\frac{I_2(4\kappa A)}{(4\kappa A)^2} - I_2(8\kappa A) - 3\frac{I_2(8\kappa A)}{(8\kappa A)^2}\Big\}\cos^2\frac{\alpha}{2}\Big] . \tag{21b}$$

The $I_n$ are the modified Bessel functions of order $n$. In Fig. 5 the amplitudes of the intensity oscillations are plotted against the absorption $\mu_0 T_e$ for the ideal and the focusing geometries. If in equations (20b) or (21b) the factor in front of $\cos^2 \alpha/2$ vanishes, the height of the oscillations and the contrast

$$C_{0,G} = \frac{R_{0,G}^{max} - R_{0,G}^{min}}{R_{0,G}^{max} + R_{0,G}^{min}} \tag{22}$$

are zero. Because of the average over the Pendellösung oscillations there are no terms proportional to $\sin\alpha$ as compared with the exact integrated powers. These terms, which vanish at

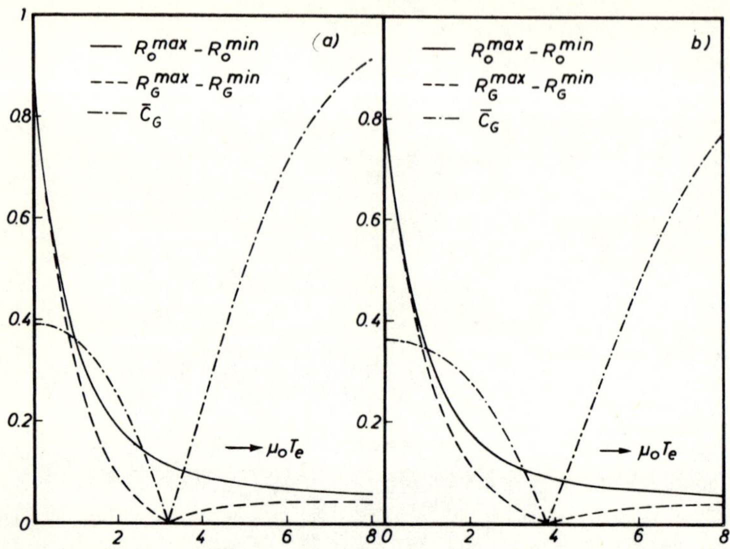


FIG. 5. Amplitudes of the intensity oscillations and the contrast calculated for the symmetric case from the averaged intensities: (a) $D_M = D$ and $\mu_0 T_e = 6\kappa A$; (b) $D_M = 2D$ and $\mu_0 T_e = 8\kappa A$.

zero absorption as well as at strong anomalous absorption, mean that in the region of medium absorption (and crystals which are not too thick) the contrast does not go down exactly to zero (Bonse and Graeff 1977) and the minimum intensity $\bar{R}_G(\alpha)$ is shifted from $\alpha = 0$ to $\alpha = \pi$.

For zero absorption, the mean integrated powers are as follows.

(a) Ideal geometry:

$$\bar{R}_0 = \frac{9\pi}{32} \qquad \bar{R}_G = \frac{7\pi}{32} \quad . \tag{23}$$

(b) Focusing geometry:

$$\bar{R}_0 = \frac{8.5\pi}{32} \qquad \bar{R}_G = \frac{7.5\pi}{32} \quad . \tag{24}$$

(c) Other cases:

$$\bar{R}_0 = \frac{7\pi}{32} \qquad \bar{R}_G = \frac{9\pi}{32} \quad . \tag{25}$$

The total integrated reflectivity is $\pi/2$ in all three cases, which follows from equation (12).

## 4. Spherical waves

Let us assume that the beam is incident through a narrow slit. Then the intensity spreads in the Borrmann fan at each crystal plate (see Fig. 6(b)). In such a situation the wave functions

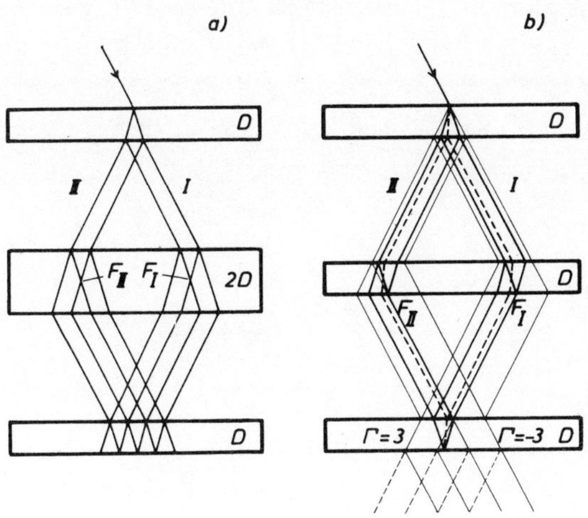


FIG. 6. Ray considerations. (a) Focusing geometry: from any incident ray there are components reaching the focusing point F. (b) Ideal geometry: the rays which contribute only to the central region are drawn with solid lines.

are described by the spherical wave theory (Kato 1968). For a skew interferometer with $\psi_i = u_0\{\exp ik(\mathbf{r}-\mathbf{r}_E)\}$ ($\mathbf{r}_E$ designates the entrance point, see Fig. 1), one obtains at the rear surface of the analyser the following wave functions:

$$\psi_0^{\,I} = \psi_0^{\,II} = c_0[\cos\{A(z^2+\nu^2)^{\frac{1}{2}}\} + \frac{iz}{(z^2+\nu^2)^{\frac{1}{2}}}\sin\{A(z^2+\nu^2)^{\frac{1}{2}}\}]$$

$$\frac{\sin\{A_M(z^2+\nu^2)^{\frac{1}{2}}\}}{(z^2+\nu^2)^{\frac{1}{2}}}\;\frac{\sin\{A(z^2+\nu^2)^{\frac{1}{2}}\}}{(z^2+\nu^2)^{\frac{1}{2}}}\exp(iA\Gamma z) \tag{26}$$

$$\psi_G{}^{I} = - c_G \frac{\sin\{A_M(z^2+\nu^2)^{\frac{1}{2}}\}}{(z^2+\nu^2)^{\frac{1}{2}}} \exp(iA\Gamma z) + \psi_G{}^{II} \tag{27}$$

$$\psi_G{}^{II} = c_G\nu^2 \frac{\sin^2\{A(z^2+\nu^2)^{\frac{1}{2}}\}\sin\{A_M(z^2+\nu^2)^{\frac{1}{2}}\}}{(z^2+\nu^2)^{3/2}} \exp(iA\Gamma z) \tag{28}$$

$c_0$ and $c_G$ are 'phase factors' which contain the normal attenuation. As usual, the paths to reach any point at the exit surface are composed of 0 and G directions. Their length $l$, which depends only on the point chosen at the exist surface, is multiplied by $\mu_0$ (Rauch and Suda 1974, Petrascheck and Rauch 1976):

$$c_0 = - \mu_0 \exp\{-i\mathbf{k}_B(\mathbf{r}-\mathbf{r}_E) - \frac{ikV(0)}{2E}\, l\} \tag{29a}$$

$$c_G = - ic_0 b^{\frac{1}{2}} \frac{V(G)}{\{|V(G)V(-G)|\}^{\frac{1}{2}}} \exp(i\mathbf{G}\cdot\mathbf{r}) \tag{29b}$$

$$l = \frac{1}{\sin 2\theta_B} \{(x-x_t)(\cos\gamma-\cos\gamma_G) + T(\sin\gamma+\sin\gamma_G)\} . \tag{29c}$$

$\mathbf{k}_B$ is the Bragg-vector and $\Gamma$ is a geometrical parameter for the position $x$ at the exit surface which is defined for the asymmetric case as (Rauch and Petrascheck 1978)

$$\Gamma = \frac{2 \cos\gamma \cos\gamma_G}{D \sin 2\theta_B} (x-x_M) . \tag{30}$$

If the origin 0 is placed as in Fig. 1, then the centre $x_M$ of the Borrmann fan is shifted with respect to 0 owing to the path difference between the crystal plates

$$x_t = t_A{}^{I} \tan\gamma_G - t_S{}^{II} \tan\gamma \tag{31a}$$

and the asymmetric reflection with the crystal plates

$$x_M = x_t + \frac{\sin(\gamma_G - \gamma)}{2\cos\gamma\cos\gamma_G} T . \tag{31b}$$

Therefore in the skew (symmetric) interferometer both the edges and the centre of the wave fan depend on the wavelength.

At the exit surface the Borrman fan lies in the range $-T/D \leqslant \Gamma \leqslant T/D$. The wave functions which describe the intensity distributions within this region are superpositions of the plane waves $\psi_{0,G}$:

$$\phi_{0,G} = \int \mathrm{d}y \psi_{0,G} . \tag{32}$$

They are taken as approximations of spherical waves. If we introduce

$$F(\nu, A, A_M, p) = \frac{\nu^2}{2} \int_{-\infty+ig}^{\infty+ig} \mathrm{d}z \exp(ipz) \frac{\sin^2\{A(z^2+\nu^2)^{\frac{1}{2}}\}\sin\{A_M(z^2+\nu^2)^{\frac{1}{2}}\}}{(z^2+\nu^2)^{3/2}} \tag{33}$$

the wave functions become

$$\phi_0 = 2c_0\left\{\frac{\partial}{\partial b} F(\nu, b, A_M, A\Gamma)\Big|_{b=A} + 2\frac{\partial}{\partial p} F(\nu, A, A_M, p)\Big|_{p=A\Gamma}\right\} \tag{34}$$

$$\phi_G{}^{I} = -\pi c_G g_0\left(\nu A_M, \frac{D\Gamma}{D_M}\right) + \phi_G{}^{II} \tag{35a}$$

$$\phi_G{}^{II} = c_G F(\nu, A, A_M, A\Gamma) . \tag{35b}$$

With $g_n(A,\Gamma)$ we denote

$$g_n(A,\Gamma) = \begin{cases} J_n\{A(1-\Gamma^2)^{\frac{1}{2}}\} & -1 < \Gamma < 1 \\ 0 & |\Gamma| > 1 \end{cases} \tag{36}$$

where $J_n$ is a Bessel function of order $n$. For thick crystals the integral (33) can be approximated as described in the appendix. Then, for the forward beam, one obtains from equations (A4) and (A6)

$$\begin{aligned}\phi_0(\alpha) = \frac{c_0\pi\nu}{4}\Big\{&(1-\Gamma_d)(1-\Gamma_d{}^2)^{\frac{1}{2}}g_1(\nu|2A-A_M|,\Gamma_d) - \\ &- 2\Gamma_M(1-\Gamma_M{}^2)^{\frac{1}{2}}g_1(\nu A_M,\Gamma_M) - \\ &- (1-\Gamma_T)(1-\Gamma_T{}^2)^{\frac{1}{2}}g_1\left(\frac{\nu TA}{D},\Gamma_T\right)\Big\}\{1+\exp(i\alpha)\} .\end{aligned} \tag{37a}$$

Equation (37a) does not hold for $D_M \approx 2D$. Therefore the focusing geometry has to be considered separately (equation (A5)):

$$\begin{aligned}\phi_0(\alpha) = \frac{c_0\pi\nu}{4}\Big\{&\exp(-\nu A|\Gamma|) - \Gamma\left(1-\frac{\Gamma^2}{4}\right)^{\frac{1}{2}}g_1\left(2\nu A,\frac{\Gamma}{2}\right) - \\ &- \left(1-\frac{\Gamma}{4}\right)\left(1-\frac{\Gamma^2}{16}\right)^{\frac{1}{2}}g_1\left(4\nu A,\frac{\Gamma}{4}\right)\Big\}\{1+\exp(i\alpha)\} .\end{aligned} \tag{37b}$$

For the deviated beam we do not have to distinguish between different geometries[†]

$$\begin{aligned}\phi_G(\alpha) = -\frac{c_G\pi}{4}\Big\{&4g_0(\nu A_M,\Gamma_M)+\Big\{\mathrm{sgn}(2D-D_M)(1-\Gamma_d{}^2)g_2(\nu|2A-A_M|,\Gamma_d) + \\ &+ 2(1-\Gamma_M{}^2)g_2(\nu A_M,\Gamma_M) - (1-\Gamma_T{}^2)g_2\left(\nu\frac{TA}{D},\Gamma_T\right)\Big\}\{1+\exp(i\alpha)\}\Big\}\end{aligned} \tag{38}$$

because, as in the case of two crystal plates (Indenbom *et al.* 1974, 1976) no diffraction focusing appears. In equations (37) and (38) the ratios

---

[†]In our approximation $g_2(A,\Gamma) = -g_0(A,\Gamma)$ holds.

$$\Gamma_T = \frac{D\Gamma}{D_T}, \quad \Gamma_M = \frac{D\Gamma}{D_M}, \quad \Gamma_d = \frac{D\Gamma}{2D-D_M} \tag{39a}$$

are introduced, which are characteristic of the different parts of the wave fan. From ray considerations we see that the Borrmann fan is divided into several parts which differ in the number of rays impinging on the exit surface.

The broken lines in Fig. 6(b) designate rays which propagate in all wafers in the same direction. They give rise to the anomalously high and the anomalously weak absorbed parts of the wave functions (terms with $g_n(\nu TA/D, \Gamma_T)$). As is displayed in Fig. 6(b), there are additional rays (solid lines) which contribute only in the central region $-1 < \Gamma < 1$. These rays change their direction of propagation in the different wafers and therefore have no extremely anomalous absorbed parts. Their contributions to the intensity vanish with increasing absorption so that they are only important in the cases of weak absorption. In the focusing points rays from any energy flow direction meet.

Our interest is now aimed at the averaged intensities. Because of the different mean values, we have again to distinguis between the three different cases. To abbreviate the following equations, we introduce

$$Cs(\Gamma_T) = \begin{cases} \cosh\left\{2\kappa A\left|\frac{\Gamma}{\Gamma_T}\right|(1-\Gamma_T^2)^{\frac{1}{2}}\right\} & -1 < \Gamma_T < 1 \\ 0 & |\Gamma_T| > 1 \end{cases} \tag{39b}$$

Then we can write the following for the intensities.

(a) Ideal geometry:

$$\overline{P}_o(\Gamma,\alpha) = \frac{\exp(-\mu_0 l)}{8}\left\{(1-3\Gamma)^2(1-\Gamma^2)^{\frac{1}{2}}Cs(\Gamma) + \frac{1}{3}\left(1-\frac{\Gamma}{3}\right)^2\left(1-\frac{\Gamma^2}{9}\right)^{\frac{1}{2}}Cs\left(\frac{\Gamma}{3}\right)\right\}\cos^2\frac{\alpha}{2} \tag{40a}$$

$$\overline{P}_G(\Gamma,\alpha) = \frac{\exp(-\mu_0 l)}{2(1-\Gamma^2)^{\frac{1}{2}}} Cs(\Gamma) - \frac{\exp(-\mu_0 l)}{8}\left\{3\frac{1+2\Gamma^2-3\Gamma^4}{(1-\Gamma^2)^{\frac{1}{2}}} Cs(\Gamma) - \right.$$

$$\left. - \frac{1}{3}\left(1-\frac{\Gamma^2}{9}\right)^{3/2} Cs\left(\frac{\Gamma}{3}\right)\right\}\cos^2\frac{\alpha}{2} \quad . \tag{40b}$$

(b) Focusing geometry:

$$\overline{P}_o(\Gamma,\alpha) = \frac{\exp(-\mu_0 l)}{16}\left\{2A\pi\exp(-2A|\Gamma|) + \Gamma^2\left(1-\frac{\Gamma^2}{4}\right)^{\frac{1}{2}} Cs\left(\frac{\Gamma}{2}\right) + \right.$$

$$\left. + \frac{1}{4}\left(1-\frac{\Gamma}{4}\right)^2\left(1-\frac{\Gamma^2}{16}\right)^{\frac{1}{2}} Cs\left(\frac{\Gamma}{4}\right)\right\}\cos^2\frac{\alpha}{2} \tag{41a}$$

$$\overline{P}_G(\Gamma,\alpha) = \frac{\exp(-\mu_0 l)}{4(1-\Gamma^2/4)^{\frac{1}{2}}} Cs\left(\frac{\Gamma}{2}\right) - \frac{\exp(-\mu_0 l)}{4}\left\{\frac{1-(\Gamma/2)^4}{(1-\Gamma^2/4)^{\frac{1}{2}}} Cs\left(\frac{\Gamma}{2}\right) - \right.$$

$$\left. - \frac{1}{8}\left(1-\frac{\Gamma^2}{16}\right)^{3/2} Cs\left(\frac{\Gamma}{4}\right)\right\}\cos^2\frac{\alpha}{2} \quad . \tag{41b}$$

(c) Other cases:

$$\overline{P}_o(\Gamma,\alpha) = \frac{\exp(-\mu_0 l)}{8}\left\{\frac{D}{|2D-D_M|}(1-\Gamma_d)^2(1-\Gamma_d^{\,2})^{\frac{1}{2}} Cs(\Gamma_d) + \right.$$

$$\left. + \frac{4D}{D_M}\Gamma_M^{\,2}(1-\Gamma_M^{\,2})^{\frac{1}{2}} Cs(\Gamma_M) + \frac{D}{T}(1-\Gamma_T)^2(1-\Gamma_T^{\,2})^{\frac{1}{2}} Cs(\Gamma_T)\right\}\cos^2\frac{\alpha}{2} \tag{42a}$$

$$\overline{P}_G(\Gamma,\alpha) = \frac{\exp(-\mu_0 l)}{2(1-\Gamma_M^{\,2})^{\frac{1}{2}}}\frac{D}{D_M} Cs(\Gamma_M) -$$

$$- \frac{\exp(-\mu_0 l)}{8}\left\{-\frac{D}{|2D-D_M|}(1-\Gamma_d^{\,2})^{3/2} Cs(\Gamma_d) + \right. \tag{42b}$$

$$+ \frac{4D}{D_M} \frac{1-\Gamma_M^{\;4}}{(1-\Gamma_M^{\;2})^{\frac{1}{2}}} Cs(\Gamma_M) - \frac{D}{T}(1-\Gamma_T^{\;2})^{3/2} Cs(\Gamma_T) \Big\} \cos^2 \frac{\alpha}{2} .$$

In the symmetric case the path length $l$ is independent of $\Gamma$. For the neutron interferometer, the zero absorption case is of main interest and is displayed in Figs. 7–9. The intensity profiles for the ideal geometry are shown in Fig. 7. The increase of the intensity of the deviated beam, varying the phase $\alpha$ from 0 to $\pi$, is shown in Figs. 7(b), 8(b), and 9(b). If $\alpha = \pi$, one obtains in any case the same intensity profile as for diffraction only at the mirror.

Figure 8 shows the spatial intensity distributions of the focusing geometry. The width $\Delta x$ of the peak is independent of the crystal thickness and is about $\Delta x = 2\pi\Delta_0$ (symmetric case). This peak has a needle structure (Indenbom *et al.* 1974, 1976) and contains about 50 per cent of the forward intensity (see equation (44)). In Fig. 9 the case with $D_M = 1.5\ D$ is displayed.

With increasing absorption, the contributions of the

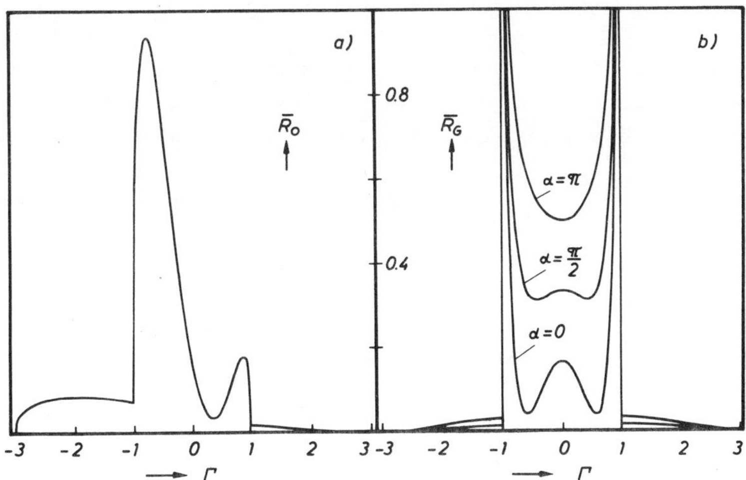


FIG. 7. Averaged intensity profiles for zero absorption and $D_M = D$: (a) forward beam; (b) deviated beam. The phase shift $\alpha = 0$, $\pi/2$, and $\pi$.

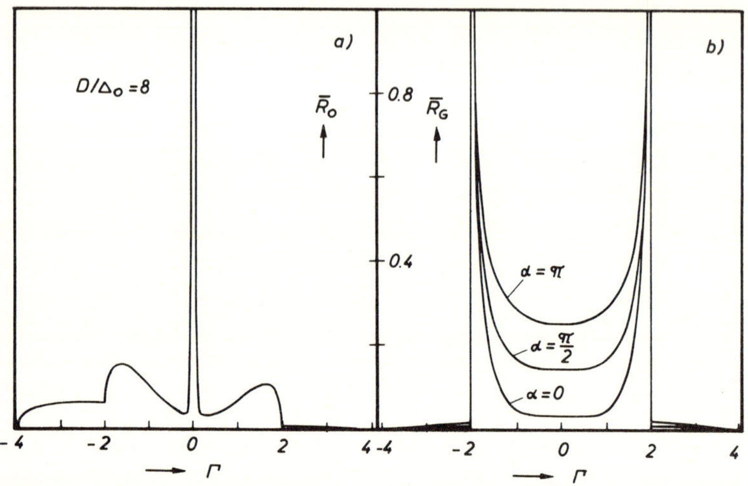


FIG. 8. Averaged intensity profiles for zero absorption and $D_M = 2D$: (a) forward beam; (b) deviated beam. The phase shift $\alpha = 0$, $\pi/2$, and $\pi$.

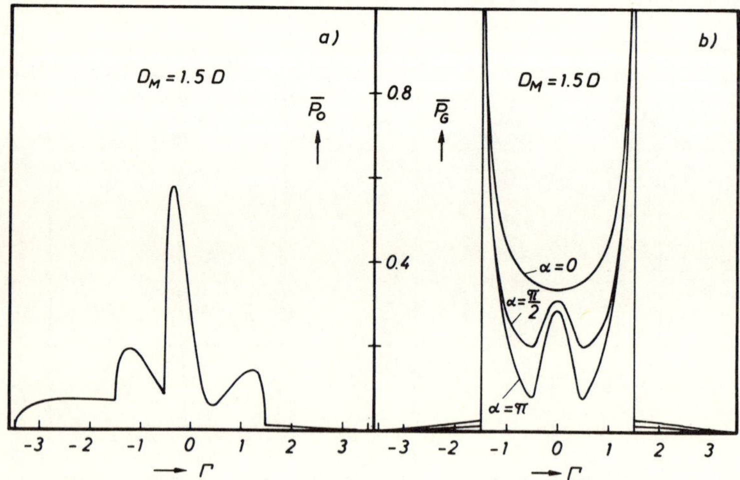


FIG. 9. Averaged intensity profiles for zero absorption and $D_M = 1.5D$: (a) forward beam; (b) deviated beam. The phase shift $\alpha = 0$, $\pi/2$, and $\pi$.

additional rays decrease faster than those of the rays which are present in the whole region. Figure 10 shows the influence of the absorption for the ideal geometry. The terms surviving the strong Borrman absorption are equal in all three cases, as is expected.

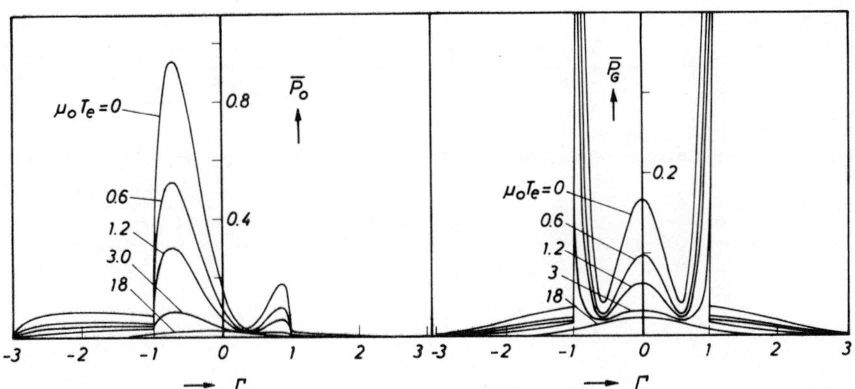


FIG. 10. Averaged intensity profiles for $D_M = D$ for different absorptions ($6\kappa A = \mu_0 T_e$): (a) forward beam; (b) deviated beam (after Petrascheck and Folk 1976).

Returning to the non-absorbing case, we compare the intensities computed for the ideal geometry (Fig. 7) with those of the LLL interferometer in Grenoble (Fig. 11 and Rauch 1978). The experimental set-up was not intended to give a precise check of the theoretical predictions. Furthermore, the contrast $C_0$ of this interferometer is about 40 per cent. Therefore we cannot expect a quantitative agreement with the idealized theoretical calculations.

The integrated powers calculated from equations (40)–(42) are the same as those obtained from plane-wave theory. They are listed in equations (43)–(45) for zero absorption.

(a) $D_M = D$, ideal geometry:

$$\overline{R}_0 = \frac{13\pi}{64} + \frac{5\pi}{64}$$

$$\overline{R}_G = \frac{\pi}{2} - \left(\frac{21\pi}{64} - \frac{3\pi}{64}\right)\cos^2\frac{\alpha}{2} \; . \tag{43}$$

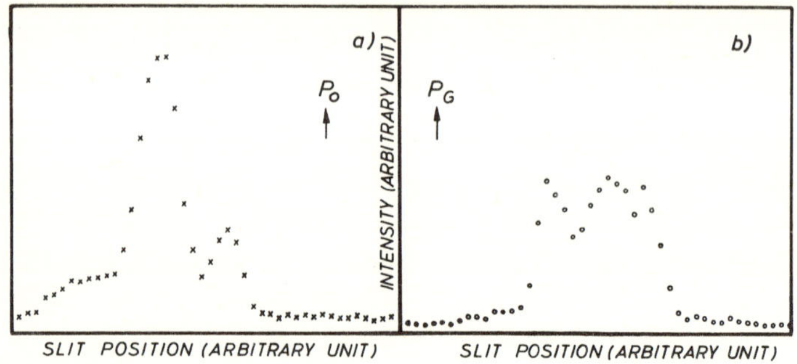


FIG. 11. Intensity profiles measured at the LLL interferometer ($D_M = D$) in Grenoble in 1977: (a) forward beam; (b) deviated beam. (After Rauch 1978.)

(b) $D_M = 2D$, focusing geometry:

$$\begin{aligned} \overline{R}_0 &= \frac{8\pi}{64} + \frac{4\pi}{64} + \frac{5\pi}{64} \\ \overline{R}_G &= \frac{\pi}{2} - \left(\frac{20\pi}{64} - \frac{3\pi}{64}\right)\cos^2\frac{\alpha}{2}\,. \end{aligned} \tag{44}$$

(c) Other cases:

$$\begin{aligned} \overline{R}_0 &= \frac{5\pi}{64} + \frac{4\pi}{64} + \frac{5\pi}{64} \\ \overline{R}_G &= \frac{\pi}{2} - \left(-\frac{3\pi}{64} + \frac{20\pi}{64} - \frac{3\pi}{64}\right)\cos^2\frac{\alpha}{2}\,. \end{aligned} \tag{45}$$

The order of the terms is the same as in equations (40)–(42). Therefore we can directly compare the integrated intensities of the different rays. As mentioned before, the intensities of the central regions give the main contributions.

## 5. The defocused interferometer

In general the contrast $C_0$ of a neutron interferometer is less than unity. Here we shall not deal with reduction due to distortions which gives rise to rotational moiré patterns (Bonse and Hart 1965c, Treimer 1975) or reduction due to vibrations, thermal instabilities, etc. Only geometrical defocusing as given by deviations from equation (2) is considered.

In X-ray interferometry there are two different defocusing effects: different distances between the wafers $t_A^{\text{I}} = t_S^{\text{II}} + \Delta t$ and/or differently thick mirrors. In neutron interferometry we have a third equation, $D_S = D_A$, to fulfil. The contrast for these three kinds of defocusing calculated from the averaged intensities of the non-absorbing ideal geometry is given by Petrascheck and Rauch (1976). Special situations for neutrons and X-rays are described by Bonse and Graeff (1977).

Here we shall restrict ourselves to the most common case, i.e. different distances between the crystal plates $t_A^{\text{I}} = t + \Delta t$. $\Delta t$ is usually called the 'defocus' (see Fig. 12).

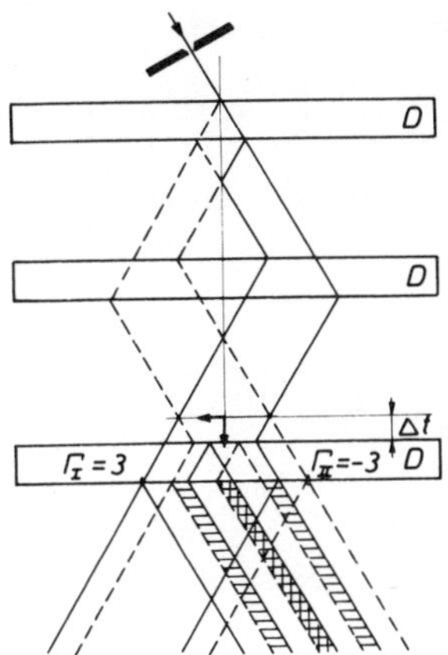


FIG. 12. Defocused interferometer. The wave fans do not cover the same region.

If we want to calculate the contrast (22) for such a defocused interferometer, we have only to replace $\alpha$ in equation (10) by $\alpha + 2\pi y \Delta t/\Delta_0$ and to perform the integration over $y$.

In the zero absorption case, the contrast is straightforwardly calculated from the averaged intensities (Petrascheck and Rauch 1976).

(a) Ideal geometry:

$$\bar{C}_0 = \left|1 + \frac{2\pi|\Delta t|}{\Delta_0} - \frac{5}{9}\left(\frac{2\pi\Delta t}{\Delta_o}\right)^2\right| \exp(-2\pi|\Delta t|/\Delta_0) \ . \quad (46a)$$

(b) Focusing geometry:

$$\bar{C}_0 = \left|1 + \frac{2\pi|\Delta t|}{\Delta_0} - \frac{5}{17}\left(\frac{2\pi\Delta t}{\Delta_0}\right)^2\right| \exp(-2\pi|\Delta t|/\Delta_0) \ . \quad (46b)$$

(c) Other cases:

$$\bar{C}_0 = \left|1 + \frac{2\pi|\Delta t|}{\Delta_0} - \frac{3}{7}\left(\frac{2\pi\Delta t}{\Delta_0}\right)^2\right| \exp(-2\pi|\Delta t|/\Delta_0) \ . \quad (46c)$$

$\bar{C}_0$ is shown in Fig. 13 for $D_M = D$ and $D_M = 2D$. To assess the validity of (46), Fig. 14 shows $C_0$ computed numerically for the ideal geometry with $D = 10\ \Delta_0$ and $D = 100\ \Delta_0$. The approximate results (46) hold if the crystals are not too thin and if $\Delta t \ll \Delta_0$.

In the same figure we see that the contrast for strong anomalous absorption is appreciably less sensitive to geometrical deviations (Petrascheck and Folk 1976). Since rays with small $y$ are only slightly influenced by the defocus, a narrow reflection curve effects only a small change of the contrast

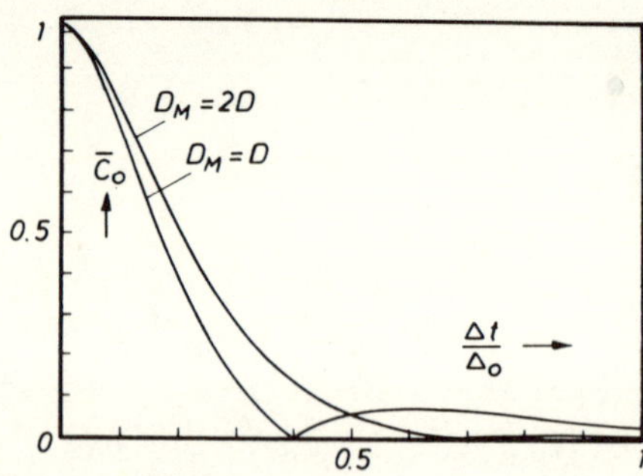


FIG. 13. The contrast $\bar{C}_0$ calculated from the averaged intensities *versus* the defocus $\Delta t$ (see Fig. 12). (After Petrascheck and Rauch 1976.)

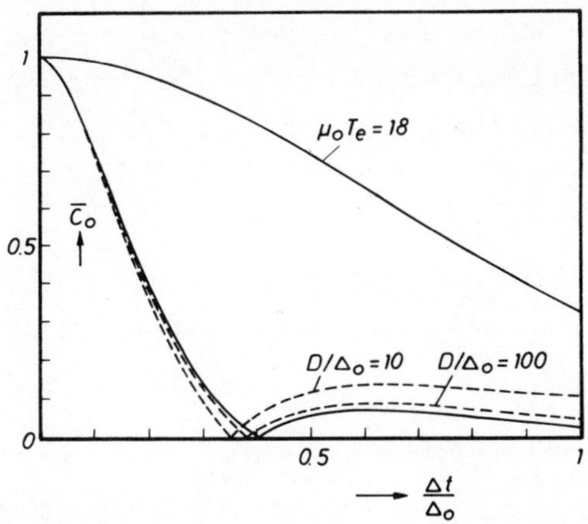


FIG. 14. The contrast $C_0$ *versus* the defocus $\Delta t$ for $D_M = D = 10\ \Delta_0$ and for $D_M = D = 100\ \Delta_0$ (broken lines) with $\mu_0 = \kappa = 0$. The solid lines show $\overline{C}_0$ in the symmetric case for $\mu_0 T_e = 6\kappa A = 0$ and 18. (After Petrascheck and Rauch 1976.)

with the defocus. Having this in mind one may compare Fig. 3 with Fig. 13.

Next we consider the intensity profile for the defocused interferometer (see Fig. 12). Since the wave fans do not cover the same region, it is convenient to introduce the parameter $\Gamma$ for each path separately (we restrict ourselves to the symmetric case sketched in Fig. 12):

$$\Gamma_{I,II} = \mp \frac{\Delta t}{D} + \Gamma = \mp \varepsilon + \Gamma \quad . \tag{47}$$

Then we develop the intensity calculated from $\phi_0^{I}(\Gamma_I)$ and $\phi_0^{II}(\Gamma_{II})$ with respect to $\varepsilon$. Averaging over the Pendellösung oscillations we obtain for the ideal geometry

$$\overline{P}_0(\Gamma) = \frac{\exp(-\mu_0 T_e)}{8}\Bigg[(1-3\Gamma)^2(1-\Gamma^2)^{\frac{1}{2}} Cs(\Gamma)\ \cos^2\left\{\frac{A\Gamma\varepsilon}{(1-\Gamma^2)^{\frac{1}{2}}}\right\} + \\ + \frac{1}{3}\left(1-\frac{\Gamma}{3}\right)^2\left(1-\frac{\Gamma^2}{9}\right)^{\frac{1}{2}} Cs\left(\frac{\Gamma}{3}\right)\cos^2\left\{\frac{A\Gamma\varepsilon}{3(1-\Gamma^2/9)^{\frac{1}{2}}}\right\}\Bigg] \ . \tag{48}$$

The fringe system extending over the whole profile shows fringe spacings known from X-ray interferometry (Bonse and te Kaat 1971). In the central region an additional fringe system is superposed with spacings which are roughly one-third of those of the fundamental system.

## References


BAUSPIESS, W., BONSE, U., and GRAEFF, W. (1976). *J. appl. Crystallog.* **9**, 68.

BAUSPIESS, W., BONSE, U., RAUCH, H., and TREIMER, W. (1974). *Z. Phys.* **271**, 177.

BONSE, U. and HART, M. (1965a). *J. appl. Phys. Lett.* **6**, 155.

—— (1965b). *Z. Phys.* **188**, 154.

—— (1965c). *Z. Phys.* **190**, 455.

BONSE, U. and TE KAAT, E. (1971). *Z. Phys.* **243**, 14.

BONSE, U. and GRAEFF, W. (1977). 'X-ray and neutron interferometry'. In *Topics in applied physics*, Vol. 22, *X-ray optics* (ed. H.J. Queisser), p. 93. Springer, Berlin.

INDENBOM, V.L., SLOBODETSKII, I. SH., and TRUNI, K.G. (1974). *Sov. Phys.—JETP* **39**, 542.

INDENBOM, V.L., SUVOROV, E.V., and SLOBODETSKII, I.SH. (1976). *Sov. Phys.—JETP* **44**, 187.

KATO, N. (1968). *J. appl. Phys.* **39**, 2225.

PETRASCHECK, D. (1976). *Acta Phys. Aust.* **45**, 217.

PETRASCHECK, D. and FOLK, R. (1976). *Phys. Status Solidi A* **36**, 147.

PETRASCHECK, D. and RAUCH, H. (1976). 'Theorie des Interferometers', AIAU 76401, Atominstitut der Österreichsichen Universitäten, Vienna.

RAUCH, H. and SUDA, M. (1974). *Phys. Status Solidi A* **25**, 495.

RAUCH, H., TREIMER, W., and BONSE, U. (1974). *Phys. Lett. A* **47**, 369.

RAUCH, H. (1978). Seminar, Atominstitut, Vienna.

RAUCH, H. and PETRASCHECK, D. (1978). 'Dynamical neutron diffraction and its application'. In *Topics in current physics*, vol. 6, *Neutron diffraction* (ed. H. Dachs). Springer, Berlin.

TREIMER, W. (1975). Thesis, Rep. AIAU 75405, Atominstitut der Österreichischen Universitäten, Vienna.

## Appendix

The integral given by equation (30) is calculated in a manner similar to the method described by Petrascheck (1976) and Petrascheck and Folk (1976). We start with

$$F(\nu,A,A_{\mathrm{M}},p) = \tfrac{1}{4}\{I(\nu,2A-A_{\mathrm{M}},p) + 2I(\nu,A_{\mathrm{M}},p) - I(\nu,2A+A_{\mathrm{M}},p)\} \tag{A1}$$

where $I$ is rewritten as follows:

$$I(\nu,A,p) = \frac{\nu^2}{2}\int_{-\infty+ig}^{\infty+ig} \mathrm{d}z\, \exp(\mathrm{i}pz)\frac{\sin\{A(\nu^2+z^2)^{\frac{1}{2}}\}}{(\nu^2+z^2)^{3/2}}$$

$$= \begin{cases} \dfrac{\pi\nu A}{2}\exp(-\nu|p|) - \displaystyle\int_0^{(A^2-p^2)^{\frac{1}{2}}} \mathrm{d}t\, \frac{t^2 J_1(\nu t)}{(t^2+p^2)^{3/2}} & |p| < |A| \\ \dfrac{\pi\nu A}{2}\exp(-\nu|p|) & |p| > |A| \end{cases} \tag{A2}$$

Assuming thick crystals, $A \gg \pi$ and $A_{\mathrm{M}} \gg \pi$, one may partially integrate (A2) and then neglect terms of order $A^{-3/2}$ and $A_{\mathrm{M}}^{-3/2}$. This leads to the following approximate expression for (A1):

$$F(\nu,A,A_{\mathrm{M}},p) \approx \frac{\pi}{8}\Big[-\mathrm{sgn}(2A-A_{\mathrm{M}})\Big\{1-\frac{p^2}{(2A-A_{\mathrm{M}})^2}\Big\}g_2\Big(\nu|2A-A_{\mathrm{M}}|,\frac{p}{2A-A_{\mathrm{M}}}\Big)$$

$$-2\Big(1-\frac{p^2}{A_{\mathrm{M}}^2}\Big)g_2\Big(A_{\mathrm{M}},\frac{p}{A_{\mathrm{M}}}\Big)$$

$$+\Big\{1-\frac{p^2}{(2A+A_{\mathrm{M}})^2}\Big\}g_2\Big\{\nu(2A+A_{\mathrm{M}}),\frac{p}{2A+A_{\mathrm{M}}}\Big\}\Big] \tag{A3}$$

where $g_2(A_{\mathrm{M}},p/A_{\mathrm{M}})$ is defined by equation (36).

For the forward beam one has to perform the derivatives of $F(\nu,A,A_{\mathrm{M}},p)$ with respect to $A$ and $p$:

$$\frac{\partial F}{\partial A} = \frac{\pi\nu}{4}\left[\left\{1-\frac{p^2}{(2A-A_M)^2}\right\}^{\frac{1}{2}} g_1\left(\nu|2A-A_M|, \frac{p}{2A-A_M}\right) + \right.$$

$$+ \left\{1-\frac{p^2}{(2A+A_M)^2}\right\}^{\frac{1}{2}} g_1\left(\nu(2A+A_M), \frac{p}{2A+A_M}\right) +$$

$$\left. + \int_{\{(2A-A_M)^2-p^2\}^{\frac{1}{2}}}^{\{(2A+A_M)^2-p^2\}^{\frac{1}{2}}} \mathrm{d}t \; \frac{t^2 J_1(\nu t)}{(t^2+p^2)^{3/2}}\right] . \tag{A4}$$

As long as $|2A-A_M| \gg \pi$, the integral on the right-hand side of equation (A4) is negligible. If, on the other hand, $A_M \approx A$, this contribution becomes important. Therefore in the case of the focusing geometry we have from equation (A4)

$$\frac{\partial F}{\partial A} \approx \frac{\pi\nu}{4}\left\{\left(1-\frac{p^2}{16A^2}\right)^{\frac{1}{2}} g_1\left(4\nu A, \frac{p}{4A}\right) + \exp(-\nu|p|)\right\} . \tag{A5}$$

Finally, one obtains for the derivative with respect to $p$

$$\frac{\partial F}{\partial p} \approx \frac{\pi\nu}{8}\left[\frac{p}{2A-A_M}\left\{1-\frac{p^2}{(2A-A_M)^2}\right\}^{\frac{1}{2}} g_1\left(\nu|2A-A_M|, \frac{p}{2A-A_M}\right) + \right.$$

$$+ 2\frac{p}{A_M}\left(1-\frac{p^2}{A_M^2}\right)^{\frac{1}{2}} g_1\left(\nu A_M, \frac{p}{A_M}\right) -$$

$$\left. - \frac{p}{2A+A_M}\left\{1-\frac{p^2}{(2A+A_M)^2}\right\}^{\frac{1}{2}} g_1\left\{\nu(2A+A_M), \frac{p}{2A+A_M}\right\}\right] . \tag{A6}$$

# 9. INTEGRATED INTENSITIES AND FLIPPING RATIOS IN NEUTRON DIFFRACTION BY PERFECT MAGNETIC CRYSTALS

J.P. GUIGAY

*Service de Physique du Solide, Département de Recherche Fondamentale, C.E.N.G., Grenoble, France*

M. SCHLENKER

*Institut Laue Langevin, BP 156 Centre de Tri, 38 Grenoble, France*

## 1. Introduction

Interest in the theory of neutron diffraction by perfect magnetic crystals is steadily growing and there appears to be at least three reasons for this. First, there is now a reasonable hope of obtaining magnetic crystals which are sufficiently perfect to test the theoretical predictions experimentally; for instance flux-grown garnets with few dislocations are always affected by growth bands, but this would be harmless if the scattering vector used is parallel to these bands and it should then be possible to use apparently perfect slices, some $10^{-1}$ mm thick. Secondly, we must know the dynamical behaviour of the scattering by a perfect crystal for a better treatment of the extinction problem (Hamilton 1958, Bonnet, Delapalme and Becker 1976). Finally, the advent of neutron topography, a technique well suited to the observation of all kinds of magnetic domains (Schlenker and Baruchel 1978), makes it desirable to investigate the behaviour of neutron beams in magnetic crystals under conditions which are unusual with respect both to the quality of the samples and the involved geometrical configurations.

Few papers have been published to date on the dynamical theory of neutron diffraction by magnetic crystals, and in most of them the incident plane wave case was treated (Stassis and Oberteuffer 1974, Sivardière 1975, Schmidt, Deimel, and Daniel 1975, Schmidt and Deimel 1975, Mendiratta and Blume 1976, Belyakov and Bokun 1976). The main properties of the dispersion surface have been studied by Sivardière, for instance. Some additional properties will be pointed out in the present paper. We shall calculate integrated intensities and flipping ratios. Such calculations have to be done numerically except for some simple cases or in the kinematic limit for the general case.

## 2. The interaction of neutrons with the atomic magnetic moments

The amplitude $A_n$ of nuclear scattering by a single atom is independent of the direction of the scattering vector $\mathbf{g}$ (isotropic scattering). The magnetic scattering amplitude is shown to be (Halpern and Johnson 1939)

$$A_m(\mathbf{g}) = \frac{2M}{\hbar^2} \mu_n \cdot \mu_{e\perp} f(\mathbf{g}) \tag{1}$$

(this is in c.g.s. units; in mks units we have the usual factor $\mu_0/4\pi = 10^{-7}$). In equation (1) $\boldsymbol{\mu}_n$ is the neutron magnetic moment, $\boldsymbol{\mu}_{e\perp}$ is the projection of the atom magnetic moment on the plane perpendicular to the scattering vector $\mathbf{g}$, and $f(\mathbf{g})$ is an angular form factor reaching its maximum value of unity when the scattering angle is zero.

The nuclear and magnetic scattering amplitudes for the unit cell of a crystal are, of course, obtained by summing $A_n$ and $A_m$ over the contributing atoms, taking into account the relative positions of these atoms in the unit cell. The total scattering amplitude is the sum of the nuclear and magnetic scattering amplitudes.

The wave equation of neutrons in a magnetic crystal is

$$\nabla^2\Psi + k^2\Psi = \frac{2M}{\hbar^2}\{V(\mathbf{r}) + \tfrac{1}{2}\mu\ \boldsymbol{\sigma}\cdot\mathbf{B}(\mathbf{r})\}\Psi(\mathbf{r})$$

in which $M$ is the neutron mass, $V(\mathbf{r})$ the nuclear potential, $\mathbf{B}(\mathbf{r})$ the microscopic magnetic field inside the crystal, $\mu_n$ the absolute value of the neutron magnetic moment and the $\boldsymbol{\sigma}$ are the Pauli matrices. $\Psi(\mathbf{r})$ is a two-component spinor wave function. $V(\mathbf{r})$ and $\mathbf{B}(\mathbf{r})$ have Fourier series expansions

$$V(\mathbf{r}) = \sum_{\mathbf{g}} V_{\mathbf{g}} \exp(i\mathbf{g}\cdot\mathbf{r})$$
$$\mathbf{B}(\mathbf{r}) = \sum_{\mathbf{g}} \mathbf{B}_{\mathbf{g}} \exp(i\mathbf{g}\cdot\mathbf{r}) \ . \tag{2}$$

When there is no Bragg reflection, the neutron wave is only affected by the mean values $V_0$ and $\mathbf{B}_0$ of $V(\mathbf{r})$ and $\mathbf{B}(\mathbf{r})$. $\mathbf{B}_0$ is of course the macroscopic magnetic field. The interaction is then interpreted in terms of an index of refraction $n$ which depends on the orientation of the polarization vector of the neutron beam.

When the Bragg condition is satisfied for one scattering vector $\mathbf{g}$ of the reciprocal lattice, we also need the coefficients $V_{\mathbf{g}}$ and $\mathbf{B}_{\mathbf{g}}$ of the Fourier expansions (2). The relation $\mathrm{div}(\mathbf{B}) = 0$ shows that $\mathbf{B}_{\mathbf{g}}$ is perpendicular to $\mathbf{g}$. More precisely, $\mathbf{B}_{\mathbf{g}}$ is collinear with the projection of $\mathbf{B}_0$ on the plane perpendicular to $\mathbf{g}$, in the case of collinear magnetic structures (we only consider this case in the present paper). This important property can be justified as follows. The magnetic field due to atoms located in $\mathbf{r}_j$ and having parallel magnetic moments $\boldsymbol{\mu}_j$ is

$$\mathbf{B}(\mathbf{r}) = \sum_j \nabla \times \left(\nabla\times\frac{\mu_j}{|\mathbf{r}-\mathbf{r}_j|}\right) = \sum_j \nabla\left(\nabla\cdot\frac{\mu_j}{|\mathbf{r}-\mathbf{r}_j|}\right) - \nabla^2\left(\frac{\mu_j}{|\mathbf{r}-\mathbf{r}_j|}\right) \ ,$$

from which it follows that

$$\nabla \times \mathbf{B}(\mathbf{r}) = -\sum_j \nabla \times \left\{\boldsymbol{\mu}_j\nabla^2\left(\frac{1}{|\mathbf{r}-\mathbf{r}_j|}\right)\right\}$$

is perpendicular to $\boldsymbol{\mu}_j$. Introducing the Fourier transform $\mathbf{B}_{\mathbf{g}}$ of $\mathbf{B}(\mathbf{r})$, we obtain $(\mathbf{g}\times\mathbf{B}_{\mathbf{g}}) \cdot \boldsymbol{\mu}_j = 0$. This shows indeed that g, $\boldsymbol{\mu}_j$ and $\mathbf{B}_{\mathbf{g}}$ are coplanar.

## 3. The dispersion surface and its asymptotic form

### *3.1. General equations*

We introduce a spinor wave function of the form

$$\Psi = \exp(\mathrm{i}\mathbf{k}_0\cdot\mathbf{r})\begin{pmatrix} a_0 \\ b_0 \end{pmatrix} + \exp(\mathrm{i}\mathbf{k}_g\cdot\mathbf{r})\begin{pmatrix} a_g \\ b_g \end{pmatrix} \tag{3}$$

with $\mathbf{k}_g = \mathbf{k}_0 + \mathbf{g}$ the spin quantification axis being along $\mathbf{B}_0$. It is convenient to introduce the functions

$$u(\mathbf{r}) = \frac{2M}{\hbar^2} V(\mathbf{r})$$

$$\mathbf{Q}(\mathbf{r}) = \frac{M}{\hbar^2} \mu_n \mathbf{B}_\perp(\mathbf{r}) \ .$$

The four amplitudes $a_0$, $a_g$, $b_0$, and $b_g$ of equation (3) are coupled by the Fourier coefficients

$$u_0,\ u_g,\ u_{-g} = u_g^*,\ \mathrm{Q}_0,\ \mathbf{Q}_g,\ \mathrm{Q}_{-g} = \mathrm{Q}_g^*.$$

We note that $u_g$ and $\mathbf{Q}_g$ are equal to the usual nuclear and magnetic structure factors $F_N$ and $\mathbf{F}_M(\mathbf{g})$ divided by the volume of the unit cell.

By a straightforward calculation we obtain the following equations which must be satisfied by $a_0$, $a_g$, $b_0$, and $b_g$:

$$a_0(k^2-k_0{}^2) = u_0a_0+u_g^*\,a_g+Q_0a_0+Q_g^*\,a_g\ \sin\beta-Q_g^*\,b_g\ \cos\beta$$

$$a_g(k^2-k_g{}^2) = u_ga_0+u_0a_g+Q_0a_g+Q_ga_0\ \sin\beta-Q_gb_0\ \cos\beta$$

$$b_0(k^2-k_0{}^2) = u_0b_0+u_g^*\,b_g-Q_0b_0-Q_g^*\,b_g\ \sin\beta-Q_g^*\,a_g\ \cos\beta$$

$$b_g(k^2-k_g{}^2) = u_gb_0+u_0b_g-Q_0b_g-Q_gb_0\ \sin\beta-Q_ga_0\ \cos\beta$$

where $\beta$ is the angle $(\mathbf{g},\mathbf{B}_0)$ defined in Fig. 1.

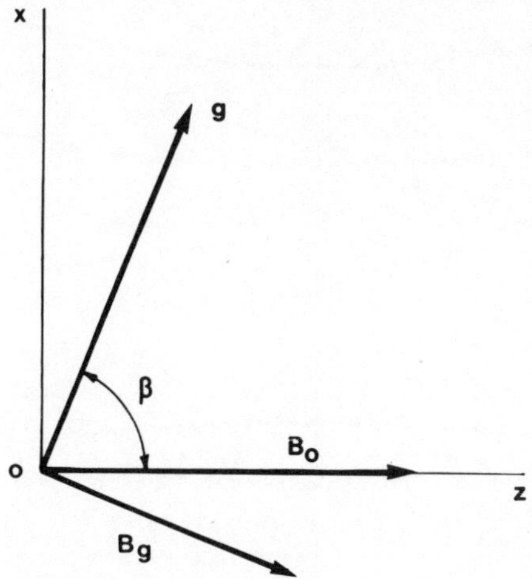


FIG. 1. Relative positions of the vectors $\mathbf{B}_0$, $\mathbf{B}_g$ and $\mathbf{g}$ in the $(\mathbf{g},\mathbf{B}_0)$ plane.

We introduce new quantities

$$K = (k^2-u_0)^{\frac{1}{2}}, \qquad X = k_0-K, \qquad Y = k_g-K$$

$$u = u_g/2K, \qquad Q = Q_g/2K, \qquad A = Q_0/2K$$

together with the usual approximation

$$k^2 - k_0{}^2 - u_0 = 2K(K-k_0)$$

$$k^2 - k_g{}^2 - u_0 = 2K(K-k_g) \;.$$

The geometrical significance of $X$, $Y$, and $A$ is indicated on Fig. 2. We then obtain our fundamental equations

$$\begin{aligned}
(X+A)a_0 + (u^*+Q^* \sin\beta)a_g - Q^*b_g \cos\beta &= 0\\
(Y+A)a_g + (u+Q \sin\beta)a_0 - Qb_0 \cos\beta &= 0\\
(X-A)b_0 + (u^*-Q^* \sin\beta)b_g - Q^*a_g \cos\beta &= 0\\
(Y-A)b_g + (u-Q \sin\beta)b_0 - Qa_0 \cos\beta &= 0.
\end{aligned} \tag{4}$$

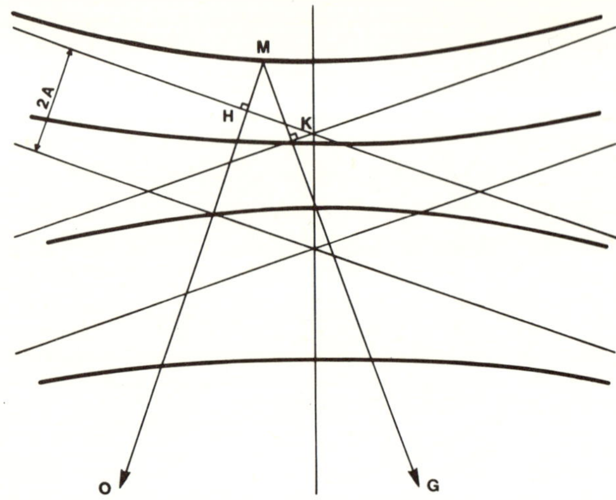


FIG. 2. Schematic representation of the four branches of the dispersion surface with their asymptotes. For any point M of the dispersion surface MH = $X - A$, and MK = $Y - A$.

The dispersion surface is defined by the condition that their determinant is equal to zero:

$$\begin{vmatrix} X+A & u^*+Q^* \sin\beta & 0 & -Q^* \cos\beta \\ u+Q \sin\beta & Y+A & -Q \cos\beta & 0 \\ 0 & -Q^* \cos\beta & X-A & u^*-Q^* \sin\beta \\ -Q \cos\beta & 0 & u-Q \sin\beta & Y-A \end{vmatrix} = 0. \tag{5}$$

*3.2. Some new properties*

From equation (4) it can be seen that the dispersion surface has four asymptotes

$$X+A = 0, \quad Y+A = 0, \quad X-A = 0, \quad Y-A = 0.$$

We can actually derive a more precise statement of this asymptotic form.

From equation (5), we obtain with some elementary calculations

$$\{(X+A)(Y+A)-|u+Q\ \sin\beta|^2-QQ^*\cos^2\beta\}\{(X-A)(Y-A)-|u-Q\ \sin\beta|^2-QQ^*\cos^2\beta\}$$

$$=\cos^2\beta\{(uQ^*+u^*Q)^2-4A^2QQ^*\}\ .$$

This is an interesting representation which shows immediately that the dispersion surface is asymptotic to the two following hyperbolic surfaces:

$$\begin{aligned}(X+A)(Y+A) &= |u+Q\ \sin\beta|^2 + QQ^*\ \cos^2\beta \\ (X-A)(Y-A) &= |u-Q\ \sin\beta|^2 + QQ^*\ \cos^2\beta\ .\end{aligned} \tag{6}$$

This asymptotic property will be useful for our numerical calculations and for our study of the kinematical approximation.

A reversal of the magnetization ($\mathbf{B}_0\rightarrow-\mathbf{B}_0$) corresponds in equations (4) to $A\rightarrow -A$ and $Q\rightarrow -Q$. We then obtain a set of four equations which can be reconverted into equation (4) if we apply the following transformation:

$$\begin{pmatrix}a_0\\ \\ b_0\end{pmatrix}\rightarrow\begin{pmatrix}b_0\\ \\ -a_0\end{pmatrix}\qquad\qquad\begin{pmatrix}a_g\\ \\ b_g\end{pmatrix}\rightarrow\begin{pmatrix}b_g\\ \\ -a_g\end{pmatrix}$$

This is exactly the reversal rule for any spinor having its polarization vector in the $(x,z)$ plane. We conclude that the wave fields corresponding to the different points of the dispersion surface are polarized in the $(\mathbf{g},\mathbf{B}_0)$ plane.

*3.3. The simple case* $\beta=\pi/2$

This is a simple and usual experimental situation. It is readily seen from equation (5) that the (±) spinor components satisfy separated equations; we obtain two hyperbolic dispersion surfaces $\Sigma_+$ and $\Sigma_-$ corresponding respectively to neutrons with polarization parallel and antiparallel to $\mathbf{B}_0$:

$$\begin{aligned}(X+A)(Y+A) &= |u+Q|^2 \qquad (\Sigma_+)\\ (X-A)(Y-A) &= |u-Q|^2 \qquad (\Sigma_-)\ .\end{aligned}$$

## 4. Integrated intensities and flipping ratios in the symmetric Laue case

### *4.1. General considerations*

According to the usual Huyghens construction, four coherent wave fields are excited by an incident plane wave. The actual incident wave cannot be considered as a plane wave. In the case of an incident spherical wave, which can be expressed as a superposition of coherent plane-wave components, the whole dispersion surface is excited.

In X-ray section topography, which requires crystals with thicknesses much larger than the beam width, the spherical wave approach takes into account the interferences of wave fields propagating along the same direction. In practical neutron topography (Schlenker and Baruchel 1978), the cross-section of the incident beam is larger than the crystal thickness. In this case (projection topographs), general arguments due to Kato (1969) indicate that the resulting intensities are the same when calculated by the spherical wave treatment or by the older method of integrating the intensities due to a uniform set of incoherent plane waves, thus disregarding the phase relations between these plane-wave components.

We shall calculate such integrated diffracted intensities in the symmetric Laue case, i.e. the entrance and exit surfaces of the crystal are parallel and the scattering vector **g** is parallel to them.

The flipping ratio is defined as the ratio of the diffracted intensities when the incident polarization $\mathbf{P}_i$ is reversed. Usually the incident polarization is parallel to $\mathbf{B}_0$, but this definition can be applied for any direction of $\mathbf{P}_i$. In general, we shall have to consider four diffracted intensities $I_{++}$, $I_{+-}$, $I_{-+}$, and $I_{++}$, in which the subscripts indicate the initial and final polarization. The flipping ratio is then

$$R = \frac{I_{++} + I_{+-}}{I_{-+} + I_{--}} \quad .$$

*4.2. Numerical calculations in the case* $\beta \neq \pi/2$

From equations (4), the amplitudes ($a_0, b_0$; $a_g, b_g$) of any wave fields are defined to within an arbitrary multiplicative constant. From the matching conditions at the entrance surface we obtain the correct amplitudes $a_{0j}$, $b_{0j}$, $a_{gj}$, $b_{gj}$ ($j$ = 1,2,3,4) of the four wave fields excited by a plane-wave component. We use the notations

$$s = X + Y, \qquad D = X - Y, \qquad t = d/\cos\theta$$

in which $d$ is the crystal thickness and $\theta$ the Bragg angle. Each plane-wave component of the incident beam is defined by a value of $D$, $D$ = 0 corresponding to the exact Bragg position. The diffracted intensities are then

$$I_+(t) = \int_{-\infty}^{+\infty} \mathrm{d}D \,|\sum_j a_{gj} \exp(\mathrm{i}tS_j)|^2$$

$$I_+(t) = \int_{-\infty}^{-\infty} \mathrm{d}D \,|\sum_j b_{gj} \exp(\mathrm{i}tS_j)|^2 \quad .$$

When $t$ is varied, variations of $I_+$ and $I_-$ (Pendellösung oscillations) will be observed. From stationary phase considerations, the dynamical limit $t \to \infty$ is

$$I_+ = \sum_j \int_{-\infty}^{-\infty} \mathrm{d}D \,|a_{gj}|^2$$

$$I_- = \sum_j \int_{-\infty}^{-\infty} \mathrm{d}D \,|b_{gj}|^2 \quad .$$

We have performed numerical calculations of these integrals and of the flipping ratio in the case of a centrosymmetric crystal ($u$ and $Q$ are then real quantities). Some results are shown in Fig. 3.

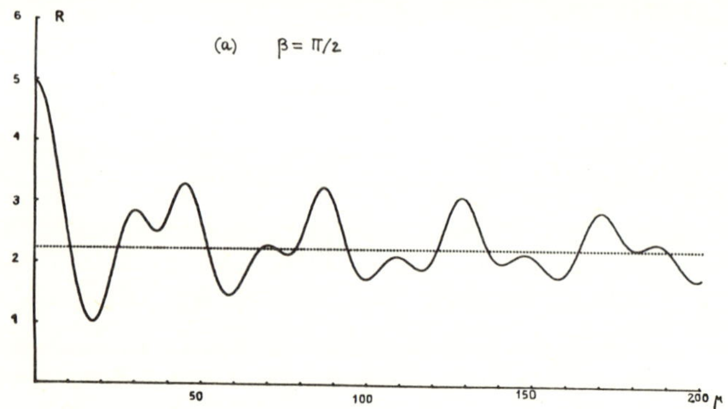


FIG. 3. Calculation of the flipping ratio as a function of the specimen thickness (in microns) for the 110 reflection of a neutron beam with wavelength $\lambda$ = 1.35 Å by a Fe–Si crystal (3% Si) of the type used in a recent topographic work (Schlenker, Linarès-Galvez, and Baruchel 1978, Linarès-Galvez 1977). In this case $u = 10.7 \times 10^4\ \text{m}^{-1}$ and $A = 6.24 \times 10^4\ \text{m}^{-1}$. The angular form factor (see formula (1)) is equal to 0.63. (a) Corresponding to $\beta = \pi/2$; $Q = 0.63A$ is due to Linarès-Galvez (1977). (b) Corresponds to $\beta = \pi/4$ and $Q = 0.63\ A \sin\beta$. In each case the dynamical limit $R_D$ is indicated by a broken line.

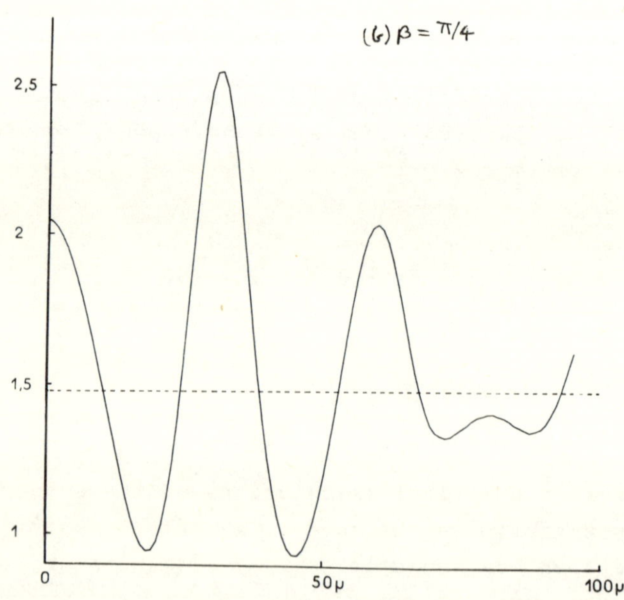

*4.3. The flipping ratio in the simple case* $\beta = \pi/2$

In this case $I_{+-}$ and $I_{-+}$ are equal to zero. According to the usual dynamical theory (Zachariasen 1967)

$$I_{\pm\pm}(t) = |u\pm Q| \int_{-\infty}^{\infty} \frac{dx}{1+x^2} \sin^2\{t|u\pm Q|(1+x^2)^{\frac{1}{2}}\} .$$

This is expressed in terms of the well-known Waller integral

$$W(A) = \int_{-\infty}^{\infty} \frac{dx}{1+x^2} \sin^2\{A(1+x^2)^{\frac{1}{2}}\} .$$

Let us recall that

$$W(A) \approx \frac{\pi}{2} - \frac{1}{2}\left(\frac{\pi}{A}\right)^{\frac{1}{2}} \cos\left(2A-\frac{\pi}{4}\right) \quad \text{for } A \to \infty$$

$$W(A) \approx \int_{-\infty}^{\infty} \frac{dx}{x^2} \sin^2(Ax) = \pi A \quad \text{for } A \to 0$$

from which we obtain simple results for the kinematical value $R_k$ $(t\to 0)$ and for the dynamical value $R_D(t\to\infty)$ of the flipping ratio

$$R_k = \frac{|u+Q|^2}{|u-Q|^2} \quad \text{and} \quad R_D = \left|\frac{u+Q}{u-Q}\right| .$$

## 5. The kinematical limit

This limit corresponds to $t \to 0$. The diffracted intensity from a plane incident wave is proportional to $t^2$, but the integrated intensity is proportional to $t$.

In the following calculations $u$ and $Q$ are real (centro-symmetric crystal with no absorption). As seen in §4 concerning the properties of the Waller integral, the kinematical limit is obtained by replacing the expression to be integrated by its asymptotic form for large deviations from the Bragg reflecting position. We shall follow the same method for our general expressions ($\beta \neq \pi/2$).

We can actually calculate the solution of equation (4) for $D = X - Y \to \infty$, by using our result (equation (6)) that $(X+A)(Y+A)$ or $(X-A)(Y-A)$ have finite limits. We find, with some calculations and keeping only the first-order terms in $1/D$, the results given in Table 1.

TABLE 1

| $S$ | $D + 2A$ | $-D + 2A$ | $D - 2A$ | $-D - 2A$ |
|---|---|---|---|---|
| $a_0$ | $\frac{Q\cos\beta}{D}$ | $\frac{uQ\cos\beta}{AD}$ | $-\frac{u+Q\sin\beta}{D}$ | 1 |
| $a_g$ | $-\frac{AQ\cos\beta}{UD}$ | $-\frac{Q\cos\beta}{D}$ | 1 | $\frac{u+Q\sin\beta}{D}$ |
| $b_0$ | $-\frac{u-Q\sin\beta}{D}$ | 1 | $\frac{Q\cos\beta}{D}$ | $-\frac{uQ\cos\beta}{AD}$ |
| $b_g$ | 1 | $\frac{u-Q\sin\beta}{D}$ | $\frac{AQ\cos\beta}{UD}$ | $-\frac{Q\cos\beta}{D}$ |

Each column of this table gives the coefficients ($a_0$, $b_0$; $a_g$, $b_g$) of a wave field corresponding to a point $(S,D)$ of the dispersion surface; these coefficients are defined to within an arbitrary multiplicative constant. We then calculate the diffracted amplitudes ($a_{gj}^+$, $b_{gj}^+$) and ($a_{gj}^-$, $b_{gj}^-$), with $j$ = 1,2,3,4, corresponding to an incident wave with polarization ±1. The results are given in Table 2.

TABLE 2

| $S$ | $D + 2A$ | $-D + 2A$ | $D - 2A$ | $-D - 2A$ |
|---|---|---|---|---|
| $a_g^+$ | 0 | 0 | $-\frac{u+Q\sin\beta}{D}$ | $\frac{u+Q\sin\beta}{D}$ |
| $b_g^+$ | $\frac{Q\cos\beta}{D}$ | 0 | 0 | $-\frac{Q\cos\beta}{D}$ |
| $a_g^-$ | 0 | $-\frac{Q\cos\beta}{D}$ | $\frac{Q\cos\beta}{D}$ | 0 |
| $b_g^-$ | $-\frac{u-Q\sin\beta}{D}$ | $\frac{u-Q\sin\beta}{D}$ | 0 | 0 |

It is then easy to obtain the integrated intensities and the flipping ratio $R_K$ in the kinematical limit:

$$R_K = \frac{(u+Q\ \sin\beta)^2 + Q^2\cos^2\beta}{(u-Q\ \sin\beta)^2 + Q^2\cos^2\beta}$$

which gives the well-known result $(u+Q)^2/(u-Q)^2$ in the simple case $\beta = \pi/2$.

We are able to express in a simple form the polarization of the diffracted wave resulting from an incident plane wave $\exp(i\mathbf{k}\cdot\mathbf{r})\binom{a}{b}$ with any state of polarization as:

$$\begin{pmatrix} u+Q\ \sin\beta & -Q\ \cos\beta \\ -Q\ \cos\beta & u-Q\ \sin\beta \end{pmatrix} \begin{pmatrix} a \\ b \end{pmatrix} .$$

This is of course valid for a large deviation from the Bragg position. We can look for the eigenvectors of the above matrix and we find polarization states along the direction of $\mathbf{B}_g$. This means that if the incident wave is polarized along this direction, this polarization will be conserved. This is in agreement with a result of Moon, Riste, and Koehler (1969).

## 6. The Pendellösung regime and the Larmor precession

It is well-known that when there is no Bragg reflection the Larmor precession of the neutron spin in the mean magnetic field $\mathbf{B}_0$ can be described in terms of interference between the + and - spinor components, which suffer different phase shifts because there are two different refractive indexes.

In the two-beam case there is a gradual shift from the Larmor precession regime to a more complicated Pendellösung regime as the conditions for Bragg reflection are approached, i.e. as the points representing the excited wave fields in the crystal move away from the asymptotes of the dispersion surface.

It may be possible to obtain a measurement of $\mathbf{B}_g$ from an investigation of the polarization behaviour of the forward transmitted beam when a perfect magnetic crystal is rotated into the vicinity of a Bragg reflecting position. This would be conceptually similar to an experiment by Graeff *et al.* (1978) reported during the present workshop (Papers I.1 and I.2), in which $V_{\mathbf{g}}$ is determined for a non-magnetic crystal plate placed across both paths of a neutron interferometer from the anomalous intensity oscillations that occur in the vicinity of a Bragg setting.

## 7. Conclusion

When the magnetization is perpendicular to the diffraction vector ($\beta=\pi/2$), the dispersion surface is made up of two hyperbolic surfaces corresponding to simple polarization states and the results of the two-beam dynamical theory for non-magnetic crystals can be directly applied. In the more general case ($\beta\neq\pi/2$) we have been able to discuss the asymptotic properties of the dispersion surface and to present an analytical treatment of the kinematical limit. Integrated intensities and flipping ratios outside this limit can only be calculated numerically. We have shown that the wave fields defined by the different points of the dispersion surface are polarized in the ($\mathbf{g}$,$B_0$) plane; this is a generalization of the fact that they are (±) states with respect to $\mathbf{B}_0$ in the simple case $\beta = \pi/2$.

## References

BELYAKOV, V.A. and BOKUN, R.CH. (1976). *Sov. Phys.—Solid State* **18**, 1399.
BONNET, M., DELAPALME, A., and BECKER, P. (1976). *Acta crystallogr. A* **32**, 945.
GRAEFF, W., BAUSPIESS, W., BONSE, U., and RAUCH, H. (1978). *Acta Crystallogr.* **A34**, S 238.
HALPERN, O. and JOHNSON, M.R. (1939). *Phys. Rev.* **55**, 898.
HAMILTON, W.C. (1958). *Acta Crystallogr.* **11**, 585.
KATO, N. (1969). *Acta Crystallogr. A* **25**, 119.
LINARES-GALVEZ, J. (1977). Thesis, Grenoble.
MOON, R.M., RISTE, T., and KOEHLER, W.C. (1969). *Phys. Rev.* **181**, 920.
MENDIRATTA, S.K. and BLUME, M. (1976). *Phys. Rev. B* **14**, 144.
SCHLENKER, M. and BARUCHEL, J. (1978). *J. appl. Phys.* 49(3), 1996.
SCHLENKER, M., LINARES-GALVES, J., and BARUCHEL, J. (1978). *Phil. Mag. B* **37**, 1.
SCHMIDT, H.H., DEIMEL, P., and DANIEL, H. (1975). *J. Appl. Crystallogr.* **8**, 128.
SCHMIDT, H.H. and DEIMEL, P. (1975). *J. Phys. C*, **8**, 1991.
SIVARDIÈRE, J. (1975). *Acta Crystallogr. A* **31**, 340.
STASSIS, C. and OBERTEUFFER, J.A. (1974). *Phys. Rev. B* **10**, 5192.
ZACHARIASEN, W.H. (1967). *Theory of X-ray diffraction in crystals*, Dover Publications, New York.

# 10. INVESTIGATION OF NEARLY PERFECT MAGNETIC SINGLE CRYSTALS BY MEANS OF NEUTRON DIFFRACTION

H.H. SCHMIDT and K. KÖNIG

*Physik-Department der Technischen Universität München, München, Germany*

## 1. Introduction

Although the dynamical diffraction theory is well developed, a number of publications dealing with dynamical effects have appeared in the last few years. One special point treated by various authors is the dynamical neutron diffraction by magnetic crystals (Stassis and Oberteuffer 1974, Sivardière 1975, Schmidt, Deimel, and Daniel 1975, Schmidt and Deimel 1975, 1976, Belyakov and Bokun 1976). One special feature of this diffraction is the dependence of the effects occurring on the relative orientation of the average magnetic field $\mathbf{B}_0$ inside the crystal to the Fourier transform $\mathbf{B}_1$ of the magnetic field with respect to the scattering vector, which is always perpendicular to the scattering vector (Ekstein 1949). This orientation affects the form of the fundamental equations of dynamical diffraction. In the most simple case $\mathbf{B}_0$ is parallel to $\mathbf{B}_1$. The problem can then be solved for each spin direction separately in a manner closely analogous to the case of pure nuclear scattering, whereas in general the equations for the two spin states (irrespective of the choice of quantization axis) are coupled to each other. Therefore the equation for the dispersion surface is of the fourth degree instead of the second degree, and there are sixteen plane waves in the crystal instead of eight.

Solutions for the general case were first given by Schmidt and Deimel (1976) for the symmetric Laue case and one of the following restrictions: (a) the angle between $B_0$ and $B_1$ is 90°; (b) the nuclear contribution to the diffraction process vanishes. In both cases, a splitting of the reflection curve arises. For polarized neutrons there are up to two peaks, and for unpolarized neutrons there are up to three peaks. Belyakov and Bokun (1976) set up the equation for the dispersion surface in the most general case and treated the symmetric Laue case and the symmetric Bragg case in more detail. For the symmetric Bragg case they show the reflection curve for unpolarized neutrons. Also in this case there is a splitting of the reflection curve because there are three peaks, whereas for $\mathbf{B}_0$ parallel to $\mathbf{B}_1$ only two peaks would appear. It is stated in both papers that the distance between the peaks increases with increasing magnetic field, whereas the width of the peaks increases with increasing magnetic or nuclear structure factor.

## 2. Properties of $DyFeO_3$

It will now be shown that crystals of rare-earth orthoferrites provide good possibilities for studying these effects. In the experiments reported here a crystal of $DyFeO_3$ was used, but such experiments could also be done with other orthoferrites. $DyFeO_3$ is antiferromagnetic and exhibits a weak ferromagnetism due to a slight misalignment of the directions of the magnetic moments in the two sublattices. Because of the magnetic structure of $DyFeO_3$ the Bragg reflections are of either pure nuclear or pure magnetic origin, and therefore the nuclear contribution to the diffraction process can be suppressed. Furthermore, the rare earth orthoferrites exhibit a strong magnetic anisotropy. Apart from temperature regions near a magnetic transition point the threshold field for rotating the magnetic moments out of the antiferromagnetic axis by applying a magnetic field perpendicular to it is at least 5T (Belov, Kadomtseva, and Levitin 1967, Belov *et al.* 1969). This provides the possibility of changing the magnitude of $\mathbf{B}_0$ and the angle between $\mathbf{B}_0$ and $\mathbf{B}_1$ by means of an external magnetic field. It should be noted that the effects to be studied here are the same whether $\mathbf{B}_0$ is an external or an internal field, because they do not arise from refraction at the crystal surface but, roughly speaking, from the change of the polarization state of the neutron caused by the scattering process and, resulting from this, the change in the magnetic potential seen by the neutrons (Schmidt and Deimel 1976). For an external field of 0.8 T, $\mathbf{B}_0$ perpendicular to $\mathbf{B}_1$, and a Bragg angle of 19° one expects a splitting of 0.6″.

## 3. Description of the experiments

A crystal of $DyFeO_3$ (grown by Crystal Tec, Grenoble) was tested at the FRM, Garching. It turned out not to be good enough to observe the magnetic line splitting. However, some peculiar effects occurred, which will be described here. A conventional double-crystal arrangement was used with parallel setting of the two crystals. The monochromator is a perfect silicon crystal. The (2-20) reflex of the silicon crystal is used with a Bragg angle of 19°, yielding a wavelength of 1.26 Å. The second crystal is rotated around an axis perpendicular to the scattering plane and the intensities of the transmitted and reflected beams are measured as a function of the rotation angle.

## 4. Properties of the double-crystal arrangement

The measured intensities depend on the reflection functions of the two crystals, as well as on the angular divergence and the wavelength distribution of the neutron beam. In the parallel setting of the double-crystal arrangement the ratio $P$ of the intensity of the reflected beam to the intensity incident upon the second crystal is given by

$$P = \frac{1}{R_1^{\theta}} \int g(\theta_{B_1}) \left\{ \int R_1(\theta-\theta_{B_1}) R_2(\theta-\theta_{B_2}-\phi)\, d\theta \right\} d\theta_{B_1} \tag{1}$$

where $R_{1,2}$ are the reflection functions and $\theta_{B_{1,2}}$ are the Bragg angles for the two crystals. $R_{1,2}^{\theta}$ is the reflecting power of the crystals, $\theta$ is the angle of incidence relative to the first crystal, and $g(\theta)$ is the beam intensity as a function of $\theta$. $\phi$ is the rotation angle of the second crystal. This expression has already been simplified by using the fact that the intensity of the neutron beam varies only slowly with the wavelength. Furthermore, only the divergence in the scattering plane is taken into consideration. This is only correct if the crystals are ideally adjusted.

In the following two cases are particularly important.

(1) The distances between the reflecting lattice planes in the two crystals are equal. This is called the non-dispersive case. Then the above expression reduces to

$$P = \frac{1}{R_1^{\theta}} \int R_1(\theta-\theta_B)R_2(\theta-\theta_B-\phi)d\theta \ . \tag{2}$$

This is simply the convolution of the reflection functions of the two crystals. For identical crystals and reflexes and the symmetric Laue case the maximum value of $P$ becomes 0.25 (neglecting absorption). Any imperfection of the crystals will decrease $P$ and increase the width of the reflection curve. The integral over the reflected intensity is proportional to the reflecting power of the crystals. It should be mentioned that the last result is valid in each case.

(2) If the lattice plane spacings are not equal to each other, the Bragg condition is no longer fulfilled for all directions of the incident beam simultaneously as in case (1). The second crystal must be rotated through a certain angle in order to fulfil the Bragg condition for each direction of the divergent beam. If this angle is much larger than the widths of the reflection curves (strong dispersive case), the above expression yields

$$P = \frac{R_2^{\theta}}{\eta}\, g\left(\frac{\phi}{\eta}\right) \tag{3}$$

where

$$\eta = \frac{\Delta d}{d\cos^2\theta_B} \ .$$

Now the intensity of the reflected beam as a function of the angular position of the second crystal reproduces the angular distribution of the beam. The width of the reflection curve is proportional to $\eta$. The integrated reflected intensity is proportional to the reflecting power of the crystals. It should be noted that these results are only valid if, as mentioned above, the intensity of the beam varies only slowly with wavelength and if the reflecting power of the crystals does not depend heavily on wavelength.

## 5. Experimental results and discussion

Measurements were carried out with the reflexes (2-20), (031), (-130), and (3-10). The relative difference $\Delta d/d$ in the lattice-plane spacing for the two crystals is about 0.025 per cent for (2-20), 6.0 per cent for (031), 9.1 per cent for (-130), and 14.0 per cent for (3-10). (031) is a pure magnetic reflex; the others are nuclear reflexes. In all cases except that of (2-20), which is a symmetric Laue reflex, the $DyFeO_3$ crystal is in an asymmetric Laue position. Figures 1–5 always show the intensity of the reflected beam as a function of the rotation angle of the second crystal. Figure 1 shows the (2-20) reflex. The theoretical full width at half maximum (fwhm) in this case is about 5″. Because $\Delta d/d$ is very small there should be a broadening of the reflection curve of only about 2″. The measured fwhm is about 25″. The maximum value of $P$ is about 0.2. The theoretical value is about 0.4. The value for the integrated intensity is about four times higher than that predicted theoretically. This is also valid for the other reflexes being measured. Because of the large half-width, it

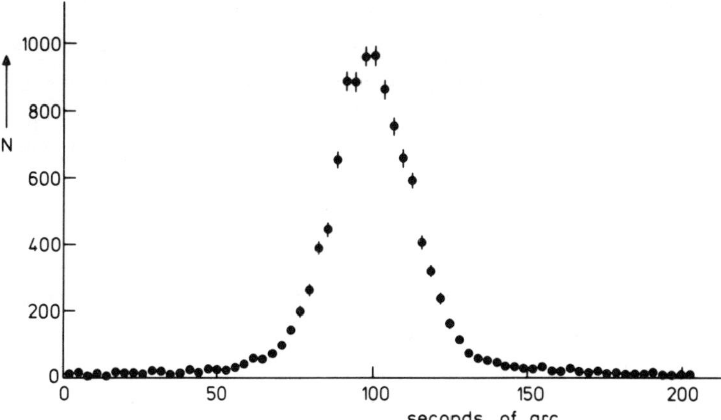


FIG. 1. Intensity of the reflected beam for the (2-20) reflex of $DyFeO_3$. Measuring time per measuring point, 300 s.

seems impossible to observe the splitting effects mentioned above with this crystal.

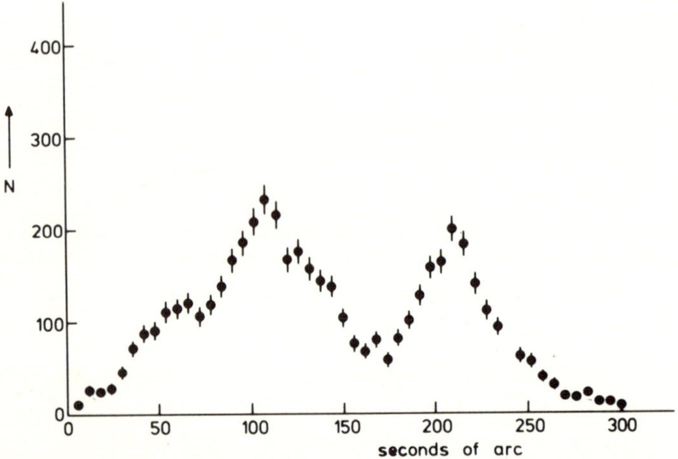


FIG. 2. Intensity of the reflected beam for the (-130) reflex of $DyFeO_3$. Measuring time per measuring point, 1000 s.

Figure 2 shows the (-130) reflex. Now one has two distinct peaks separated by about 100″. Furthermore, some other line structures can be seen. Figure 3 shows the (3-10) reflex in which the separation has increased to about 160″. All these effects could be reproduced in a series of experiments.

To discuss this we should remember the fact that in the dispersive case only a part of the beam is reflected by the second crystal, depending on its angular position. Therefore a particular angle of incidence and a particular wavelength are associated with each point of the reflection curve. Because of this the double peaks can be explained as special properties of either the neutron beam or the crystal. In the first case, one should assume that the wavelength distribution of the neutron beam varies rapidly for wavelength changes of the order of magnitude $\Delta\lambda/\lambda = 10^{-2}$, or that the neutron intensity varies strongly with the point of origin of the neutrons. In the second case, one must assume that the reflecting power of the crystal varies rapidly with wavelength or with the angle

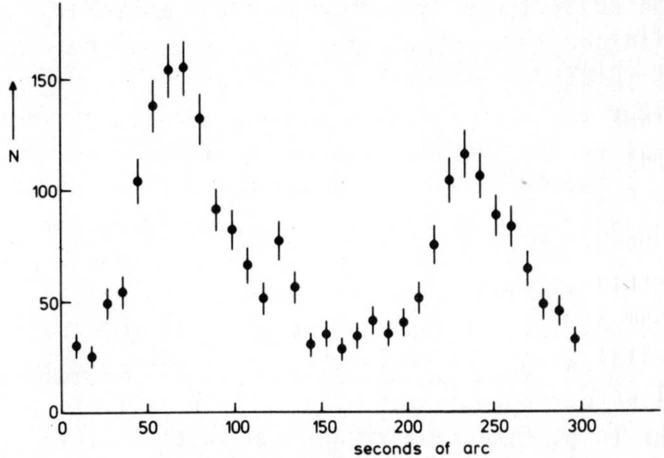


FIG. 3. Intensity of the reflected beam for the (3-20) reflex of $DyFeO_3$. Measuring time per measuring point, 1800 s.

of incidence. To decide which explanation is true, an experiment was carried out with two silicon crystals using different reflexes, namely (2-20) for the first crystal and (311) for the second crystal. The difference in lattice-plane spacing is 17.3 per cent. The result of this experiment is shown in Fig. 4.

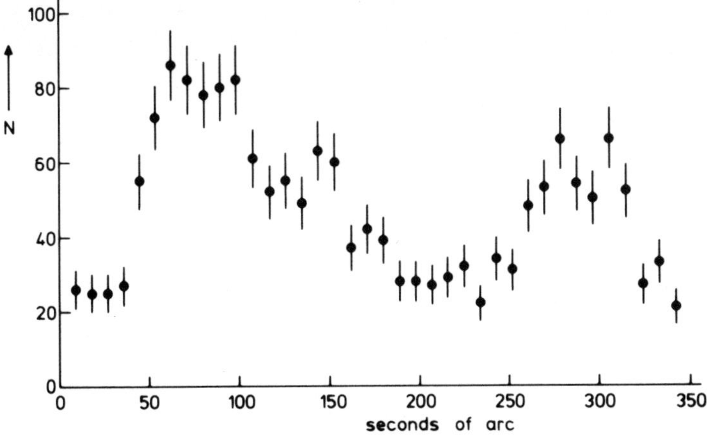


FIG. 4. Intensity of the reflected beam for the (311) reflex of silicon. Measuring time per measuring point, 1800 s.

Also in this case there are two peaks, and therefore it must be assumed that the double-peak structure arises from properties of the beam. Figure 5 shows the difference in angular position for the two peaks as a function of $\Delta d/d$ for the asymmetric $DyFeO_3$ reflexes and for the silicon experiment. One can see that the linear relation between $\Delta d/d$ and the separation of the peaks which follows from equation (3) is in good agreement with the experimental results. From the slope of the straight line the difference between the angles of incidence for the two peaks can be deduced. This yields a value of $(4.61 \pm 0.08) \times 10^{-3}$. A possible explanation for these effects is the fact that the neutron beam of the beam tube P2 used here is filtered by a single crystal of bismutium (Bauer and Seyfarth 1971). Neutrons of special wavelengths are scattered out of the beam by Bragg reflections in the crystal. Thus, the beam may have the structure described above in the wavelength.

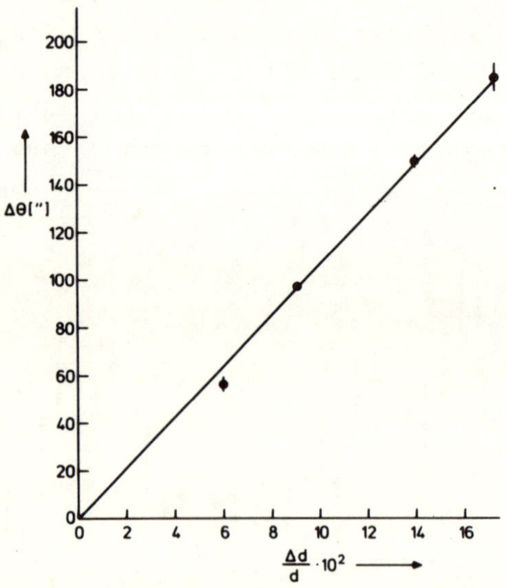


FIG. 5. Difference in angle between the two peaks as a function of $\Delta d/d$.

## Acknowledgments

We are indebted to Professor H. Daniel for his continuous interest in this work and to P. Stoeckel, H. Hagn, and H. Angerer for technical assistance.

## References

BAUER, G. and SEYFARTH, H. (1971). *Atomkernenergie* **18**, 236.
BELOV, K.P., KADOMTSEVA, A.M., and LEVITIN, R.Z. (1967). *Sov. Phys.—JETP* **24**, 878.
BELOV, K.P., KADOMTSEVA, A.M., LEVITIN, R.Z., TIMOFEEVA, V.A., USKOV, V.S., and KHOKHLOV, V.A. (1969). *Sov. Phys.—JETP* **28**, 1139.
BELYAKOV, V.A. and BOKUN, R.CH. (1976). *Sov. Phys.—Solid State* **18**, 1399.
EKSTEIN, H. (1949). *Phys. Rev.* **76**, 1328.
SCHMIDT, H.H. and DEIMEL, P. (1975). *J. Phys. C* **8**, 1991.
—— (1976). *Phys. Status Solidi B* **73**, 87.
SCHMIDT, H.H., DEIMEL, P., and DANIEL, H. (1975). *J. appl. Crystallogr.* **8**, 128.
SIVARDIÈRE, J. (1975). *Acta Crystallogr. A* **31**, 340.
STASSIS, C. and OBERTEUFFER, J.A. (1974). *Phys. Rev. B* **10**, 5192.

# PART II APPLICATION OF NEUTRON INTERFEROMETRY

# 1. SCOPE OF NEUTRON INTERFEROMETRY

H. RAUCH
*Atominstitut der Österreichischen Universitäten, A-1020 Wien, Austria*

## 1. Introduction

The great success of optical interferometry has motivated the extension of this technique to X-rays and particle radiation such as electrons and neutrons. Because of the much shorter wavelengths of such radiation the usual optical systems cannot be used. The interference effects within crystals giving the well-known Bragg reflections form a separate field of research. Here we deal with the interferometry of well-separated coherent beams, where the phase of the beams can be manipulated individually.

The electron microscope was used by Möllenstedt and Düker (1954) to construct the first electron interferometer. It is analogous in operation to a Fresnel biprism interferometer. The action of the electrostatic lenses allows well-separated beams, even though the wavelengths of such electrons are rather short (about 0.05 Å). Up to 2200 interference fringes could be observed in the arrangement described by Keller (1961) produced by a variation of the optical path length by the electrostatic field within the microscope. With such electron interferometers Boersch *et al.* (1961, 1962) and Möllenstedt and Bayh (1962) demonstrated the Aharonov–Bohm (1959) effect, i.e. the coupling of the electron to the vector potential. Lischke (1969) used this method to investigate the action of the quantized magnetic flux in superconductors. Another type of interferometer, where crystal laminae cause separation and superposition of the coherent beams, were proposed by Marton (1952) for electron beams and tested later (Marton, Simpson, and Suddeth 1953). The interference fringes can only be observed microscopically because of various stability problems due to the extremely short wavelength. The coherent action of the crystal is based on the dynamical diffraction theory developed for particle radiation by Bethe (1928) and von Laue (1948).

Bonse and Hart (1965) constructed the first powerful X-ray interferometer in the ångström range of wavelengths. These interferometers are of the Marton type with perfect crystal plates as beam splitter, bender, and analyser. There are many possibilities for research in the field of crystallography and for precision measurements in the ånström region, which have been summarized recently by Bonse and Graeff (1977).

The first attempt to construct a neutron interferometer by Maier-Leibnitz and Springer (1962) was based on classical optical components, i.e. single slit diffraction and biprism deflection. Owing to the rather small beam separation of such systems (about 60 μm) the range of application is rather limited (Landkammer 1966). Therefore proposals for a Marton-type neutron interferometer with well-separated beams appeared in the literature (Bonse and Hart 1966, Rauch 1971, Shull 1973). The first construction of such a perfect crystal neutron interferometer was achieved by Rauch, Treimer, and Bonse (1975) at the 250 kW TRIGA reactor in Vienna, where the expected intensity modulation for various optical path lengths could be observed (Fig. 1). Afterwards the interferometer set-up was moved to the Institut Laue-Langevin to utilize the higher flux for more sophisticated experiments. The features of this set up are described by Bauspiess (1977) and Bauspiess, Bonse, and Rauch (1978).

## 2. Basic equations

The perfect crystal interferometer is based on dynamical diffraction theory, which was formulated for neutrons by Goldberger and Seitz (1947) and has recently been reviewed by Rauch and Petrascheck (1978) and Sears (1978). Most of the important theoretical relations obtained from dynamical diffraction theory and applied to the case of a neutron interferometer are given by Rauch and Suda (1974), Bauspiess, Bonse, and Graeff (1976), Petrascheck (1976), and Petrascheck and Folk (1976).

To obtain an appropriate description of the wave fields within an interferometer the spherical theory developed for

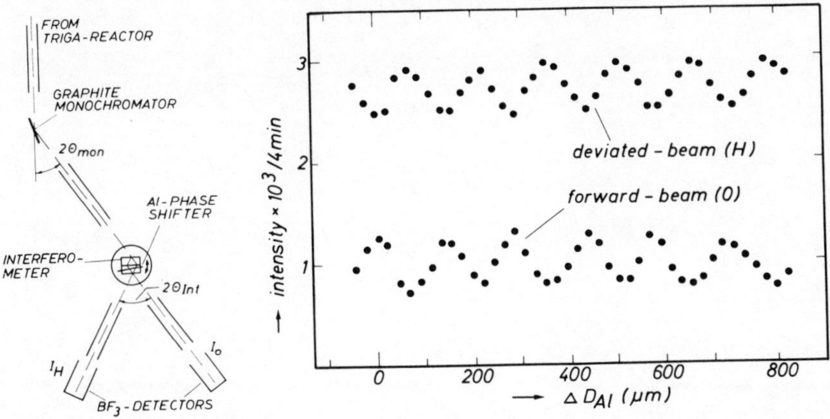


FIG. 1. First neutron interferogram obtained with a perfect-crystal interferometer and a test set-up at a TRIGA reactor (Rauch *et al.* 1974).

X-rays by Kato (1961, 1968) has to be applied. For thick crystals a ray consideration is justified and a coherent superposition of plane waves $\psi(y)$ gives the correct result:

$$\phi = (2\pi)^{-3} \int F(k,k')\exp(i\mathbf{k}'\cdot\mathbf{r})dk' \simeq \int_{-y_0}^{+y_0} dy\psi(y) \ .$$

The values for $y_0$ are determined by the beam collimation and can be replaced in most cases by $\pm\infty$ since the contribution to the reflection decreases drastically for large $y$ values. For the focused interferometer the plane wave solutions for the forward (0) and deviated (H) direction are as follows:

$$\psi_0 = \psi_0^{\,I} + \psi_0^{\,II} = \{v_0(y)v_H(v)v_{-H}(-y)+v_H(y)v_{-H}(-y)v_0(y)\} \times \tag{1}$$

$$\times \exp\frac{-2\pi iy(T+t)}{\Delta_0}\,\psi_e$$

$$\psi_H = \psi_H^{I} + \psi_H^{II} = \{v_H(y)v_H(y)v_0(-y)+v_H(y)v_{-H}(y)v_H(y)\}$$
$$\times \exp\left\{\frac{-2\pi i y(T+2t)}{\Delta_0}\right\}\exp(i\mathbf{G}\cdot\mathbf{r})\psi_e$$

with

$$v_0(y) = \left[\cos\left\{(A(1+y^2)^{\frac{1}{2}}\right\} + i\frac{y}{(1+y^2)^{\frac{1}{2}}}\left\{\sin A(1+y^2)^{\frac{1}{2}}\right\}\right]\exp(iPt)$$

$$v_H(y) = -i\frac{\sin\{A(1+y^2)^{\frac{1}{2}}\}}{(1+y^2)^{\frac{1}{2}}}\exp(iPt)$$

$$P = -\frac{\pi y}{\Delta_0} - \frac{2\pi}{D_\lambda \cos\theta_B} = -\frac{\pi}{\Delta_0}(1+y)$$

$$\Delta_0 = \frac{\pi t}{A} = \frac{2\pi\cos\theta_B}{k|V(H)/E|}$$

$$D_\lambda = \frac{4\pi}{k(V(0)/E} .$$

where t is the thickness and T the distance of the crystal plates. The Fourier components of the interaction potential $V(H)$ are determined by the coherent scattering length $b_c$, the particle density $N$, and the crystal structure. For the most frequently used reflection in a silicon crystal the following relation holds:

$$V(0) = |V(220)| = \sqrt{2}|V(111)| = \frac{2\pi\hbar^2 b_c N}{m}$$

(without the Debye—Waller correction). The incident wave is given by $\psi_e = n_0 \exp(i\mathbf{k}\cdot\mathbf{r})$. For the symmetrical Laue reflection, which is used for our interferometers, we have

$$y = \frac{(\theta_B-\theta)k^2 \sin 2\theta_B}{4\pi b_c N} .$$

For an idealized system we obtain for all components of the beam in the forward direction and therefore for the whole beam

$$\psi_0{}^{I} = \psi_0{}^{II} \qquad \phi_0{}^{I} = \phi_0{}^{II} \tag{2a}$$

and for the deviated beam

$$\psi_H{}^{I} = \psi_H{}^{II}\left\{1 - \frac{1+y^2}{\sin^2 A(1+y^2)^{\frac{1}{2}}}\right\} \qquad \phi_H{}^{I} = \phi_H{}^{II} R_H(A) \tag{2b}$$

where $R_H(A)$ is an integral quantity describing the dependence on crystal thickness.

For a real system we will have a complex relation between the wave functions of beam path I and II:

$$\frac{\phi^{I}}{\phi^{II}} = a e^{i\varphi} \ . \tag{3}$$

Therefore, the intensity behind the interferometer is

$$I = |\phi^{I} + \phi^{II}|^2 = |\phi^{II}|^2 (1 + a^2 + 2a \cos\varphi). \tag{4}$$

Here we define (similar to optics) a mutual coherence function $\gamma$

$$\gamma = \frac{2|\phi^{I}\phi^{II}|}{|\phi^{I}|^2 + |\phi^{II}|^2} = \frac{2a \cos\varphi}{1 + a^2} \ . \tag{5}$$

The quantities $a$ and $\varphi$ depend on all the imperfections of the instrument, i.e. variation of the thickness ($\Delta t$) and the distances ($\Delta T$), small lattice distortions in the ångström range (causing moiré fringes), and vibrations of the instrument which cause ångström range shifts in the coherent (parallel) lattice planes during the time of flight of the neutrons through the interferometer (about 30 μs). According to the Pendellösung structure of the intensity profile the influence of the imperfections is enhanced at large $y$ values or near the edges of the Borrmann fan. This gives an angular or a spatial variation of $a$ and $\varphi$ and therefore an appropriate average has to be taken for the intensities, which are measured over a large area.

The way to measure the mutual coherence is to introduce a

phase-shifting material with a thickness $\Delta D$ and an index of refraction $n$ into the coherent beams. This produces an additional phase factor for the quantities $v$ of equation (1) and therefore for equation (3) also. We assume that this phase shift is independent of $y$ and obtain an intensity modulation behind the interferometer, which is described by the λ thickness ($D_\lambda = 2\pi/Nb_c\lambda$) of the material:

$$n = \frac{K}{k} \approx 1 - \frac{V(0)}{2E} = 1 - \lambda^2 \frac{Nb_c}{2\pi}$$

$$\phi \to \phi \exp\{-ik(1-n)\Delta D\} = \phi \exp(-2\pi i \Delta D/D_\lambda)$$

$$I_{0,H} = |\phi_{0,H}^{II}|^2\{1+a^2+2a \cos(2\pi\Delta D/D_\lambda)\cos\varphi\} \quad . \qquad (6)$$

The quantities $a$ and $\varphi$ vary as a function of angle or of position within the Borrmann fan and the mutual coherence function is also position dependent. Usually the value for $\gamma$ increases for small beams. Figure 2 shows the measured intensity modulation at the centre of the Borrmann fan in a region with a high mutual coherence of the beams ($\gamma \sim 0.8$). $V(0)$ is equivalent to

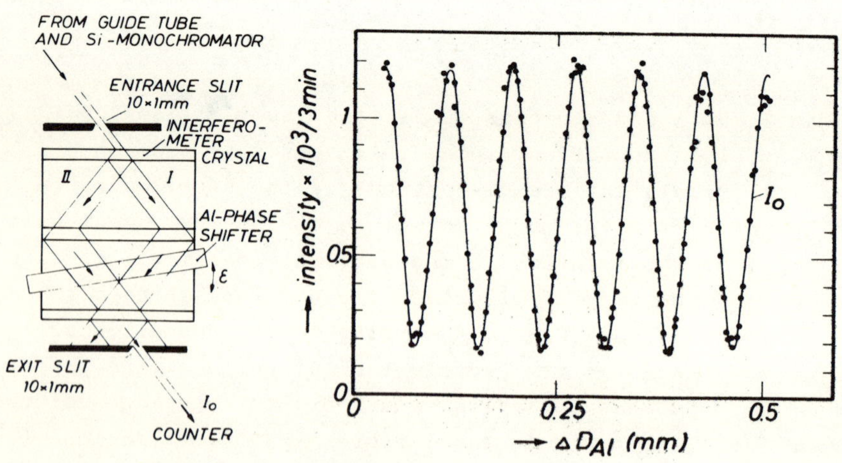


FIG. 2. Beam modulation obtained at the interferometer set-up at ILL in a region with a high degree of mutual coherence.

the mean potential of the neutrons in the sample and is independent of the internal structure. The phase shift

$$\rho_0 = k(1-n)\Delta D = 2\pi\Delta D/D_\lambda = \lambda N b_c \Delta D$$

due to the material depends strongly on the wavelength of the neutron. An 'achromatic' phase-shifting system with an appropriate $\Delta D$ variation would be restricted to a rather small range of wavelengths.

## 3. Neutron interaction

The interaction of a slow neutron with its surrounding contains several contributions:

$$V = V_N + V_m + V_{em} + V_i + V_g + V_W . \tag{7}$$

where the terms on the right-hand side denote the nuclear, magnetic, electromagnetic, intrinsic, gravitational, and weak interactions respectively.

The nuclear part $V_N$ can be written as a Fermi pseudo potential (Fermi 1936):

$$V_N = \frac{2\pi\hbar^2}{m}\left(b_c - ib_c' + \frac{b_i}{\{I(I+1)\}^{\frac{1}{2}}}\boldsymbol{\sigma}\cdot\mathbf{I}\right)\delta(\mathbf{r}-\mathbf{R}) . \tag{8}$$

The real part of the coherent scattering length is obtained from the average phase shift $\langle\delta\rangle$ for the scattering of neutrons by the various nuclei of the target ($\langle\delta\rangle = -b_c k$). As the nuclear forces are spin dependent we have various scattering lengths $(b^+, b^-)$ for the interaction in the state $I + \frac{1}{2}$ and $I - \frac{1}{2}$ and statistical weights $g_+ = (I+1)/(2I+1)$ and $g_- = I(2I+1)$ and therefore

$$b_c = \langle b\rangle = g^+b^+ + g^-b^- . \tag{9}$$

The average over various isotopes with abundances $\rho_j$ yields $b_c = \sum_j \rho_j b_c^{\,j}$. The nuclear part produces the most significant contribution to the phase shift, which can be measured with the interferometer. The coherent scattering length is of the order of $10^{-13}$ cm and the related mean potential $\langle V_N \rangle = 2\pi\hbar^2 N b_c / m$ is of the order $10^{-7}$ eV. While $b_c$ is tabulated for all elements (e.g. Koester 1977) and $b_c^j$ is known for about 10% of the stable isotopes the values of $b^+$ and $b^-$, which are relevant for an interpretation of the nuclear interaction, are known for a few isotopes only. To obtain the values of $b^+$ and $b^-$ the coherent scattering length $b_c$ and either the incoherent scattering length $b_i$ or the total free scattering cross-section have to be measured. The quantity $b'$ can be derived from the total cross-section since $b' = \sigma_t / 2\lambda$. The incoherent contribution $b_i$ follows from the variance of the individual phase shifts $b_i^2 = (\langle \delta^2 \rangle - \langle \delta \rangle^2)/k^2 = g_+ g_- (b^+ - b^-)^2$. The incoherent part of eqn (8) vanishes for unpolarized neutrons and/or unpolarized targets $(\langle \sigma \rangle = \langle I \rangle = 0)$ as the phase shift measured with the interferometer represents an average phase shift due to many particles. For polarized neutrons a small contribution may arise if the sample is within a magnetic field $B$ which produces a small nuclear polarization. For $\mu_K B / k_B T \ll 1$ we can write

$$\Delta b_i = \pm \frac{b^+ - b^-}{3} \frac{\mu_K B}{k_B T} \frac{I+1}{2I + 1} \tag{10}$$

where $\mu_K$ denotes the magnetic moment of the nucleus and $k_B$ the Boltzmann constant. The contribution is small even for materials with a pronounced spin incoherence. For room temperature and fields $B \leqslant 2 \times 10^4$ G we obtain $\Delta b \sim 10^{-17}$ cm, increasing at lower temperatures with the nuclear polarization. The associated spin precession is treated by Forte (1973).

The *magnetic contribution* $V_M$ has for magnetic substances the same order of magnitude as the nuclear part. $V_M$ corresponds to a dipole interaction between the magnetic moment of the neutron $\mu$ and a magnetic field $B$ (Halpern and Johnson 1939):

$$V_M = -\mu \boldsymbol{\sigma} \cdot \mathbf{B} \ . \tag{11}$$

The magnetic field of an atom is produced by the outer electrons and therefore the magnetic form factor $f_m(\mathbf{Q})$ enters the scattering amplitude $p(Q)$ (e.g. Marshall and Lovesey 1971):

$$p(\mathbf{Q}) = -\frac{\gamma e^2}{m_e c^2} f_m(\mathbf{Q}) \langle \mathbf{S} \rangle (\boldsymbol{\sigma} \cdot \mathbf{q}) \quad . \tag{12}$$

where $\gamma$ is the neutron magnetic moment in units of the nuclear magneton ($\mu=\gamma\mu_k$), $m_e$ the mass of the electron, $\langle \mathbf{S} \rangle$ denotes the mean magnetization of the magnetic atoms and $\mathbf{q} = \hat{\mathbf{h}} - (\hat{\mathbf{h}} \cdot \hat{\mathbf{e}})\hat{\mathbf{e}}$ is the magnetic interaction vector expressed by the unit vectors $\hat{\mathbf{h}}$ in the direction of $\langle \mathbf{S} \rangle$ and $\hat{\mathbf{e}}$ in the direction of $\mathbf{Q}$. For $\mathbf{Q} = 0$ only the mean potential enters $p(0)$. The magnitude of $p(0)$ follows from the magnetic field $B$ within the sample and we have for the two spin states ($f_m(0)=1$)

$$p(0) = \pm \frac{\mu m B}{2\pi N \hbar^2} \quad . \tag{13}$$

The same result applies for a stationary magnetic field. For comparison it is more convenient to use the mean potential of the neutrons in the field region $\langle V_M \rangle = \pm \mu B = 2\pi\hbar^2 p(0) B/m$, which gives, for example, for fields of 10 kG a potential of $\pm 6 \times 10^{-8}$ eV. As the interferometer allows a precise measurement of $p(0)$ the $B$ value can be determined. The accuracies of such measurements are in the range of 1% and are comparable to the measurements of the prism deflection made by Schneider and Shull (1971), but at present conventional measurements achieve higher accuracies.

For paramagnetic and diamagnetic substances $B$ in equation (13) has to be replaced by $H(1+4\pi\chi)$, where $\chi$ is the magnetic susceptibility. Within a field of about 10 kG the paramagnetic or diamagnetic contribution is in the range $\Delta p \sim 10^{-16}$ cm, which can be measured in appropriate cases (Stassis 1970). An accurate separation from the direct magnetic term seems to be rather difficult for interferometer measurements. The same formalism but with different form factors can be used for a Kondo system (e.g. Dickens *et al.* 1975, Stassis and Shull 1972).

The electromagnetic coupling $V_{em}$ arises from the movement of a magnetic dipole within an electric field. It has been calculated from the non-relativistic Dirac equation by Schwinger (1948) and Foldy (1951, 1958):

$$V_{em} = -\frac{\mu\sigma\hbar}{mc}(\mathbf{E}\times\mathbf{k}) - \frac{\hbar\mu}{2mc}\nabla\mathbf{E} \quad . \tag{14}$$

The first term is known as the spin-orbit or Schwinger contribution and the second term is the Foldy contribution. The $E$ field of interest is the Coulomb field of the positive charge $Ze$ located within the nucleus and the negative charge of the electron cloud. This causes a form factor of the type $Z\{1-f(Q)\}$, where $f(Q)$ is the atom form factor known from X-rays with $f(0) = 1$. According to this form factor both contributions vanish for $Q = 0$, but the Pendellösung length $\Delta_0$ has contributions due to $V(\mathrm{H})$ and therefore accurate and independent measurements at $Q = 0$ and for $Q \neq 0$ can isolate these terms (Graeff *et al.* 1978).

The Fourier transformation of the spin-orbit term gives an imaginary scattering length (e.g. Lovesey 1969)

$$b_{s1} = i\frac{m_e}{2m}\left(\gamma\frac{e^2}{m_e c^2}\right)Z\{1-f(Q)\}\cot\frac{\theta}{2}(\boldsymbol{\sigma}\cdot\mathbf{n}) \tag{15}$$

where $\theta$ is the scattering angle and $\mathbf{n}$ is a unit vector perpendicular to the scattering plane. For intermediate $Q$ and $Z$ values this contribution reaches values of $\langle b_{s1}\rangle \sim 10^{-14}$ cm (Obermaier 1967). As well as the contribution of the Coulomb field of the atom a stationary field $E$ perpendicular to $\mathbf{k}$ produces a mean potential $\langle V_e\rangle = \pm\ \mu\hbar Ek/mc$ and therefore a contribution even for $Q = 0$. For $E = 10^5$ V cm$^{-1}$ and 2 Å neutrons this potential is, however, only $1.3 \times 10^{-14}$ eV, which is equivalent to the action of a 2.2 mG magnetic field.

The Foldy contribution yields a scattering length of

$$b_F = -\frac{m_e}{2m}\left(\gamma\frac{e^2}{m_e c^2}\right)Z\{1-f(Q)\} = 1.468 \times 10^{-16}\ Z\{1-f(Q)\} \tag{16}$$

Accurate measurements at various $Q$ values allow a separation from the much larger nuclear scattering length (Krohn and Ringo 1966, Koester, Nistler and Waschkowski 1976).

The intrinsic interaction $V_i$ can be split into a term $V_{iF}$

due to a virtual charge separation within the neutron (Foldy 1958), a term $V_{ip}$ due to an electric polarizability $\alpha$ of the neutron (Thaler 1959), and a contribution from a speculative static electric dipole moment (Shull and Nathans 1967). These terms can be written as

$$V_i = -\varepsilon_2 \nabla \mathbf{E} - \tfrac{1}{2}\alpha E^2 - d\boldsymbol{\sigma}\cdot\mathbf{E} \,. \tag{17}$$

A Fourier transformation yields the related scattering length for the interaction with atoms:

$$b_i = -\frac{2m}{\hbar^2} e\varepsilon_2 Z\{1-f(Q)\} - \frac{me^2}{\hbar^2 R_c}\alpha Z^2\left(1-\frac{\pi}{2}kR_c \sin\frac{\theta}{2}\right)$$

$$-\, i\frac{Z\{1-f(Q)\}}{\hbar^2 k}(\boldsymbol{\sigma}\cdot\mathbf{e}) \,. \tag{18}$$

The first term $b_{iF}$ is due to the charge separation and is determined by the second moment of the charge distribution within the neutron $\varepsilon_2 \approx \int \mathbf{r}^2 \rho(\mathbf{r}) d\mathbf{r}/6 = \langle r^2 \rangle e/6$. Recently Koester *et al.* (1976) have separated such a contribution from the larger Foldy term $b_F$. They obtained a value of $-(9\pm2) \times 10^{-18}\ Z\{1-f(Q)\}$ cm, which gives a r.m.s. charge radius of the neutron of $(\langle r^2\rangle)^{\frac{1}{2}} \approx (8\pm5) \times 10^{-15}$ cm, i.e. much smaller than the magnetic r.m.s. radius of about $8 \times 10^{-14}$ cm.

The second term $b_{ip}$ arises from an induced electric dipole moment $d_i$, which is proportional to the electric field ($d_i = \alpha E$). For an atom the main contribution arises from the region near to the nucleus, where the potential can be written as $V_{ip} = -\tfrac{1}{2}\alpha Z^2 e^2/r^4$. This relation is valid for the region $r > R_c$, where $R_c$ is the Coulomb cut-off radius. Therefore another form factor with a characteristic left–right asymmetry appears and a residual contribution remains for the forward direction (Thaler 1959). The electric polarizability can be calculated by using various models for the intrinsic hadron structure (Ericson 1975). For various quark models Dattoli, Matone, and Prosperi (1977) obtained neutron polarizability values of (4–20) $\times 10^{-43}$ cm$^3$, which are in rough agreement with estimates obtained from photoproduction in deuterium and comparable with the $\alpha$ value for the proton determined from Compton scattering. With these values the contribution may be quite large for heavy elements, (1.4–8) × $10^{-15}$ cm, but separation from the nuclear part $b_c$ is, at present, not possible and the angular dependence is not pronounced for thermal neutrons. The influence of the electric polarizability on the neutron–neutron scattering length is discussed by Arnold (1973). A static electric field yields a further contribution. The mean potential is $-\frac{1}{2}\alpha E^2$, and corresponds for a field of $10^5$ V $cm^{-1}$ to values of $-(1.3\text{–}5.5) \times 10^{-26}$ eV, i.e. much smaller than the value of $V_{s1}$ discussed earlier.

The third term in equation (18) exists only if the static electric dipole moment is different from zero. This is predicted from various theories and would provide a sensitive test for time invariance (Golob and Pendlebury 1972). A new experimental limit, $d < 3 \times 10^{-24}$ $e$ cm, was recently given by Dress *et al.* (1977). The interaction is again linear in $E$ and an imaginary polarization-dependent contribution to the scattering length arises. For intermediate $Q$ and $Z$ values one has $|b_i| < 8 \times 10^{-20}$ cm and a stationary electric field of $10^5$ V $cm^{-1}$ produces a mean potential of less than $3 \times 10^{-19}$ eV.

The gravitational interaction $V_g$ can be written as

$$V_g = mgH - \hbar\mathbf{\Omega}(\mathbf{r}\times\mathbf{k}). \tag{19}$$

The first term is the potential of a particle with mass $m$ at a height $H$ in a gravitational field with a gravitational acceleration $g$. The potential difference for a neutron at sea level due to a $\Delta H$ of about 5 cm, which can be realized within the interferometer, is about $\langle \Delta V_g \rangle \approx 5 \times 10^{-9}$ eV. The possibility of observing this effect by neutron interferometry was suggested by Overhauser and Colella (1974) and has been successfully performed by Colella, Overhauser, and Werner (1975).

The second term is due to the earth's rotation, where $\mathbf{\Omega}$ is the angular frequency. We choose the angle $\phi$ for the latitude and $\psi$ for the angle with the east–west direction and obtain for this contribution $V_{gr} = -\hbar k H\omega \cos\phi \cos\psi$. This leads to a difference between beams in the east and the north direction and for a beam separation of 5 cm $\langle \Delta V_{gr} \rangle = 7.5 \times 10^{-11}$ eV. This additional phase shift within an interferometer was pointed out by Page (1975) and treated in detail by Anandan (1977), who also discussed relativistic effects and a relativistic coupling of particles with spin to the space–time curvature. These effects are extremely small and the gravity refractometer may have distinct advantages for related experiments, because the interaction time can be made much larger than within a perfect crystal interferometer (e.g. Koester 1967, McReynolds 1967, Koester 1976).

The *weak interaction* potential $V_w$ may be caused by parity non-conserving forces, which give a correlation between the spin and the momentum of the neutron $(\boldsymbol{\sigma}\cdot\mathbf{k})$ and an optical rotation as pointed out by Michel (1964). The potential can be written in the form

$$V_w = G'(\rho\boldsymbol{\sigma}\cdot\mathbf{k}+\sigma\mathbf{k}\cdot\rho) \tag{20}$$

where $G'$ is related to the weak coupling constant and $\rho$ denotes the proton and neutron interaction densities within the nucleus. This gives the following contribution to the scattering length:

$$b_w \simeq \frac{9\times 10^{-9}}{4\pi L} mCZ(\boldsymbol{\sigma}\cdot\mathbf{k}) \tag{21}$$

where $L$ is the Avogadro number and $C$ accounts for the wave distortion within the nucleus (approximately unity in our case). For heavy elements this expression leads to $b_w \sim 10^{-47}$ cm and mean potentials of only about $10^{-43}$ eV and is therefore orders of magnitude smaller than the other contributions. Near resonances a marked increase can be expected (Forte 1978), but to obtain measurable effects even in this case thick samples would have to be used.

A further term was introduced by Eckstein (1953) and may arise from multiple scattering and radiation damping within condensed matter. The corresponding mean potential is $\langle V_E\rangle = -Nb^2\hbar^2/2dm$, where $d$ is the nearest-neighbour distance. For bismuth this contribution is about $\langle V_E\rangle \approx 7\times 10^{-12}$ eV.

Figure 3 shows a comparison of the various terms for properly chosen conditions. So far nuclear, magnetic, and gravitational effects have been detected by neutron interferometry. The main problem in detecting the smaller contributions

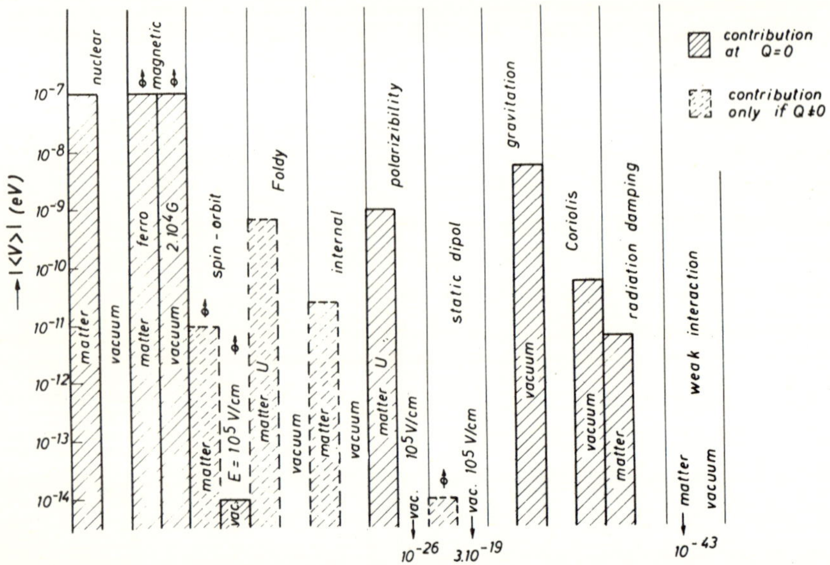


FIG. 3. Comparison of the mean interaction potential for various types of interactions

lies in their separation from the larger terms. Some of these effects are sensitive to the neutron polarization, which may serve for identification.

## 4. Applications to fundamental physics

The first experiments in this field were the observation of interference fringes of two widely separated beams by Rauch *et al.* (1974) and the gravitational experiment by Colella *et al.* (1975) which showed the action of a gravitational field on the phase of the wave function.

The extension $\Delta x$ of the wave packet has also received some interest in the literature (Shull 1969, 1973, Scheckenhofer and Steyerl 1977). There is a distinct difference to the electromagnetic radiation owing to the dispersion of matter waves even in a vacuum (e.g. Messiah 1965):

$$(\Delta x)^2 = (\Delta x_0)^2 + \left\{\frac{(\hbar/2m)\,t}{\Delta x_0}\right\}^2 . \qquad (22)$$

It is assumed that the initial length of the wave packet $\Delta x_0$ is defined by the experimental conditions, i.e. the $\Delta k$ value, via the Heisenberg uncertainty principle ($\Delta k_0 \Delta x_0 \geqslant \frac{1}{2}$). In our experimental arrangement we have $\Delta k/k \approx 10^{-3}$ and therefore $\Delta x_0 \approx 1.6 \times 10^{-6}$ cm. The $\Delta k$ value is given by various crystal reflections or by the selection of waves within the Borrmann fan. For a slow neutron beam the second term of equation (22) dominates for flight paths available within an interferometer. At our interferometer set-up at ILL (Bauspiess *et al.* 1978) the interferometer crystal is about 30 cm behind the low dispersive silicon–germanium arrangement which defines $\Delta k$. Therefore $\Delta x$ is about $3 \times 10^{-2}$ cm. In the vertical direction $\Delta k_v$ is usually much larger and therefore $\Delta x_v$ is smaller.

Considering the volume of the wave packet, the volume of the interferometer, and the counting rates available we see that only self-interference is of importance. In spite of this discussion the realisation of a Hanbury-Brown–Twiss (1951) experiment for neutrons can be discussed. For fermion fields it is expected that the coincidence counting rates of two counters I and II detecting the coherence volume of the wave packet decrease for times below the flight time of the packet through the coherence volume (here $\Delta t \sim 1$ μs) (Ledinegg 1967, Haken 1977). Although the requirements for the resolution and for the detector are rather stringent, such experiments may be possible in future. A natural limit for $\Delta x$ can be found from the live time of the neutron ($\Delta E \Delta \tau \geqslant \hbar/2$ which gives $\Delta x_c = 1.3 \times 10^8$ cm, i.e. much larger than can be observed with thermal neutrons).

We made some observations of high-order interferences with the set-up at the high flux reactor (Fig. 4). The decrease of the mutual coherence at high orders is visible. It is influenced partly by the finite $\Delta k$ value and partly by the beam attenuation due to the complex interaction potential of equation (8). Assuming a Gaussian wavelength distribution with a halfwidth $\Delta\lambda$ the interference pattern at high orders ($n = D/D_\lambda$) obeys the following relation (Petrascheck and Rauch 1976)

$$I = \frac{I(0)}{2} \exp\left(\frac{-\Sigma_t D}{2}\right)\left\{\cosh\left(\frac{\Sigma_t D}{2}\right) + \cos\left(\frac{2\pi D}{D_\lambda}\right)\right\} \times \exp\left\{-\left(\frac{1.886\, D\Delta\lambda}{D_\lambda \lambda}\right)^2\right\} . \tag{23}$$

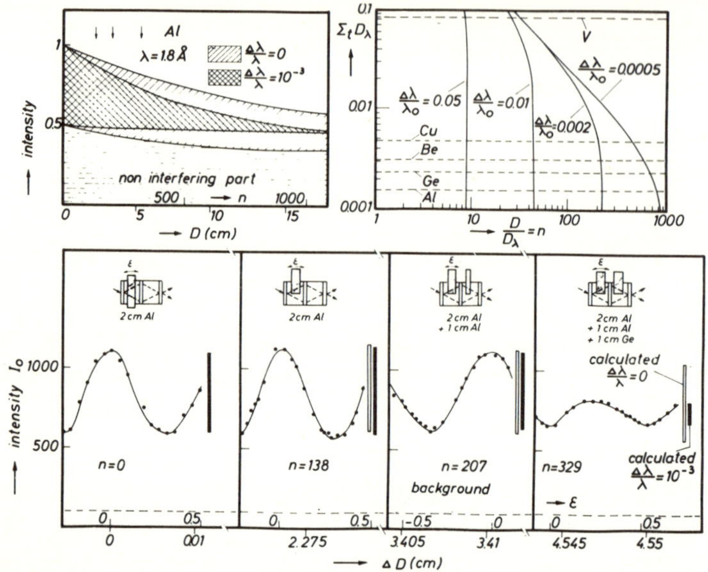

FIG. 4. Observed high-order interferences (below) and the calculated decrease of the visibility for finite and infinite resolution and the beam attenuation within aluminium (above, left). Decrease of the visibility function by a factor of 2 as a function of the interference order and various beam attenuations (above, right).

The non-interfering part, on the other hand, follows the usual attenuation law

$$I_n = I_n(0)\{C+(1-C)\exp(-\Sigma_t D)\},$$

where $C$ is the probability for the neutron to be within beam I. The insert in Fig. 4 shows for various substances the order ($n = D/D_\lambda$) for which the mutual coherence is reduced by a factor of 2. This shows that in addition to the limited space $D$ between the interferometer plates and the finite wavelength resolution there is a further limitation on the visibility of the interference fringes due to beam attenuation. In our case a coherence length $\Delta x \approx 5 \times 10^{-5}$ cm can be obtained from these measurements. In principle such measurements can serve for a determination of $\Sigma_t$ in the forward direction, but the accuracy depends rather strongly on other variations ($\Delta D$, $\Delta N$, etc.).

In 1975 the neutron interferometer was used to observe the

$4\pi$ periodicity of a spinor wave function (Rauch *et al.* 1975, Werner *et al.* 1975). Since these measurements this effect has also been observed by a Fresnel diffraction interferometer (Klein and Opat 1976), for a molecular beam system (Klempt 1976), and for a nuclear magnetic resonance transition (Stoll, Wolff, and Mehring 1978). The theoretical treatment of this subject is given for neutrons by Eder and Zeilinger (1976), Zeilinger (1976), and Balcar (1978) and showed that $4\pi$ periodicity can even be observed with unpolarized neutrons. The unpolarized beam is represented by an incoherent mixture of two oppositely polarized sub-beams. By choosing the axis of quantization perpendicular to **B** a 'rotation picture' can be applied, where the polarization of the individual sub-beams obey the Bloch equation $d\langle \mathbf{s}\rangle/dt = g\langle \mathbf{s}\rangle \times \mathbf{B}$. This gives the well-known Larmor precision angle $\alpha = gB\tau$, where $\tau$ is the flight time within the field $B$ ($g = 2\mu/\hbar$). For measurements with the interferometer the difference $\Delta\int B\mathrm{d}s$ between the effects of the field on beams I and II has to be taken. The spinor wave function transforms as (Messiah 1965, Vol. 2)

$$|\pm\rangle' = \exp(-\mathrm{i}\boldsymbol{\sigma}\cdot\boldsymbol{\alpha}/2)|\pm\rangle = \begin{pmatrix} \exp(-\mathrm{i}\alpha/2) & 0 \\ 0 & \exp(\mathrm{i}\alpha/2) \end{pmatrix} |\pm\rangle \quad (24)$$

and the intensity modulation behind the interferometer is

$$I = \frac{I(0)}{2}\left\{1 + \cos\left(\frac{\alpha}{2}\right)\right\}. \quad (25)$$

For an air-gap magnet the large stray fields hinder an accurate determination of the periodicity. Therefore we repeated the experiment with magnetized sheets of mu-metal which produce well-defined magnetic fields (Rauch *et al.* 1978a). The sheets are magnetized in opposite direction and rotated in the same direction (Fig. 5). In this case the nuclear phase shift is cancelled, while the magnetic part is enhanced. The most accurate value we obtained for the periodicity is $\alpha_0 = 716.8 \pm 3.8°$, which is in agreement with the theoretical predictions.

Even for unpolarized incident neutrons additional interference and polarization effects occur for simultaneous nuclear

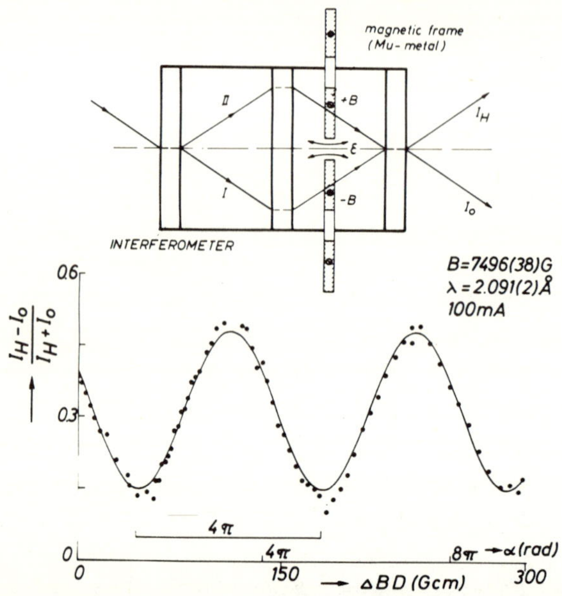


FIG. 5. Precise measurement of the 4π periodicity factor of a spinor wave function by the rotation of oppositely magnetized sheets of mu-metal. (Rauch *et al.* 1978a).

and magnetic phase shifts as observed by Badurek *et al.*(1975, 1976). For polarized incident neutrons further effects are predicted by Eder and Zeilinger (1976) and by Zeilinger (1976) and will be discussed in detail by other contributions to this workshop.

It should be mentioned that according to the self-interference properties of the neutrons within the interferometer the periodicity for a pure state described by a wave function

$$\chi_c = \sqrt{I}\begin{bmatrix}\cos \tfrac{1}{2}\alpha & \exp(-i\beta/2) \\ \sin \tfrac{1}{2}\alpha & \exp(i\beta/2)\end{bmatrix}_c \qquad (26)$$

is the same as for a mixed state to which the density matrix formalism is applied (Robson 1974). In this case α and β define the orientation of the polarization vector to some reference axis $c$.

## 5. Applications to nuclear physics

The low-energy neutron—nucleus interaction is described by a Fermi pseudo-potential (equation (8)), and the mean potential defines an index of refraction, and therefore a transmitted wave suffers a phase shift $\chi = (1-n)kD = \lambda Nb_c D$. For an interpretation of nuclear forces only the scattering lengths for the interaction in well-defined states are of interest. Therefore the spin and isotope incoherence has to be separated, which requires in most cases at least the measurement of the coherent scattering length and of the free cross-section or of the incoherent cross-section (e.g. Koester 1977).

A phenomenological interpretation of scattering lengths starts from the Breit—Wigner formalism with a separation into a potential ($b_p$) and resonance ($b_r$) scattering length, where only the resonance part depends on the total spin ($I\pm\frac{1}{2}$):

$$b = b_p + g_+ b_r^{\,+} + g_- b_r^{\,-}. \tag{27}$$

For $b_p$ one uses the potential scattering radius, which corresponds to the effective nuclear radius (Seth *et al.* 1958) and may also depend on the spin value. For well-separated resonances ($E_r^{\,j} \gg \Gamma_n^{\,j}$) the resonance scattering length for low-energy neutrons can be written as a sum of contributions arising from various resonances $j$:

$$b_r^{\,\pm} \approx -\hbar \sum_j \Gamma_n^{\,j} \left(E_r^{\,j}\right)^{-3/2} (8m)^{-1} \;. \tag{28}$$

Using this method one obtains only a rather rough estimate of $b_r^{\,\pm}$ because at least the parameters for the resonances below the binding energy ($E_r < 0$) are unknown for most nuclei. Recently Koester, Knopf, and Waschkowski (1978) pointed out this situation in connection with measurements of $b_c$ for the chromium isotopes.

A very reliable microscopic calculation of low-energy scattering lengths has been carried out on the basis of a shell-model formalism by Normand (1977). He obtained values for nuclei with closed or nearly closed shells which depend strongly on the parameters chosen for the one- and two-body potential.

A more fundamental treatment can be done for very light nuclei where few-body theory can be applied to describe the neutron–nucleus interaction. Here the zero-energy scattering length can provide a very sensitive test of various theories besides the binding energy which gives only one parameter. As we discuss nuclear problems it is more appropriate to use the free scattering length $a = bA/(A+1)$, where $A$ is the nucleus/neutron mass ratio. The singlet scattering length ($I$-½) for the proton–neutron system of $a_s = -23.749(8) \times 10^{-13}$ cm (Koester and Nistler 1975) is quite different from the proton–proton and the neutron–neutron value, which is about $-17(1) \times 10^{-13}$ cm. This indicates a charge dependence of nuclear forces, while charge symmetry may be conserved. The recent values given for the neutron–deuteron interaction are satisfactory with respect to the doublet and quartet scattering lengths ($a_{\frac{1}{2}} = 0.65(4) \times 10^{-13}$ cm and $a_{3/2} = 6.35(2) \times 10^{-13}$ cm) and are in reasonable agreement with three-body calculations which also fit the triton binding energy (see Koester 1977).

For interferometer measurements we have started a programme for a more precise determination of the neutron–$^3$He and the neutron–triton scattering length to obtain a better understanding of the four-body system. Together with the related low-energy proton scattering data and the well-known binding energy of $^4$He the four-body system can serve as a sensitive test of fundamental properties of nuclear forces, because the whole spin and isospin set can be measured. For the interactions with $^3$He and tritons, which have a nuclear spin $I$ = ½, we deal with the singlet and triplet scattering length and have the following relations (Sears and Khanna 1975):

$$a_c = \frac{3}{4}a_t + \frac{1}{4}a_s$$

$$a_i = \frac{\sqrt{3}}{4}(a_t - a_s) \qquad (29)$$

$$\sigma_{s,\text{free}} = 4\pi(|a_c|^2 + |a_i|^2) = \pi(3a_t{}^2 + a_s{}^2 + a_s'^2)$$

According to the large absorption cross-section of $^3$He ($\sigma_a$ = 5327(10)$b$) the imaginary part of the scattering length (equation (8)) has to be taken into account ($a_s \to a_s - ia_s'$). The absorption is almost entirely due to (n,p) reactions, which occur in the singlet state. The imaginary part ($a_s' = 4.4440(3) \times 10^{-13}$ cm) causes a reduction of the visibility function described by equation (23) and has to be taken into account if the measured value $\sigma_{s,free} = 3.16(20)b$ is used to obtain the results for $a_s$ and $a_t$ respectively (Alfimenkov *et al.* 1977). For this clarification we are very grateful to Dr. E. Sharapov from Dubna, USSR, and to Dr. G.R. Plattner from the University of Basel, Switzerland. Characteristic measurements with $^3$He using the interferometer set-up at ILL are shown in Fig. 6 (Kaiser *et al.* 1977). An optimal fit procedure according to equation (23) yields the bound coherent scattering length $b_c = 5.70(7) \times 10^{-13}$ cm, which is much more accurate than previous measurements by Kitchens *et al.* (1974) done with mirror reflections.

The present situation for the determination of $a_s$ and $a_t$ from available experimental data is shown in Fig. 7. It is extremely unsatisfactory for tritium where no common intersection of the experimental curves within their error bars exists. For $^3$He we have the rather accurate value for $b_c$ (or $a_c$) given

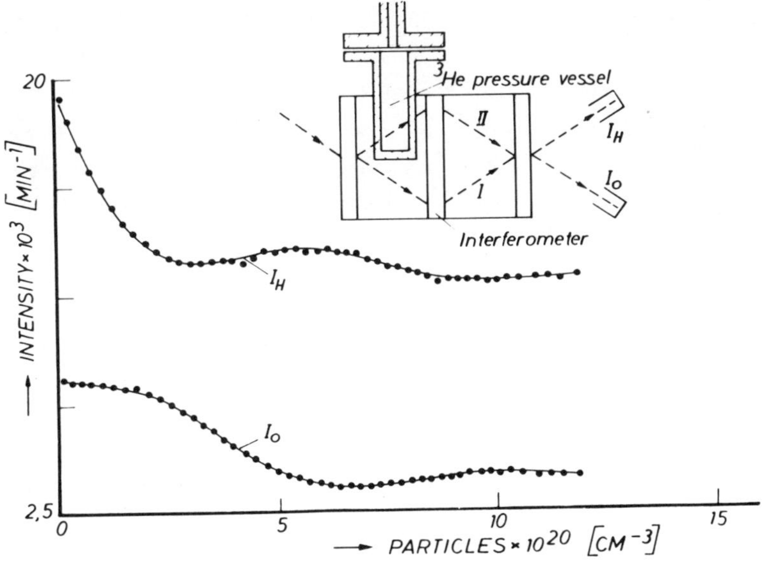


FIG. 6. Measured beam modulation for $^3$He (Kaiser *et al.* 1977).

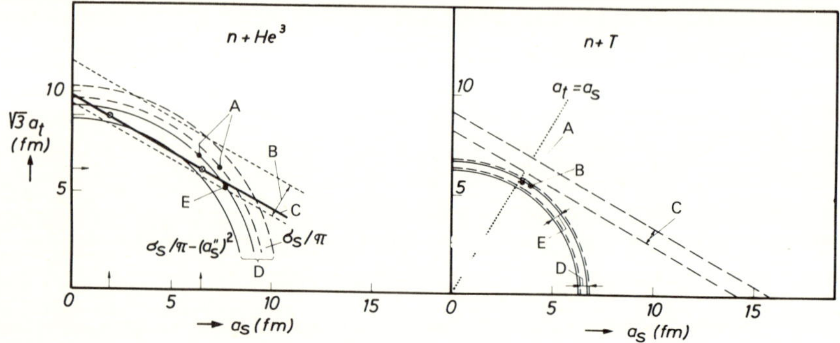


FIG. 7. Present situation for the determination of the singlet and triplet scattering length for $^3$He and T from experimental available data: (a) A, Sears and Khanna (1975); B, Kitchens *et al.* (1974) ($a_c$); C, Kaiser *et al.* (1977) ($a_c$) D, Alfimenkov *et al.* (1977); E, Kharchenko and Levashev (1976). (b) A, Szydlik and Werntz (1965); B, Kharchenko and Levashev (1976); C, Donaldson *et al.* (1972) ($a_c$); D, Battat *et al.* (1959); E, Kirilyuk (1972

earlier but a more precise determination of $\sigma_{s,free}$ or $a_i$ would be required to obtain sufficiently small error bars for $a_s$ and $a_t$. Using Gaussian error law, the new value for $a_c$ and the $\sigma_{s,free}$ of Alfimenkov *et al.* (1977) we obtain the following data sets:

(a) $a_s = 6.6(7) \times 10^{-13}$ cm $\quad$ (b) $a_s = 1.95(70) \times 10^{-13}$ cm
$\quad\;\; a_t = 3.5(4) \times 10^{-13}$ cm $\quad\quad\;\; a_t = 5.1(4) \times 10^{-13}$ cm.

A comparison with theoretical values indicates that data set (a) is the most probable.

The theoretical treatment of the four-body problem starts from the Schrödinger equation (Iotti and Donnert 1976)

$$(T+\Sigma V_{ij}-E)\psi(1,2,3,4) = 0 \tag{30}$$

where 1 and 3 denote the two neutrons and 2 and 4 the two protons. $E$ is the total energy, i.e. the sum of the binding energy and the kinetic energy,

$$E = E_{^3\text{He}} + E_n = E_T + E_p \tag{31}$$

and the kinetic energy operator $T$ can be split for the $^3$He–n system into

$$T = -\frac{\hbar^2}{M}\Delta_{34} - \frac{3}{4}\frac{\hbar^2}{M}\Delta_{2-(34)} - \frac{2}{3}\frac{\hbar^2}{M}\Delta_{1-(342)} \,. \qquad (32)$$

$V_{ij}$ is the internuclear interaction which is assumed to have a central form, exchange properties, and the Coulomb term. For $^3$He–n the complete antisymmetrized wave function is

$$\psi(1234) = (1-P_{13})\phi(234)\boldsymbol{\sigma}_s\{1-(234)F(1-(234)\} \qquad (33)$$

where $P_{13}$ is the exchange operator for particles 1 and 3, $\phi(234)$ is the wave function of the $^3$He cluster, $\boldsymbol{\sigma}_s$ corresponds to the total spin ($S$=0,1) and $F(1-(234))$ describes the motion of the neutron relative to the $^3$He cluster. Equation (33) represents a reduction to the three-body problem, which is treated by Levinger (1974) and Phillips (1977). Unfortunately the paper of Iotti and Donnert (1976) is devoted to the scattering of neutrons with energies of the order of MeV and the results for low-energy neutrons are therefore not very representative.

Another approach was chosen by Kharchenkov and Levashev (1976) who used Fadeev–Yacubovski equations with a charge-independent nucleon–nucleon interaction of the Yukawa type with spin and isospin exchange character. Therefore scattering lengths $a_{ST}$ for distinct spin ($S$) and isospin ($T$) values are obtained. While for tritium ($T_z$=1) only $T$ = 1 and $S$ = 0,1 are allowed ($a_s = a_{01}$ and $a_t = a_{11}$), for $^3$He ($T_2$=0) $T$ = 0,1 and $S$ = 0,1 are allowed which gives $a_s = (a_{00}+a_{01})/2$ and $a_t = (a_{10}+a_{11})/2$. For tritium the calculated values are $a_s = 3.76 \times 10^{-13}$ cm and $a_t = 3.07 \times 10^{-13}$ cm, which is near data set (a) (Fig. 7). In accordance with the Pauli principle for both interactions in tritium and for the triplet interaction in $^3$He only an effective repulsion exists and the calculated values are therefore not very sensitive to the details of the nuclear potential. The effective attraction in the singlet state of $^3$He is very sensitive to the detailed form of the potential. It is proposed to extend the experiments to tritium, where accurate values for $a_s$ and $a_t$ are relevant to the question of whether the repulsion is stronger in the singlet or triplet state ($a_s \gtrless a_t$).

Other new measurements of scattering lengths of various metals with the interferometer are reported elsewhere (Bauspiess *et al.* 1978). The accuracy obtained is in the range $\Delta b/b \sim 10^{-3}$ and can be increased to $10^{-4}$ or even better if the experimental parameters, i.e. the wavelength and other corrective parameters, are known with a high precision. Such accurate values can help to separate the additional interaction terms discussed in Chapter 3.

Recently we measured the scattering length of various gases like $H_2$, $D_2$, He, $O_2$, $N_2$, Ne, Ar, Xe, Kr (Kaiser *et al.* 1979). Figure 8 shows characteristic oscillations for light gases as a

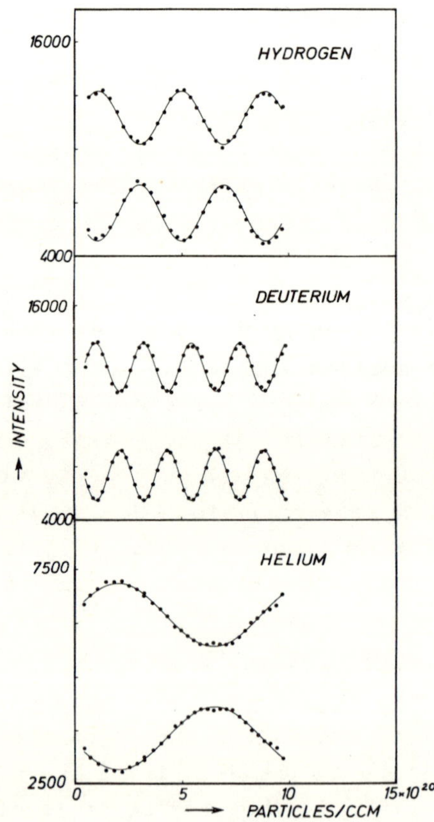


FIG. 8. Measured beam modulation for hydrogen, deuterium, and helium (Kaiser *et al.* 1979).

function of the particle density. Even for gases the index of refraction is given by the bound coherent scattering length (Kleinman and Snow 1951) and any temperature effects enter only via the density. Precise measurements can serve to obtain information about the virial coefficients.

## 6. Applications to solid state physics

Measurement with the interferometer yields the forward scattering lengths, and therefore the averaged interaction of the neutron with the sample is observed and not the microscopic structure of the sample. There might be some special cases where the influence of the microscopic structure appears, i.e. near a Bragg reflection of a perfect sample crystal where a rapid change of the index of refraction ($n = K/k$) occurs (Graeff *et al.* 1978) or in the presence of large precipitates which give rise to a pronounced forward scattering.

The relative abundances for a homogeneous mixture of elements (A and B) or isotopes can be determined if the scattering lengths are known. The period of beam modulation is given by the $\lambda$ thickness $D_\lambda = 2\pi/N\bar{b}_c\lambda$, where in the case of a mixture $N\bar{b}_c = N_A b_A + N_B b_B$. We performed such measurements with the metal–hydrogen systems V–H, V–D and Nb–H in the $\alpha$ and $\beta$ phases (Rauch *et al.* 1978b). Figure 9 shows the result of the measurement for the system V–H and V–D, where the variation of the $\lambda$ thickness due to the H(D) content is demonstrated. According to the large coherent scattering length and the high accuracy of $D_\lambda$ ($\Delta D_\lambda/D_\lambda < 10^{-3}$) a high sensitivity is achieved. In our case we obtain an accuracy for hydrogen detection of ±0.04 at.% or ±10 ppm by weight. This method is therefore more sensitive than the usual weighing method. Our experiences indicate that an increase of the sensitivity by a factor of 5 can be achieved. The interferometer method is not sensitive to specific elements and therefore a combination with other methods is necessary if the sample composition is unknown.

For a higher hydrogen concentration in the sample, near to the phase transition lines, a drastic reduction of the mutual coherence is observed (Fig. 9). This reduction of the contrast is caused by precipitation of hydrides within the sample. These

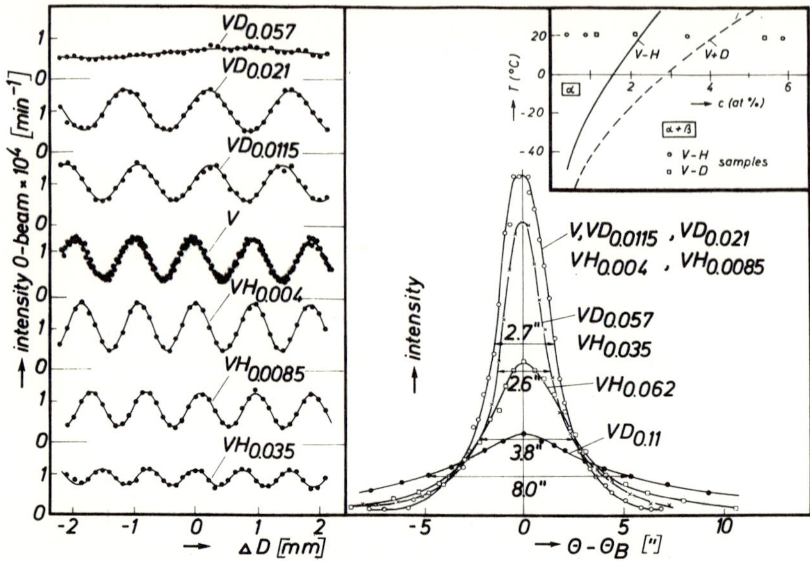


FIG. 9. Beam modulation (left) observed for the V–H and V–D systems; the related small-angle scattering (right) measured with a perfect-crystal arrangement and the related phase diagram (above). (Rauch *et al.* 1978b)

precipitates cause a spatial variation of the scattering density $N\bar{b}$ and give rise to small-angle scattering. Such small-angle scattering can be observed with the ILL interferometer set-up since a non-dispersive arrangement of a perfect-crystal monochromator and the interferometer exists with a rocking curve half-width of 2″. Characteristic results for small-angle scattering are shown in Fig. 9. It is obvious that the contrast of the interferometer is more sensitive to precipitates than to small angle scattering. A quantitative understanding of this reduction of coherence within the interferometer due to small-angle scattering can serve to create a new powerful method for the observation of sample inhomogeneities. Figure 10 shows another example, where additional precipitates are created during the annealing of an Al–(1.5% MgZn) alloy.

An exact description of the loss of coherence due to an inhomogeneous sample has to account for both the deflection of neutrons and the relative phase shift ($\Delta\varphi = 2\pi\Delta(Nb)\delta/k$), which appears in the forward direction as a result of crossing a region of thickness $\delta$ with a different scattering length density $\Delta(Nb)$. The angle of acceptance of the interferometer must

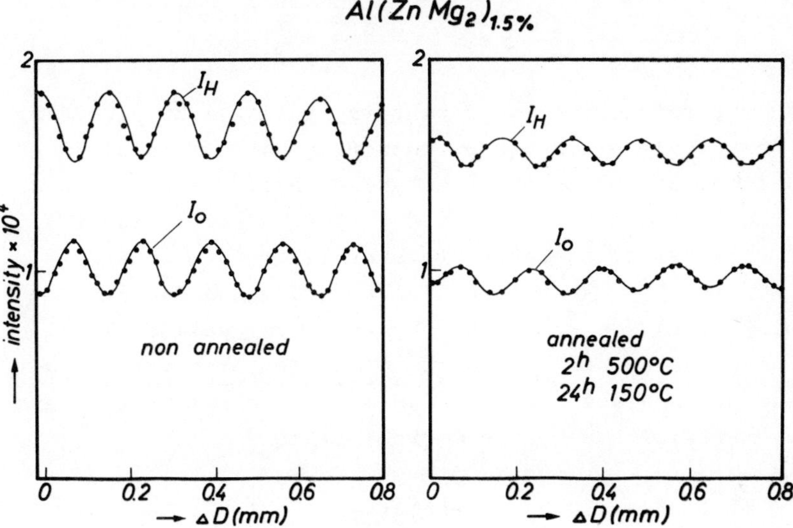


FIG. 10. Loss of contrast and of the mutual coherence due to precipitates in an Al–Mg–Zn alloy formed during the annealing of the sample.

therefore be considered. This angular region is given by the Pendellösung structure of the Laue reflection curve of the interferometer plates with a thickness $t$. For our interferometer ($t \sim 4$ mm) the characteristic period of the Pendellösung oscillations in the angular scale is about $\Delta\Theta_c = 0.06''$, which is much smaller than the half-width of the reflection curve. The $\Delta\Theta_c$ value corresponds to a diffraction object with a dimension $\delta_c \sim 750$ μm.

The general expression for small-angle scattering can be written as (e.g. Schmatz *et al.* 1974)

$$\frac{d\sigma}{d\Omega} = \frac{1}{N}\left|\int \exp(i\mathbf{Q}\cdot\mathbf{r})N(\mathbf{r})b(\mathbf{r})d\mathbf{r}\right|^2 = \frac{1}{N}|a(\mathbf{Q})|^2 \tag{34}$$

where $a(\mathbf{Q})$ denotes the scattering length of an atom within a precipitate. Corresponding to $\Delta\varphi \ll 2\pi$ or $\Delta\varphi \gg 2\pi$ we have the diffraction or refraction range of small-angle scattering. For the diffraction range we can use the approximation given by Guinier (1963)

$$\frac{d\sigma}{d\Omega} \approx \frac{[\Delta(\mathrm{Nb})]^2}{N} \exp - \frac{Q^2\delta^2}{3} = \frac{1}{N}|a(Q)|^2 \tag{35}$$

and for the refraction range other approximations according to the geometrical optics can be used (von Nardoff 1926).

Equation (35) shows that the phase shift $\varphi$ depends on $Q$, which can be related to a $y$ value with respect to the following reflections at an interferometer plate ($Q \approx k(\Theta-\Theta_B) = 4\pi y/D_\lambda^{\mathrm{Si}} \sin 2\Theta_B$). Therefore we obtain a phase factor $\exp\{i\varphi(Q)\}$ in addition to the $v$ values of equation (1) which now depends on $\Delta y$ and has to be integrated over various $Q$ transfers and weighed according to the mean path within the homogeneous material and within a precipitate

$$\begin{aligned}\varphi(Q) = \varphi(y) &= b_c\lambda N(D-n\delta)\delta(y-y') + \lambda\Delta(Nb_c)n\delta \\ &\times \exp\left[-\frac{1}{6}\left\{-\frac{4\pi(y-y')\delta}{D_\lambda^{\mathrm{Si}}\sin 2\Theta}\right\}^2\right]\end{aligned} \tag{36}$$

and, for example

$$\begin{aligned}\psi_0^{\mathrm{I}} = v_0(y) \int &\exp\left\{i\varphi(y-y')\right\}v_H(y-y')v_{-H}(y'-y) \\ &\times \exp\left\{-\frac{2\pi i(y-y')(T+t)}{\Delta_0}\right\}dy' .\end{aligned}$$

Such an expression yields a damping of the contrast and a decrease of the mutual coherence depending on $\delta$, the mean path length within a precipitate, and $n$, the mean number of precipitates, crossed by the neutron beam. For a rather crude estimate we can omit the angular dependence and obtain a dephasing factor $D_\varphi$

$$D_\varphi = \exp\left\{-\frac{4h^2(\Delta Nb)^2\delta^2 n}{mv^2}\right\} \tag{37}$$

similar to the depolarization factor discussed later.

A ferromagnetic sample with magnetic domains represents again an inhomogeneous medium for neutrons. The index of refraction varies through the sample according to the magnetization within the domain ($n=1\mp\lambda^2 m\mu B_\parallel^i/h^2$). $B_\parallel^i$ is the magnetic induction within the domain $i$ parallel to the axis of quantization chosen for the sub-beams which form the unpolarized incident beam. It is further assumed that a non-adiabatic transition of the neutron spins occurs between the various domains. This is justified as long as the Larmor frequency $\omega_L = gB = 2\mu B/\hbar$ is smaller than the frequency $\omega = 2\pi v/\xi$ seen by the neutron during transmission through a Bloch wall of thickness $\xi_0$ at a glancing angle $\Theta$ ($\xi=\xi_0/\sin\Theta$) (Rauch, Rockenbauer, and Rupar 1970). The mean angle of deflection from large and arbitrarily oriented domains is much smaller than the critical angle for total reflection. Hughes *et al.* (1949) calculated the mean angle of deflection for iron to be $\beta_0 = 0.029'$, while the half-width $\Delta\beta$ of small-angle scattering for a sample of thickness $D$ and mean domain size $\delta$ is $\Delta\beta = \beta_0(D/\delta)^{\frac{1}{2}}$. For a more rigorous treatment we use the cross-section for magnetic small-angle scattering (Schmatz *et al.* 1974)

$$\frac{d\sigma}{d\Omega} = \frac{1}{N}\left(\frac{\gamma e^2}{4\pi\hbar c}\right)^2\left|\int \exp(i\mathbf{Q}\cdot\mathbf{r})B^{\perp}(\mathbf{r})d\mathbf{r}\right|^2 = \frac{1}{N}|a_M(\mathbf{Q})|^2 \tag{38}$$

where $B^{\perp}$ is the magnetic induction perpendicular to $\mathbf{Q}$ and $a_M$ is the magnetic scattering length of an atom within a magnetic domain. Equation (38) allows us to proceed in the same way as for nuclear small-angle scattering. The case of ferromagnetic precipitates within a non-magnetic material where nuclear and magnetic small-angle scattering occur simultaneously is treated by Ernst, Schelten, and Schmatz (1971).

Another approach neglecting small-angle scattering is the depolarization method developed for polarized neutrons by Halpern and Holstein (1941) and generalized by Rekveldt (1973). To use this theory for unpolarized neutrons within the interferometer it is assumed that the polarization of the sub-beams ($|P_s|=1$) obeys the Bloch equation ($g=2\mu/\hbar$)

$$\frac{d\mathbf{P}_s}{dt} = g\mathbf{P}_s \times \mathbf{B} \tag{39}$$

which connects the polarization before entering and the polarization after leaving the domain by a rotation dyad

$$\mathbf{P}_s^{i} = \{\mathbf{b}_i\cdot\mathbf{b}_i\bullet + (1-\mathbf{b}_i\cdot\mathbf{b}_i\bullet)\cos\int_0^{\tau_i} gB_i dt - \sin\int_0^{\tau_i} gB_i dt\,\mathbf{b}_i \times\}\mathbf{P}_s^{i-1} \tag{40}$$

where $b_i$ is a unit vector in direction of $B_i$, $\tau_i$ is the time the neutron spends in the domain $i(\tau_i=\delta_i/v)$, and

$$\int_0^{\tau_i} gB_i \mathrm{d}t = gB_i\delta_i/v = \chi$$

is the angle of rotation within this domain. For arbitrary oriented domains one obtains

$$\mathbf{P}_{\mathrm{s}} = \prod_{i=1}^{n}\left(1-\frac{4}{3}\sin^2 g\frac{B_{\mathrm{i}}\delta_{\mathrm{i}}}{2v}\right)\mathbf{P}_{\mathrm{s}}^{0} \tag{41}$$

which have to be averaged over many possible neutron paths through the sample $(n=D/\delta)$. For $\chi \ll 2\pi$ we obtain

$$\frac{P_{\mathrm{s}}^{z}}{P_{\mathrm{s}}^{z0}} = D_{\mathrm{M}} = \exp\left(-\frac{g^2B^2\delta D}{3v^2}\right) \tag{42}$$

and for $\chi \gg 2\pi$

$$\frac{P_{\mathrm{s}}^{z}}{P_{\mathrm{s}}^{z0}} = D_{\mathrm{M}} = \exp\left(-\frac{D\ln 3}{\delta}\right) .$$

For the measurements with the interferometer $D_{\mathrm{M}} \neq 1$ corresponds to a loss of coherence of the spinor wave function. For a special field arrangement an alternative description is given by Schwink and Schärpf (1975), who give a solution to the Pauli equation in varying fields, and by Balcar (1978).

We performed such measurements with various thick iron foils in both beams of the interferometer and measured the mutual coherence with an aluminium phase shifter in the second part of the interferometer (Fig. 11). The decrease of the mutual coherence roughly agrees with the value $D_{\mathrm{M}}^2$, which was determined independently from measurement of the depolarization of a polarized beam transmitted through these foils. We thank Dr. M.Th. Rekveldt from I.R.I. Netherland, who made a three-dimensional depolarization analysis of the iron foils in 1975 and indicated a preferred axis of magnetization for non-annealed samples.

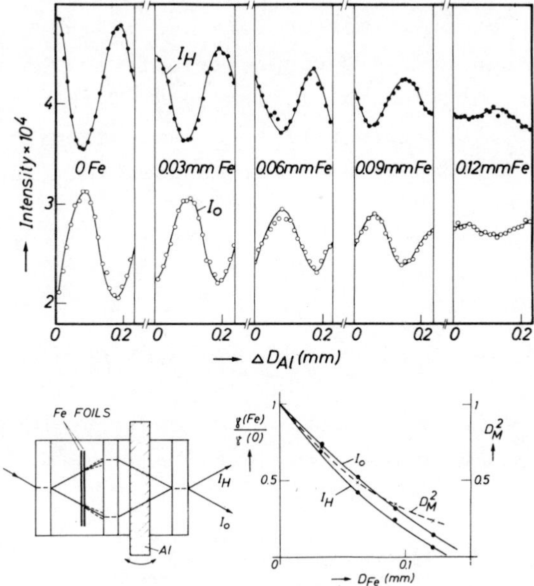


FIG. 11. Loss of contrast and mutual coherence due to the ferromagnetic domain structure in iron foils.

## 7. Conclusion

The development of neutron interferometry with perfect-crystal interferometers increased the capacities of particle radiation interferometry markedly. Instead of cross-sections the phase shifts caused by the samples are measured. Various characteristic interference effects predicted by quantum theory can be performed on a macroscopic scale. It will become a standard method for the measurement of coherent scattering lengths and their behaviour near resonances will gain further interest. Neutron interferometry can also be used for an extremely sensitive test of perfect crystals and for measurements near critical points where large-range fluctuations of the density or the magnetization occur. The sensitive detection of hydrogen or other gases in metal and the observation of precipitates may lead to further applications. The sensitivity for magnetic investigations can be increased for a polarized interferometer set-up. The measurements are usually made at $Q = 0$ and $\omega = 0$. In special cases the dispersion properties of the wave vectors near a reciprocal lattice point ($\bar{Q}\approx 2\pi\tau$) can be investigated and with an interferometer moving relative to the neutron beam a certain range of $\omega \neq 0$ can contribute to the reflection. Even though one recognizes the limitations of this technique, neutron interferometry represents a new and widening field of neutron optics which can be used for fundamental, nuclear, and solid state physics.

**Acknowledgments**

The author wishes to thank Doz. Dr. Ewald Balcar for many helpful discussions and for help during the preparation of the manuscript. This work was supported by Fonds zur Förderung der Wissenschaftlichen Forschung – project 3185.

**References**

AHARONOV, Y. and BOHM, D. (1959). *Phys. Rev.* **115**, 485.
ALFIMENKOV, V.P., AKOPIAN, G.G., WIERZBICKI, J., GOVOROV, A.M., PIKELNER, L.B., and SHARAPOV, E.I. (1977). *Sov. J. nucl. Phys.* **25**, 1145.
ANANDAN, J. (1977). *Phys. Rev. D* **15**, 1448.
ARNOLD, L.G. (1973). *Phys. Lett. B* **44**, 401.
BADUREK, G., RAUCH, H., ZEILINGER, A., BAUSPIESS, W., and BONSE, U. (1975). *Phys. Lett. A* **54**, 425.
—— (1976). *Phys. Rev. D* **14**, 1177.
BATTAT, M.E., BONDELID, R.O., COON, Y.H., CRANBERG, L., DAY, R.B., EDESKUTY, F., FRENTROP, A.M., HENKEL, R.L., MILLS, R.L., NOBLES, R.A., PERRY, Y.E., PHILLIPS, D.D., ROBERTS, T.R., and SYDORIAK, G.S. (1959). *Nucl. Phys.* **12**, 291.
BALCAR, E. (1978). This book, page 252.
BAUSPIESS, W. (1977). Ph.D. Thesis, University of Dortmund.
BAUSPIESS, W., BONSE, U., and GRAEFF, W. (1976). *J. appl. Crystallogr.* **9**, 68
BAUSPIESS, W., BONSE, U., and RAUCH, H. (1978). *Nucl. Instrum. Meth.* **157**, 49
BETHE, H. (1928). *Annln Phys.* **87**, 55.
BOERSCH, H., HAMISCH, H., GROHMANN, K., and WOHLLEBEN, D. (1961). *Z. Phys.* **165**, 79.
—— (1962). *Z. Phys.* **169**, 263.
BONSE, U. and GRAEFF, W. (1977). *Topics appl. Phys.* **22**, 92.
BONSE, U. and HART, M. (1965). *Appl. Phys. Lett.* **6**, 154; **7**, 99.
—— (1966). *Z. Phys.* **194**, 1.
COLELLA, R., OVERHAUSER, A.W., and WERNER, S.A. (1975). *Phys. Rev. Lett.* **34**, 1472.
DATTOLI, G., MATONE, G., and PROSPERI, D. (1977). Rep. LNF-77/9 (R), Istituto Naz. Fisica Nucleare, Frascati.
DICKENS, M.H., SHULL, C.G., KOHLER, W.C., and MOON, R.M. (1975). *Phys. Rev. Lett.* **35**, 595.
DONALDSON, R.E., BARTOLINI, W., and OTSUKI, H. (1972). *Phys. Rev. C* **5**, 1952.
DRESS, W.B., MILLER, P.D., PENDLEBURY, J.M., and PERRIN, P. (1977). *Phys. Rev. D* **15**, 9.
ECKSTEIN, H. (1953). *Phys. Rev.* **89**, 490.

EDER, G. and ZEILINGER, A. (1976). *Nuovo Cim. B* **34**, 76.
ERICSON, T.E.O. (1975). In *Interaction studies in nuclei* (eds. H. Jochin and B. Ziegler). North-Holland, Amsterdam.
ERNST, M., SCHELTEN, J., and SCHMATZ, W. (1971). *Phys. Status Solidi A* **7**, 477.
FERMI, E. (1936). *Ric. Sci.* **1**, 13.
FOLDY, L.L. (1955). *Phys. Rev.* **83**, 688.
—— (1958). *Rev. mod. Phys.* **30**, 471.
FORTE, M. (1973). *Nuovo Cim. A* **18**, 726.
—— (1979). In *Fundamental physics with reactor neutrons and neutrinos* (ed. T. von Egidy). Institute of Physics, London/Bristol.
GOLDBERGER, M.L. and SEITZ, F. (1947). *Phys. Rev.* **71**, 294.
GOLOB, R. and PENDLEBURY, J.M. (1972). *Contemp. Phys.* **13**, 519.
GRAEFF. W., BAUSPIESS, W., BONSE, U., and RAUCH, H. (1978). *Acta Cryst.* **A34**, S 238.
GUINIER, A. (1963). *X ray diffraction*. Follman, London.
HAKEN, H. (1977). *Acta Phys. austriaca* **47**, 59.
HALPERN, O. and HOLSTEIN, R. (1941). *Phys. Rev.* **59**, 960.
HALPERN, O. and JOHNSON, M.H. (1939). *Phys. Rev.* **59**, 898.
HANBURY-BROWN, R. and TWISS, R.Q. (1957). *Proc. R. Soc. A* **242**, 300; **243**, 291.
HUGHES, D.J., BURGY, M.T., HELLER, R.B., and WALLACE, J.M. (1949). *Phys. Rev.* **75**, 565.
IOTTI, R.C. and DONNERT, H.J. (1976). *Acta Phys. austriaca* **44**, 7.
KAISER, H., RAUCH, H., BADUREK, G., BAUSPIESS, W., and BONSE, U. (1979). *Z. Physik* **291**, 231.
KAISER, H., RAUCH, H., BAUSPIESS, W., and BONSE, U. (1977). *Phys. Lett. B* **71**, 321.
KATO, N. (1961). *Acta Crystallogr.* **14**, 526, 627.
—— (1968). *Appl. Phys.* **39**, 2225, 2231.
KELLER, M. (1961). *Z. Phys.* **164**, 292.
KHARCHENKO, V.F. and LEVASHEV, V.P. (1976). *Phys. Lett. B* **60**, 317.
KIRILYUK, A.L. (1972). Thesis, Inst.Nucl.Research, Acad.Sci. Ukrain.SSR., Kiev.
KITCHENS, T.A., OVERHUISEN, T., and PASSEL, L. (1974). *Phys. Rev. Lett.* **32**, 791.
KLEIN, A.G. and OPAT, G.I. (1976). *Phys. Rev. Lett.* **37**, 238.
KLEINMAN, D. and SNOW, G. (1951). *Phys. Rev.* **82**, 952.
KLEMPT, E. (1976). *Phys. Rev. D* **13**, 3125.
KOESTER, L. (1967). *Z. Phys.* **198**, 187.
—— (1976). *Phys. Rev. D* **14**, 907.
—— (1977). Neutron scattering lengths and fundamental neutron interaction. In *Neutron physics* (ed. G. Höhler), *Springer Tracts Mod. Phys.* **80**.
KOESTER, L. and NISTLER, W. (1975). *Z. Phys. A* **272**, 189.
KOESTER, L., NISTLER, W., and WASCHKOWSKI, W. (1976). *Phys. Rev. Lett.* **36**, 1021.
KOESTER, L., KNOPF, K., and WASCHKOWSKI, W. (1978). *Z. Physik* A287, 61.
KROHN, V.E. and RINGO, G.K. (1966). *Phys. Rev.* **148**, 1303.
LANDKAMMER, F.J. (1966). *Z. Phys.* **189**, 113.
von LAUE, M. (1948). *Materiewellen und Interferenzen*. Akad. Verlagsges, Leipzig.
LEDINEGG, E. (1967). *Z. Phys.* **205**, 25.
LEVINGER, J.S. (1974). *Springer Tracts mod. Phys.* **71**, 88.
LISCHKE, B. (1969). *Phys. Rev. Lett.* **22**, 1366.
LOVESEY, S.W. (1969). *J. Phys. C* 2, 981.
McREYNOLDS, A.W. (1967). *Bull. am. phys. Soc.* **12**, 105.
MAIER-LEIBNITZ, H. and SPRINGER, T. (1962). *Z. Phys.* **167**, 386.
MARSHALL, W. and LOVESEY, S.W. (1971). *Theory of thermal neutron scattering*. Clarendon Press, Oxford.
MARTON, L. (1952). *Phys. Rev.* **85**, 1057.

MARTON, L., SIMPSON, J.A. and SUDDETH, Y. (1953). *Phys. Rev.* **90**, 410.
MESSIAH, A. (1965). *Quantum Mechanics*, Vol. 1, p. 222. North-Holland, Amsterdam.
MICHEL, F.C. (1964). *Phys. Rev.* **133**, B 329.
MÖLLENSTEDT, G. and BAYH, W. (1962). *Naturwissenschaften* **49**, 81; *Z. Phys.* **169**, 492.
MÖLLENSTEDT, G. and DÜCKER, H. (1954). *Naturwissenschaften* **42**, 41.
von NARDOFF, R. (1926). *Phys. Rev.* **28**, 240.
NORMAND, J.N. (1977). *Nucl. Phys. A* **291**, 126.
OBERMAIER, G. (1967). *Z. Phys.* **204**, 215.
OVERHAUSER, O.W. and COLELLA, R. (1974). *Phys. Rev. Lett.* **33**, 1237.
PAGE, L.A. (1975). *Phys. Rev. Lett.* **35**, 543.
PETRASCHEK, D. (1976). *Acta Phys. Austriaca* **45**, 217.
PETRASCHEK, D. and FOLK, R. (1976). *Phys. Status Solidi A* **36**, 147.
PETRASCHEK, D. and RAUCH, H. (1976). *Theorie des interferometers*. Rep. AIAU 76 401, Atominstitut, Vienna.
PHILLIPS, A.C. (1977). *Rep. Prog. Phys.* **40**, 905.
RAUCH, H. (1971). *Acta Phys. Austriaca* **33**, 50.
RAUCH, H. and PETRASCHEK, D. (1978). In *Topics in current physics* (ed. H. Dachs), Vol. 6, Chap. 9. Springer-Verlag, Berlin.
RAUCH, H., ROCKENBAUER, G., and RUPAR, H. (1970). *Atomkernenergie* **15**, 275.
RAUCH, H., SEIDL, E., ZEILINGER, A., BAUSPIESS, W., and BONSE, U. (1978b). *J. appl. Phys.* **49**, 2731.
RAUCH, H. and SUDA, M. (1974). *Phys. Status Solidi A* **24**, 495.
RAUCH, H., TREIMER, W. and BONSE, U. (1974). *Phys. Lett. A* **47**, 369.
RAUCH, H., WILFING, A., BAUSPIESS, W., and BONSE, U. (1978a). *Z. Phys. B* **29**, 281.
RAUCH, H., ZEILINGER, A., BADUREK, G., WILFING, A., BAUSPIESS, W. and BONSE, U. (1975). *Phys. Lett. A* **54**, 425.
REKVELDT, M.TH. (1973). *Z. Phys.* **259**, 391.
ROBSON, A.B. (1974). *The theory of polarization phenomena*. Clarendon Press, Oxford.
SCHECKENHOFER, H. and STEYERL, A. (1977). *Phys. Rev. Lett.* **39**, 1310.
SCHMATZ, W., SPRINGER, T., SCHELTEN, J., and IBEL, K. (1974). *J. appl. Crystallogr.* **7**, 96.
SCHNEIDER, C.S. and SHULL, C.G. (1971). *Phys. Rev. B* **3**, 830.
SCHWINGER, J. (1948). *Phys. Rev.* **73**, 407.
SCHWINK, CH. and SCHÄRPF, O. (1975). *Z. Phys. B* **21**, 305.
SEARS, V.F. (1978). *Can. J. Phys.* **56**, 1261.
SEARS, V.F. and KHANNA, F.C. (1975). *Phys. Lett. B* **56**, 1.
SETH, K.K., HUGHES, D.J., ZIMMERMANN, R.L., and GARTH, R.C. (1958). *Phys. Rev.* **110**, 692.
SHULL, C.G. (1969). *Phys. Rev.* **179**, 752.
—— (1973). *J. appl. Crystallogr.* **6**, 257.
SHULL, C.G. and NATHANS, R. (1967). *Phys. Rev. Lett.* **7**, 384.
STASSIS, C. (1970). *Phys. Rev. Lett.* **24**, 1415.
STASSIS, C. and SHULL, C.G. (1972). *Phys. Rev. B* **25**, 1040.
STOLL, M.E., WOLFF, E.K. and MEHRING, M. (1978). *Phys. Rev. A* **17**, 1561.
SZYDLIK, P. and WERNTZ, C. (1965). *Phys. Rev. B* **138**, 866.
THALER, R.M. (1959). *Phys. Rev.* **114**, 827.
WERNER, S.A., COLELLA, R., OVERHAUSER, A.W., and EAGEN, C.F. (1975). *Phys. Rev. Lett.* **35**, 1053.
ZEILINGER, A. (1976). *Z. Phys. B* **25**, 97.

# 2. SCATTERING LENGTHS

L. KOESTER
*Technische Universität München, Reaktorstation Garching, Germany*

## 1. Introduction

A plane neutron wave (wave vector **k**, neutron mass $m$) when scattered by a single fixed particle (spin $I = 0$, mass $M$) appears in the range $r$ from the scatterer with a relative amplitude $f(\Theta,k)/r$ in the direction given by $\Theta$. The quantity $f(\Theta,k)$ is the scattering amplitude. Its value depends on the neutron—particle interaction potential which may be characterized by depth, range and shape. The interaction causes phase shifts $\delta_l$ of the neutron wave.

At low neutron energies (with $k \ll R$) only $s(l=0)$-wave scattering occurs. The interaction characterized by $f_0$ or $\delta_0$ is independent of the shape of the potential but is slowly dependent on the range.

With the assumption of an effective range $r_{eff}$ the scattering theory leads to the relation

$$\frac{1}{f_0} = k \cot\delta_0 = -\frac{1}{b}\left(1-k^2 \frac{r_{eff}}{2}|b|\right).$$

Hence, as $k \to 0$ $f_0$ approaches

$$\lim |f_0| = -b .$$

$b$ is denoted as the Fermi scattering length. It is a fundamental constant for the specific particle. At thermal neutron energies the correction term due to $r_{eff}$ amounts to $10^{-6}$ $\Delta b/b$ and therefore may still be neglected. Normally the scattering length $b = R' + b_R$ is written in terms of the energy-independent potential scattering term $R'$ and the resonance scattering which contributes the term $b_R$ depending on the resonance structure of the neutron—nucleus scattering. $b_R$ may be a function of neutron energy if broad resonance lie in or near the thermal region.

For particles with spin $I \neq 0$ as scatterer there are two spin states: $J_+ = I + \frac{1}{2}$ and $J_- = I - \frac{1}{2}$. The statistical weights are $g_+$ and $g_-$ respectively. Scattering on an assembly of spin particles leads then to a coherent scattering with $b_c = g_+b_+ + g_-b_-$ and to an incoherent part with $b_i = (g_+g_-)^{\frac{1}{2}}(b_+-b_-)$. Here the spin state scattering lengths $(b_+,b_-)$ are the fundamental quantities.

At thermal energy the point-like Fermi potential for a nucleus at $\mathbf{R}$ and a neutron at $\mathbf{r}$

$$V(\mathbf{r}) = 2\pi\hbar^2 m^{-1} b\delta(\mathbf{r}-\mathbf{R})$$

is an exact solution of the wave equation in the first Born approximation. For an assembly of $N$ atoms cm$^{-3}$ the averaged potential becomes

$$\langle V\rangle = 2\pi\hbar^2 m^{-1} Nb$$

wherein $Nb$ is the so-called scattering length density. Matter of a scattering density $Nb$ has a refractive index $n$ for neutron radiation of the kinetic energy $E = \hbar^2/(2m\lambda^2)$ according to

$$n^2 - 1 = -\frac{\langle V\rangle}{E} = -\frac{1}{\pi}\lambda^2 Nb \ .$$

## 2. Scattering lengths of atoms

The neutron is a particle with mass $m$, spin $\frac{1}{2}$, a magnetic moment ($\gamma$=-1.91) nuclear magnetons), and an intrinsic charge structure, hypothetically with electric dipole moment $d$ and electric polarizability $\alpha$. All these properties play a role in the interaction of the neutron with an atom, or more exactly with the nucleus, the charge of nucleus and electrons, and the magnetic moment of the shell.

Specific interactions are described in detail below.

(1) *The neutron–charge interaction:* the dominating process is the relativistic interaction of the magnetic moment of the neutron with a weak electromagnetic field of charge $Q$. From the Dirac equation for the neutron Foldy (1958) derived the

corresponding scattering length to be

$$b_F = Q\left(\frac{\gamma e}{2mc^2}\right) = -1.468 \times 10^{-3} \text{ fm} \quad \text{for } Q = -e.$$

If there is an intrinsic charge distribution within the neutron an additional contribution $b_e$ arises so that the neutron–electron interaction can be written as $b_{ne} = b_F + b_e$.

(2) *The magnetic interaction* of unpolarized neutrons with a magnetic moment of the atomic shell is described by

$$b_{mag}(\mathbf{q}) = -\gamma\left(\frac{e^2}{m_e c^2}\right) fm(\mathbf{q}) \langle \mathbf{S} \rangle (\boldsymbol{\sigma} \cdot \mathbf{m})$$

where $fm(\mathbf{q})$ is the magnetic form factor of the atom depending on the momentum transfer $\mathbf{q}$, $\langle \mathbf{S} \rangle$ the mean magnetization, $\boldsymbol{\sigma}$ the Pauli spin matrix, and $\mathbf{m}$ the magnetic interaction vector.

(3) *The spin–orbit interaction* (Schwinger 1948) of the neutron occurs in the Coulomb field of the nucleus $(Z,A)$. The corresponding scattering amplitude $b_{LS}(\mathbf{q})$ is a purely imaginary quantity which depends on the momentum transfer $\mathbf{q}$ and the scattering angle $\Theta$:

$$b_{LS}(\mathbf{q}) = i[Zb_F\{1-f(\mathbf{q})\}\cot(\Theta/2)](\boldsymbol{\sigma},\hat{\mathbf{n}})$$

where $f(\mathbf{q})$ is the electric form factor of the atomic shell and $\hat{\mathbf{n}}$ the unit vector of the scattering plane. The Schwinger scattering has, like the Foldy interaction, its origin in the Dirac equation of the neutron.

(4) *The interaction of the hypothetical dipole moment $d$* of the neutron with the atom also gives rise to an imaginary scattering amplitude (Shull and Nathans 1967):

$$b_{ed}(\mathbf{q}.\Theta) = i\left[\frac{dZe\{1-f(\mathbf{q})\}m}{\hbar^2 k \cos(\Theta/2)}\right](\boldsymbol{\sigma}\cdot\hat{\mathbf{q}})$$

where $\hat{\mathbf{q}}$ is the unit vector in the direction of momentum transfer.

(5) *The contribution of the hypothetical electric polarizability* $\alpha_n$ to the elastic neutron scattering can be written as (i.e. Aleksandrov 1969)

$$b_P = \frac{(m/\hbar^2)\alpha_n(Ze)^2}{R} = b_\alpha Z^2 \leqslant 10^{-5}Z^2\text{fm}$$

where $R$ denotes the electric radius of the nucleus.

(6) *The neutron—nucleus interaction:* the total scattering length $b_N$ for the nucleus can be considered as the sum of contributions due to the intrinsic nuclear force interaction $b_{NF}$, the nuclear charge $Zb_{ne}$, and the nuclear electric field $b_P$. The effect of neutron reactions (absorption etc.) is described by an imaginary part i $\sigma_{abs}/2\lambda_n$ which in most cases is small and can be neglected.

Summarizing, we can present the total *neutron—atom scattering length* in the lower energy region as

$$b_c = \left\{R'+b_R(E)+Z^2b_\alpha+Zb_{ne}\right\}_{\text{nucleus}} - Z\overline{f}(E)b_{ne}$$

$$+\,\text{i}\{b_{LS}(E)+b_d(E)+\sigma_a(E)/2\lambda_n\}\ .$$

The neutron—electron contribution $Z\overline{f}(E)b_{ne}$, the Schwinger term, $b_{LS}$ and the dipole term $b_d$ can be experimentally separated because of their typical behaviour concerning energy or angle dependence. On the other hand it is impossible at present to extract the contribution of the polarizability by s-wave scattering experiments only. All quantities concerned are of fundamental interest for neutron and particle physics. Moreover, exact values of nuclear scattering lengths play an important role for the theories of nuclear forces, few-body forces, and nuclear models. Also in the wide field of applications of neutron scattering, the coherent and the incoherent scattering lengths are the basic quantities.

## 3. Methods

The separation of small contributions (i.e. $b_{ne}$) from the atomic scattering length as well as the determination of the fundamental nuclear-scattering parameters require a high accuracy of the experimental determination of the scattering lengths. Good accuracy, of course, is of advantage in all other problems in the application of neutron scattering. Consequently in the last decade new high-precision techniques have been developed. The advanced techniques and the conventional ones will be briefly discussed emphasizing the problem of accuracy.

All physical phenomena involved in the interaction of neutron waves with matter, such as extinction, reflection and refraction, diffraction and direct interference, and gravitational acceleration, have been used to determine scattering lengths. All techniques need samples which exhibit a common source of more or less serious uncertainties. The more exact the technique the more serious the uncertainties of the samples become. They concern density, dimensions, and structure as well as purity and stoichiometry.

The precision and the accuracy of a technique are limited by number and nature of the parameters which must be measured to obtain the scattering length. Brief characteristics of the various techniques are given below.

### 3.1. *Transmission method*

The total cross-section is measured with neutrons of energies in the electronvolt region. Liquids, polycrystals, powders, gases, and in special cases separated isotopes are suitable as samples. To obtain $\sigma_{sc} = 4\pi b_c^2$ from the total cross-section the absorption, the incoherent, and the neutron—electron cross-sections and a possible resonance contribution must be known. Sometimes solid state corrections are necessary. The uncertainty amounts to $\pm\ 0.01\ \Delta b/b$, but in favourable cases ($I = 0$ single isotope elements without broad resonances) an error of about $\pm\ 10^{-4}\ \Delta b/b$ can be obtained. For instance the old standard value for carbon has been determined with an uncertainty of $\pm\ 3 \times 10^{-3}\ \Delta b/b$.

### 3.2. *Diffraction techniques*

In the pioneer work (Shull and Wollan 1951) the Bragg diffraction method was used to determine the coherent scattering lengths for most of the elements and for several separated isotopes. Monochromatic neutrons are scattered from a polycrystalline sample (powder) and each set of reflecting lattice planes produces a reflex with a relative neutron intensity $I_{hkl}/I_0$ which is proportional to $b^2$. Furthermore, data for the neutron wavelength $\lambda$, the Debye–Waller factor DWF, the absorption factor, and the lattice spacing $d$ are necessary. In general, the uncertainty is restricted to $\pm\ 10^{-2}\ \Delta b/b$.

A particularly high accuracy corresponding to $\pm\ 3 \times 10^{-4}\ \Delta b/b$ was obtained by Shull (1968) who used the dynamical diffraction of neutrons on perfect crystals. An almost monochromatic neutron beam travels through the perfect crystal in Laue geometry. The diffracted wavelength can be altered by varying the Bragg angle under exact conservation of the Bragg condition. Such a change leads to a variation of the intensity of the emanating beams, the Pendellösung fringes. The period of the fringe pattern, which can be determined with high accuracy, is directly related to the scattering length $b_c$. The neutron wavelength, Debye–Waller factor and crystal thickness must be accurately known. This high precision technique is restricted to those few elements of which perfect crystals can be produced.

### 3.3. *Refraction and reflection*

In the mirror-reflection technique a monochromatic or filtered beam of slow neutrons is reflected from the plane surface of a mirror. The principal advantage of this method lies in the direct connection between the critical angle $\Theta_c$ and the mean coherent scattering length $\bar{b}$ of the mirror material: $N\bar{b}/\pi = (\Theta_c/\lambda)^2$.

The measured scattering length is an average over the atoms of the mirror and is independent of the crystalline state and of the temperature diffuse scattering. The Debye–Waller factor is unity. Since the critical angles are small and the wavelength distribution in the beam can never be known exactly, the accuracy is restricted to the order of $\pm\ 10^{-2}\ \Delta b/b$ for absolute determinations and $\pm\ 10^{-3}\ \Delta b/b$ for relative determinations (Dickinson, Passel and Halpern 1962). All kinds of plane mirrors and interfaces are suitable for the reflection measurements.

The very small bending of a monochromatic beam by *prism refraction* can be exactly measured with a double-crystal spectrometer as introduced by Shull, Billmann and Wedgwood (1967). The relation between the angular deflection $\Theta$, the coherent scattering length $b$, and the apex angle $2\alpha$ of the prism is given by

$$Nb = \pi(\Theta/tn\alpha)\lambda^{-2} .$$

This technique has the same advantages as the mirror-reflection method. However, the wavelength $\lambda$ can be exactly determined. Hence an uncertainty of only $\pm$ (2–9) $\times 10^{-4}$ $\Delta b/b$ could be obtained (Schneider 1976).

Diffraction, refraction, and reflection are the processes which lead to small-angle scattering when a beam of slow neutrons passes through a powder immersed in another substance such as a liquid or a gas. The broadening of the beam of the initial width $w_0$ (full width at half-maximum) is proportional to the difference $\Delta Nb$ between the scattering densities of powder and embedding (Weiss 1951):

$$\Delta Nb \sim (w^2 - w_0{}^2)^{\frac{1}{2}}$$

Thus the scattering density of the powder or the embedding can be obtained by measuring the broadenings if one of the $Nb$ values is already known. Because of the smallness of the broadening the error amounts to several times $10^{-2}$ $\Delta b/b$. However, this technique was the first for direct measurements on small amounts of gases such as tritium (Donaldson, Bartolini, and Otsuki 1972).

### 3.4. *Gravity refractometer*

In the gravity refractometer described by Maier-Leibnitz (1962) and Koester (1965) the fact is used that in the gravitational field freely falling neutrons gain an energy $mg_f h$ which is of the same magnitude as the potential energy $\langle V\rangle = 2\pi\hbar^2 m_i{}^{-1}\overline{Nb}$ of neutrons in matter containing $N$ atoms cm$^{-3}$. $m$ is the gravitational mass and $m_i$ the inertial mass of the neutron, $g_f$ denotes the effective gravitational acceleration acting on the free neutron, and $h$ is the height of fall. A falling neutron beam (or neutron wave) is reflected from a horizontal mirror if $mg_f h$ is smaller than $\langle V\rangle$; otherwise the neutrons can penetrate the mirror. The critical height for total reflection is reached (with $m = m_i$) at $h_c$ according to

$$mg_f h_c = 2\pi\hbar^2 m^{-1}\overline{Nb} .$$

The mean scattering density $\overline{Nb}$ of the mirror substance is directly proportional to the critical fall height, which is the only quantity to be measured. All other quantities are very well known fundamental constants. Thus a very high accuracy for the scattering density becomes possible. In practice, uncertainties of only $(\pm 1 – \pm 3) \times 10^{-4}$ $\Delta b/b$ for the scattering length could be obtained (Koester and Nistler 1975). This good accuracy can be obtained only for liquid mirrors with a high content of the element under investigation.

### 3.5. *Christiansen filter technique*

Small-angle scattering of neutrons by mixtures of powders with liquids (Christiansen 1884) can be used as an indicator for a balance method, for the scattering does not arise when the scattering densities $Nb$ of powder and liquid are equal. In this technique a well-collimated beam of cold neutrons passes through the filter. The unscattered part is completely stopped by a shield in front of the neutron counter, so that only the scattered part is detected. In this arrangement the point of balance $Nb$ (liquid) = $Nb$ (powder) can be experimentally found by changing the scattering density of the liquid phase and by measuring the disappearance of the scattering (Koester and Knopf 1971). Liquids of various scattering densities over a wide $Nb$ range are prepared by mixing different parts of two compounds of very different $Nb$ values (e.g. $H_2O/D_2O$) which themselves can be determined very exactly in the gravity refractometer. The Christiansen filter technique is also applicable to filters of thin wires in liquids or of powders in gases under various pressures. The method is useful and reliable for determinations of scattering lengths with medium accuracy to within $\pm 1 \times 10^{-3}$ and $1 \times 10^{-2}$ $\Delta b/b$. It is particularly applicable to small amounts of separated isotopes since the filters can be kept rather small (Koester, Knopf, and Waschkowski 1977).

### 3.6. *Perfect crystal interferometer*

The application of neutron interferometry (Rauch, Treimer, and Bonse 1974) to measure the index of refraction exactly is the most recent procedure for obtaining highly accurate values for scattering lengths (Bauspiess *et al.* 1974). The discussion in this method is one object of this workshop and therefore the technique is well known to the participants.

In the perfect-crystal interferometer two coherent neutron partial beams (wavelength $\lambda$) appear side by side in a rather large lateral distance of few centimetres.

Any material with thickness $D$ and refractive index $n$ in one partial beam causes a phase shift of $(1-n)2\pi D/\lambda$, which leads in the interference zone (emanating 0 beam) to a periodic intensity modulation $I(n,\lambda) = I_0 \cos^2|\pi(1-n)D/\lambda|$ if $D$ is changed. The period $D_0$ of the modulation is connected with the scattering density of the material as $D_0 = 2\pi/(\lambda Nb)$. In proper cases $D_0$ can be measured with an uncertainty of only $\pm 1 \times 10^{-4}$ $\Delta D_0/D_0$. To obtain a corresponding accuracy for the scattering density the wave length $\lambda$ must be determined with similar precision.

All homogeneous material can be used as samples, as can small volumes of gases ($^3$H) or liquids (Kaiser *et al.* 1977). The smallest uncertainty of $\pm 3 \times 10^{-4}$ $\Delta b/b$ was obtained for the scattering length of tin (Rauch *et al.* 1976).

## 4. Some experimental results

### 4.1. *Scattering lengths*

So far we have seen the considerable effort made to determine scattering lengths with high accuracy. The techniques are different in both physical principle and experimental performance so that the question of how good the agreement of the various results really is arises. Data obtained by different techniques are compiled in Table 1.

The results listed in column 1 were obtained from diffraction works. They are given in terms of the carbon or nickel scattering lengths. In column 2 the data obtained by reflection, refraction and gravity acceleration are shown. The reflection (R) data are based on the carbon scattering length whilst the other data are determined on an absolute scale. The values given in column 3 (interferometer and dynamical diffraction) and in column 4 (transmission) are also absolute data. Comparing the given figures one finds good agreement between most data, especially between the transmission data and the results of the advanced methods (e.g. Koester 1977). This fact

TABLE 1

*Compilation of scattering lengths (in Fermis) for bound atoms obtained by different techniques*

| | 1 | 2 | 3 | 4 |
|---|---|---|---|---|
| Element | Diffraction | Reflection (R)<br>Prism refraction (Pr)<br>Gravity refraction (Gr)<br>Christiansen filter (Ch) | Interferometer (J)<br>Dynamic diffraction (DD) | Transmission |
| $^{1}H$ | - 4.0(1) | R: - 3.74(2)<br>Gr: - 3.7409(11) | | - 3.733(4) |
| $^{2}H$ | 6.4(3) | R: 6.21(4)<br>Ch: 6.70(5)<br>Gr: 6.674(6) | | |
| $^{3}He$ | | R: 6.1(6) | J: 5.73(5) | |
| $^{16}O$ | 5.81(20) | R: 5.80(5)<br>Pr: 5.830(2)<br>Ch: 5.804(3)<br>Gr: 5.801(6) | | 5.804(7) |
| $^{27}Al$ | 3.5(2) | Ch: 3.449(9) | J: 3.449(5) | 3.449(5) |
| $S_i$ | | Pr: 4.1478(16)<br>Ch: 4.159(6) | DD: 4.1491(10) | 4.147(6) |
| $^{93}Nb$ | 7.12(8) | Ch: 7.11(4) | J: 7.054(3) | 7.032(4) |
| $^{208}Bi$ | 8.9(3) | Gr: 8.5256(15) | J: 8.503(12) | 8.5267(16) |

demonstrates the reliability of the errors of the order of $10^{-3}$–$10^{-4}$. The smallest uncertainty of $\pm 0.8 \times 10^{-4}$ $\Delta b/b$ has been reported by Koester and Nistler (1975) for the scattering length of chlorine ($b_c = 9.5792 \pm 0.0008$ fm). This magnitude of accuracy makes it possible that small differences of the order of $10^{-2}$ $\Delta b/b$ can be determined with an uncertainty of 1 per cent. We shall see in the following some examples of the importance of highly accurate scattering lengths.

### 4.2. *Neutron and gravity*

Considering the gravity refractometer we have learned that the critical fall height $h_c$ depends on the gravitational acceleration $g_f$ for free neutrons and the gravitational mass $m$ and the inertial mass $m_i$ of the neutrons. Because of the universality of free fall, experimentally verified to 1 part in $10^{11}$, the local gravitational acceleration $g_\ell$ for the bulk material can also be assumed for the free neutron. Taking into account the centrifugal and coriolis acceleration by a correction term $(1+\alpha)$ the product $g_f h_c$ can be written as

$$g_f h_c = g_\ell (1+\alpha) h_c = g_\ell H_c \ .$$

Then the basic equation for the determination of the actual scattering length $b_0$ becomes:

$$b_0 = \frac{\gamma m^2 g_\ell H_c}{2\pi\hbar^2 N} = \gamma b_g$$

with the equivalence factor $\gamma = m_i/m$. Consequently, if a determination of the actual $b_0$ could be performed by another experiment independent of gravity the measurement of the critical fall height leads to a determination of the equivalence as $\gamma = b_0/b_g$ (Koester 1976).

The $b_0$ value can be derived from measurements of the total cross-section and the absorption cross-section with epithermal neutrons. The necessary correction for neutron–electron scattering may be made using the 1973 average for the (n,e) scattering length $b_{ne} = -(1.40 \pm 0.05) \times 10^{-3}$ fm (Koester 1977). Experiments (Koester and Nistler 1975, Koester 1976) have been performed on carbon, lead and bismuth. The results (carbon,

$\gamma = 1.0040 \pm 0.0070$; lead, $\gamma = 0.9997 \pm 0.0003$; bismuth, $\gamma = 0.9999 \pm 0.0004$) yield an average $\overline{\gamma} = 0.99984 \pm 0.00025$ from which the equivalence is verified with an uncertainty of 1 part in 4000.

It should be emphasized that this result holds in the quantum limit since the neutron in the gravity refractometer experiences gravity while simultaneously behaving as a matter wave.

4.3. *Neutron—electron interaction*

The history of the fundamental neutron—electron interaction appears as a story of contradictions. The unexpected smallness of the intrinsic scattering length $b_e = b_{ne} - b_F$ derived from the measured $b_{ne}$ could never be explained by meson models for the nucleon, such as the $\pi$ meson cloud, nucleon recoil, and K meson effects, each of which individually is much larger than the observed effect. Only the first more exact result $b_{ne} = -1.59(5) \times 10^{-3}$ fm reported by Melkonian, Rustad, and Havens (1959) can still be discussed in the framework of the vector meson ($\rho$) model. However, the more recent and most exact value due to Krohn and Ringo (1973) of $b_{ne} = -1.30 \pm 0.03 \times 10^{-3}$ fm contradicts all predictions of meson theory. Moreover, both values are in contrast to the theoretical prediction $b_{ne} = 0$ of the quark model SU(6) which works so well for the ratio of the magnetic properties of neutron and proton. Recently a new exact value for $b_{ne}$ was derived from very precise measurements of the scattering lengths for bismuth and lead (Koester, Nistler, and Waschkowski 1976). In this experiment the scattering lengths were measured at zero energy, $b(0) = b_N + Zb_{ne}$, and at epithermal energies $E$, $b(E) = b_N + \overline{f}b_{ne}$, where the atomic form factor $\overline{f}$ is very small.

For lead, for instance, $b(0)$ was derived from gravity refractometer experiments as

$$b_N + Zb_{ne} = 9.4003 \pm 0.0014 \text{ fm}.$$

Transmission experiments at 5.2 eV result in

$$b(5.2) = b_N + 0.083\, Zb_{ne} = 9.5031 \pm 0.0020 \text{ fm} \quad .$$

From these results it follows that $b_{ne} = -(1.37\pm0.04) \times 10^{-3}$ fm. The average of many measurements of this kind

$$b_{ne} = -(1.38\pm0.02) \times 10^{-3} \text{ fm}$$

is in fair agreement with the value of Krohn and Ringo (1973). So far, both results are contrary to the predictions of meson theory and ordinary quark theory.

Recently, however, Carlitz, Ellis, and Savit (1977) have treated the problem in the quark model but have taken account of a spin dependence of the quark–quark forces. They obtained $b_{ne} = -1.3 \times 10^{-3}$ fm which for the first time is in agreement with the experimental results.

## 5. Final remark

In the last decade a drastic improvement in the accuracy of coherent and incoherent scattering lengths has been achieved by the use of new methods having their roots and some examples in the oldest physical fields of mechanics and optics. In practice, the precision of the advanced techniques is restricted to an uncertainty of 1 part in $10^4$. Contributions to fundamental problems of physics could be obtained from the high precision values of scattering lengths, for instance for neutron interactions with the electron, proton and deuteron and with gravity. The spin-state scattering lengths of the four-particle systems neutron–tritium and neutron–$^3$He will be the next contributions to fundamental questions.

In the wide field of interactions with many-particle systems which are not treated here, exact scattering lengths for separated isotopes will be of special interest for the optical nuclear model and for numerous applications in the physics of matter.

## References

ALEKSANDROV, YU.A. (1969). Communications of the Joint Institute for Nuclear Research, P3-4783, Dubna, USSR.

BAUSPIESS, W., BONSE, U., RAUCH, H., and TREIMAR, W. (1974). *Z. Phys.* **271**, 177.

CARLITZ, R.D., ELLIS, S.D., and SAVIT, R. (1977). *Phys. Lett. B* **68**, 443.
CHRISTIANSEN, C. (1884). *Wied. Annln.* **23**, 298.
DICKINSON, W.C., PASSEL, L., and HALPERN, O. (1962). *Phys. Rev.* **126**, *632*.
DONALDSON, R.E., BARTOLINI, W., and OTSUKI, H. (1972). *Phys. Rev. C* **5**, 1952.
FOLDY, L.L. (1958). *Rev. mod. Phys.* **30**, 471.
KAISER, H., RAUCH, H., BAUSPIESS, A., and BONSE, U. (1977). *Phys. Lett.* **71**, 321.
KOESTER, L. (1965). *Z. Phys.* **182**, 328.
—— (1976). *Phys. Rev. D* **14**, 907.
—— (1977). *Springer Tracts mod. Phys.* **80**, 1–52.
KOESTER, L. and KNOPF, K. (1971). *Z. Naturforsch.* **26a**, 391.
KOESTER, L., KNOPF, K., and WASCHKOWSKI, W. (1977). *Z. Phys. A* **282**, 371.
KOESTER, L. and NISTLER, W. (1975). *Z. Phys. A* **272**, 189.
KOESTER, L., NISTLER, W., and WASCHKOWSKI, W. (1976). *Phys. Rev. Lett.* **36**, 1021.
KROHN, V.E. and RINGO, G.E. (1973). *Phys. Rev. D* **8**, 1305.
MAIER-LEIBNITZ, H. (1962). *Z. angew. Phys.* **14**, 738.
MELKONIAN, E., RUSTAD, B.M., and HAVENS, W.M. (1959). *Phys. Rev.* **114**, 1571.
RAUCH, H., BADUREK, G., BAUSPIESS, W., BONSE, U., and ZEILINGER, A. (1976). *Proc. Int. Conf. on Interactions of Neutrons with Nuclei, Lowell,* USA CONF-760715-P2, p. 1027. University of Lowell, Lowell, Massachusetts, USA (1976).
RAUCH, H., TREIMER, W., and BONSE, U. (1974). *Phys. Lett. A* **47**, 369.
SCHNEIDER, C.S. (1976). *Acta Crystallogr. A* **32**, 375.
SCHWINGER, J. (1948). *Phys. Rev.* **73**, 407.
SHULL, C.G. (1968). *Phys. Rev. Lett.* **21**, 1585.
SHULL, C.G., BILLMANN, K.W., and WEDGWOOD, F.A. (1967). *Phys. Rev.* **153**, 1415.
SHULL, C.G. and NATHANS, R. (1967). *Phys. Rev. Lett.* **19**, 384.
SHULL, C.G. and WOLLAN, E.O. (1951). *Phys. Rev.* **81**, 527.
WEISS, R.J. (1951). *Phys. Rev.* **83**, 379.

# 3. GRAVITATIONAL AND ROTATIONAL EFFECTS ON THE NEUTRON PHASE

S.A. WERNER and J.-L. STAUDENMANN

*Physics Department and Research Reactor Facility, University of Missouri-Columbia, Columbia, Mo., U.S.A.*

R. COLELLA and A.W. OVERHAUSER

*Department of Physics, Purdue University, West Lafayette, Ind., U.S.A.*

## 1. Introduction

The phenomenon of interference is a fundamental difference between classical and quantum physics. Neutron interferometry experiments must therefore directly and strongly involve quantum mechanics. The experiments carried out here probe the simultaneous effect of *gravity* and *quantum mechanics* on the neutron's motion in a *non-inertial* frame of reference.

We usually think of quantum mechanics as being important in the realm of small things, i.e. atoms, nuclei, and elementary particles, and gravity as being important in the realm of large things, for example in the motion of the planets and in various celestial affairs. The reason for this is easy to understand since, for example, in atoms the electrical forces are 40 orders of magnitude larger than the gravitational forces. However, for uncharged celestial bodies we are left with the much weaker gravitational force. This is also the case for a neutron if we are able to segregate it from nuclear matter.

We have previously reported results of preliminary experiments (Colella, Overhauser, and Werner 1975, Werner *et al.* 1976) in which our first observation of the effect of the earth's gravitational potential on the neutron phase was described. We have now carried out a series of precision experiments utilizing a LLL interferometer of the type first developed for X-rays by Bonse and Hart (1965) which provide an experimental test of the principle of equivalence in the quantum limit at the 1.0 per cent level. Since these experiments are necessarily carried out on the surface of the rotating earth, we have an opportunity to test the rules for transforming a quantum-mechanical Hamiltonian into a non-inertial frame. The phase shift observed as a result of the rotation of the earth is in principle the quantum-mechanical analog of the optical interferometry experiment carried out by Sagnac (1913) and the subsequent experiment of Michelson, Gale, and Pearson (1925) in which the rotating earth was the source of rotation.

## 2. Theory and geometry of the experiment

The basic geometry of the interferometer–detector system is shown in Fig. 1. Details of the overall geometry of the experiment are given in §3. A nominally monoenergetic neutron beam ($\Delta\lambda/\lambda \approx 0.01$) is directed along the line AB (horizontal) and is coherently split near point A by Bragg reflection in the first

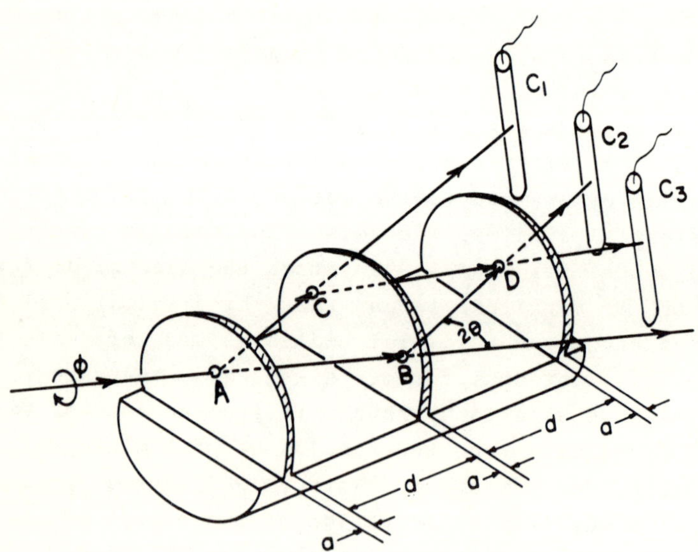


FIG. 1. Schematic diagram of the LLL interferometer. The detector $C_1$ counts the non-interfering beam, while $C_2$ and $C_3$ count the interfering beams. The Bragg angle is $\theta$ for neutrons of wavelength $\lambda$ reflected by the (220) planes in silicon. The angle $\phi$ is defined to be positive as the path CD is raised above the path AB; when the parallelogram ABDC is horizontal $\phi = 0$.

perfect silicon crystal slab. These two coherent beams are again split at points B and C in the second slab. Two of these beams are directed toward point D in the third slab where they overlap and interfere. The outgoing interfering beams are then detected using two high-pressure (40 atm) $^3$He neutron detectors of diameter 0.25 in. labelled $C_2$ and $C_3$. We also detect the non-interfering beam directed along the line AC with the detector $C_1$.

The experimental procedure involves turning the interferometer, including an entrance slit and the three detectors, about the incident beam line AB. At each angular setting $\phi$ neutrons are counted for a preset length of time (actually based on an incident beam monitor). This procedure allows the neutron on the beam trajectory CD to be somewhat higher above the surface of the earth than for the beam path AB. The difference in the earth's gravitational potential between these two levels causes a quantum-mechanical phase shift of the neutron on the trajectory ACD relative to the trajectory ABD. The phase shift on the rising path AC is exactly equal to thephase shift on theopposite rising path BD, as can be shown by applying Huygens' principle.

Because the experiment is done on the surface of the rotating earth, which is a non-inertial frame, the Hamiltonian governing the neutron's motion will involve a third term in addition to the kinetic energy and the gravitational potential energy. Using the methods of standard classical mechanics, we find that the appropriate Hamiltonian is

$$H = \frac{p^2}{2m_\mathrm{i}} - \mathrm{G}\frac{Mm_\mathrm{g}}{r} - \boldsymbol{\omega}\cdot\mathbf{L} \ . \tag{1}$$

where $\mathbf{p}$ is the canonical momentum of the neutron, $\mathbf{L} = \mathbf{r} \times \mathbf{p}$ is the angular momentum of the neutron's motion about the center of the earth ($\mathbf{r} = 0$), $\boldsymbol{\omega}$ is the angular rotation velocity of the earth, $M$ is the mass of the earth, $m_\mathrm{i}$ is the inertial mass of the neutron, and $m_\mathrm{g}$ is the gravitational mass of the neutron. For macroscopic phenomena, we know experimentally that the principle of equivalence ($m_\mathrm{i}=m_\mathrm{g}$) is correct to very high accuracy (Eötvös 1890, Eötvös, Pekár, and Fekete 1922, Roll, Krotkov, and Dicke 1967). For microscopic phenomena, which are governed by the rules of quantum mechanics, there appears to be no way to be certain from existing experimental data that the principle of equivalence is rigorously a fact of nature. However, it is well known that in the gravitational field of the earth, neutrons fall following a classical parabolic trajectory (Dobbs *et al.* 1965). The experiments described here can be viewed as an attempt to push the verification of this principle into a strongly quantum limit. The Hamiltonian above describes correctly the motion of a classical (macroscopic) particle. It cannot be taken for granted that it correctly describes quantum-mechanical phenomena, especially those involving interference. However, for lack of evidence to the contrary, we shall also assume that it is the correct quantum-mechnical Hamiltonian and then see if the predictions agree with experiment.

Since the distances involved in the interferometer are very small compared with the radius $R$ of the earth, we can write equation (1) as

$$H = \frac{p^2}{2m_i} + m_g g_0 z - \boldsymbol{\omega}\cdot\mathbf{L} + V_0 \tag{2}$$

where $V_0$ is the gravitational potential energy at some reference height $z = 0$ (say the center of the interferometer) and

$$g_0 = \frac{GM}{R^2} \,. \tag{3}$$

The classical equations of motion are Hamilton's equations

$$\dot{\mathbf{r}} = \frac{\partial H}{\partial \mathbf{p}} \qquad \text{and} \qquad \dot{\mathbf{p}} = -\frac{\partial H}{\partial \mathbf{r}} \tag{4}$$

where the dot implies a time derivative. Using the Hamiltonian (2), the first of these equations gives

$$\mathbf{p} = m_i\dot{\mathbf{r}} + m_i\boldsymbol{\omega}\times\mathbf{r} \tag{5}$$

and the second gives

$$\dot{\mathbf{p}} = -\, m_g g_0 \hat{\mathbf{z}} - \boldsymbol{\omega}\times\mathbf{p} \,. \tag{6}$$

The unit vector $\hat{\mathbf{z}}$ points normally outward along the earth's radius vector at the local point of the experiment. Combining equations (5) and (6) we obtain the well-known equation of motion for a classical particle in a rotating frame:

$$m_i \ddot{\mathbf{r}} = - m_g g_0 \hat{\mathbf{z}} - m_i \boldsymbol{\omega} \times (\boldsymbol{\omega}\times\mathbf{r}) - 2m_i \boldsymbol{\omega}\times\dot{\mathbf{r}} \quad . \tag{7}$$

We now see that the term $-\boldsymbol{\omega}\cdot\mathbf{L}$ in the Hamiltonian gives rise to both the centrifugal acceleration and the Coriolis acceleration. We again make use of the fact that the dimensions of the interferometer are very small compared with the earth's radius $R$ and define an effective gravitational acceleration in the usual way:

$$\mathbf{g} = g_0 \hat{\mathbf{z}} + \frac{m_i}{m_g} \boldsymbol{\omega} \times (\boldsymbol{\omega}\times\mathbf{R}) \quad . \tag{8}$$

If $\mathbf{g}$ is assumed to be independent of position, we can solve equation (7) for the local motion of the neutron in the frame of the rotating earth. To leading order in $\omega$ the solution is

$$\mathbf{r} = \mathbf{r}_0 + \mathbf{v}_0 t - \tfrac{1}{2}\mathbf{g}t^2 + \tfrac{1}{3}\omega t^3 \hat{\boldsymbol{\omega}}\times\mathbf{g} \quad . \tag{9}$$

For the thermal neutrons used in our experiments, the transit times through the interferometer are of order $5 \times 10^{-5}$ s. Since $\omega = 7.27 \times 10^{-5}\ \mathrm{s}^{-1}$, we see that the term involving $\omega$ is smaller than $\tfrac{1}{2}gt^2$ by a factor of about $10^{-9}$. Thus the Coriolis force has a negligible effect on the trajectory over these small distances. However, its effect on the neutron phase is not negligible as we shall now see.

In order to use the above classical results in calculating thephase shift $\beta$ in the neutron interferometer due to gravity and rotation, we assume that we can associate a de Broglie wave of wave vector $\mathbf{k}$ with the neutron which has momentum $\mathbf{p}$, i.e.

$$\mathbf{p} = \hbar\mathbf{k} \tag{10}$$

where $\hbar = h/2\pi$. The difference in phase accumulated on the path ACD relative to the path ABD is then obtained via the methods of classical optics:

$$\beta = \oint \mathbf{k}\cdot d\mathbf{r} = \frac{1}{\hbar} \oint \mathbf{p}\cdot d\mathbf{r} \quad . \tag{11}$$

The momentum **p** appearing in this line integral on the path ABDCA around the interferometer must be the canonical momentum given in equation (5). The phase shift β thus involves two terms:

$$\beta = \frac{m_i}{\hbar} \oint \dot{\mathbf{r}} \cdot d\mathbf{r} + \frac{m_i}{\hbar} \oint (\omega \times \mathbf{r}) \cdot d\mathbf{r} \ . \tag{12}$$

The velocity $\dot{\mathbf{r}}$ is obtained by differentiating equation (9). Working these integrals out we find

$$\beta = -2\pi m_i m_g \frac{g}{h^2} \lambda_0 A' \sin\phi + \frac{4\pi m_i}{h} \boldsymbol{\omega} \cdot \mathbf{A} \ . \tag{13}$$

The normal area **A** enclosed by the beam paths is

$$A = (2d^2 + 2ad)\tan\theta_B \tag{14}$$

and

$$A' = (2d^2 + 2ad \cos\theta_B)\tan\theta_B \ . \tag{15}$$

The dimensions $a$ and $d$ are shown in Fig. 1 and $\theta_B$ is the Bragg angle for neutrons of laboratory wavelength $\lambda_0$ reflecting from the reciprocal lattice point **G**, i.e.

$$\lambda_0 = \frac{h}{m_i v_0} \quad \text{and} \quad \sin\theta_B = \frac{\lambda_0 G}{2} \ . \tag{16}$$

The second term in equation (13) resulting from the Coriolis force depends on the latitude and on the interferometer orientation with respect to the local north–south longitude line. We call this term $\beta_{Sagnac}$. Working out the geometry involved in the dot product, we find

$$\beta_{Sagnac} = \frac{4\pi m_i}{h} \omega A(\cos\phi \cos\theta_L + \sin\phi \sin\Gamma \sin\theta_L) \tag{17}$$

where $\theta_L$ is the colatitude, Γ is the angle of the incident neutron beam west of due south, and ϕ is the angle of rotation of the interferometer as shown in Fig. 1 ($\phi = 0$ when the interferometer is horizontal).

It is not apparent at the outset, nor from what has been

said here, that the Sagnac effect for neutrons is directly analogous to the effect observed with light. The analogy was first pointed out by Page (1975). It can most easily be seen by a simple argument. Consider the two beam paths ABD and ACD to be the two halves of a circle of radius $R$ with the axis of rotation $\hat{\boldsymbol{\omega}}$ perpendicular to the plane of the circle. Owing to rotation the beam path ABD is shortened and the beam path ACD is lengthened, giving rise to a path-length difference

$$\Delta S = R\omega t \ . \tag{18}$$

Here $t$ is the transit time from A to D

$$t \approx \frac{\pi R}{v_{\mathrm{p}}} \tag{19}$$

and $v_{\mathrm{p}}$ is the phase velocity of the wave. The phase shift due to rotation is then seen to be

$$\beta_{\mathrm{Sagnac}} = k\Delta S = k\frac{A\omega}{v_{\mathrm{p}}} \ . \tag{20}$$

For photons the phase velocity is the velocity of light

$$v_{\mathrm{p}} = c \qquad \text{(photons)} \tag{21}$$

but for neutrons it is

$$v_{\mathrm{p}} = \frac{\hbar}{2m}k \ . \qquad \text{(neutrons)} \tag{22}$$

Thus, we see that for the optical case

$$\beta_{\mathrm{Sagnac}} = \frac{2\pi}{\lambda c}A\omega \qquad \text{(photons)} \tag{23}$$

and for neutrons

$$\beta_{\mathrm{Sagnac}} = \frac{2m}{\hbar}A\omega \qquad \text{(neutrons)} \tag{24}$$

which agrees with equation (13). It is interesting to note that the optical Sagnac effect is wavelength dependent, whereas the neutron Sagnac effect is wavelength independent.

Anandan (1977) has shown that there are three additional relativistic terms that should be added to expression (13) for the phase shift:

$$\beta_{rel} = \frac{\hbar\omega\cdot A}{mc^2}k^2 - \frac{gAk}{c^2}\sin\phi - \frac{GMS_n\omega A}{mc^2R^3}\ . \qquad (25)$$

The first term is the relativistic correction to the neutron Sagnac effect. The second term is the relativistic correction to β resulting from gravity. The last term arises from the coupling of the neutron spin to the space–time curvature. ($S_n$ is the component of the spin normal to the velocity in the plane containing **v** and **S**.) For thermal neutrons all of these terms are too small to be observable in our experiments.

## 3. Experimental

The experiments described here were carried out at the 10 MW University of Missouri research reactor. A schematic diagram of the experimental apparatus is shown in Fig. 2. A nominally collimated polychromatic thermal neutron beam is brought out of the reactor at beam port B. A monochromatic neutron beam is produced in a double-crystal monochromator assembly which utilizes two pyrolytic graphite crystals. The beam leaving the second monochromator passes through a secondary collimator and

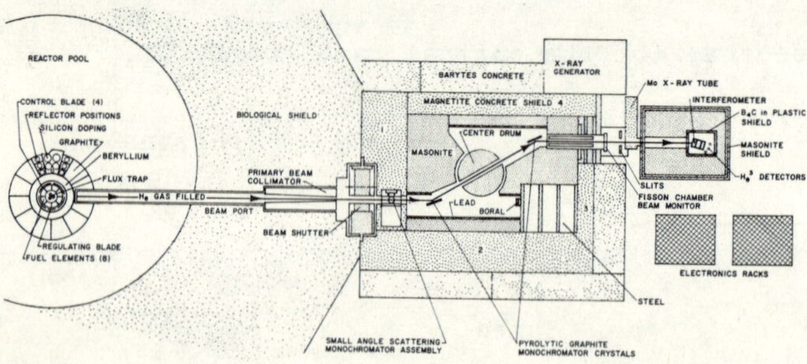


FIG. 2. Schematic diagram of the overall geometry of the experiment.

is then directed through a slit onto the interferometer. The slit size used for the data presented in this paper was 2.8 mm × 5.7 mm. The angular divergence of the neutrons striking the interferometer is approximately 0.7° in both the horizontal and vertical planes. The interferometer itself, along with the $^3$He detectors, is mounted inside a 0.5 in thick aluminium box which provides thermal isolation of the interferometer. This box is rigidly fastened to a rotator assembly which allows the interferometer to be turned about the incident beam direction with a computer-controlled stepping motor. The rotator and interferometer assembly are mounted rigidly inside a large ($\frac{2}{3}$m × $\frac{2}{3}$m × 1m) heavy (about 500 kg) masonite neutron shielding box which rests on four air-filled vibration isolation pads. We have found that both thermal stability and vibration isolation from the reactor hall are essential in performing high-contrast neutron interferometry experiments.

### 3.2. *The interferometer*

Over the past four years we have fabricated from large perfect single crystal silicon ingots five interferometers of the LLL type first developed by Bonse and Hart (1965) for X-rays. We have progressively used more sophisticated diamond-cutting techniques and we have also improved our etching techniques. The interferometer which we are currently using is by far the best one we have been able to produce. It is cut from a 5 cm diameter high-purity vacuum float zone ingot (resistivity = 3600 Ω cm) obtained from the Silicon Division of Monsanto Corporation. The dimensions of this interferometer are (see Fig. 1)

$$a = 2.464 \pm 0.002 \text{ mm}, \qquad d = 34.518 \pm 0.002 \text{ mm} . \qquad (26)$$

The precision in these dimensions is very important for a neutron interferometer as has been discussed in detail by Bonse and Graeff (1977). We have used a high-quality milling machine equipped with an optical dimensional readout, 600 grit diamond cutting wheels of diameter 4 in and thickness 1/16 in. Our etching techniques remove approximately 5 μm of work-damaged silicon after machining.

### 3.3. *Experimental parameters*

Beam port B is oriented nearly directly along a north–south line such that the monochromatic beam incident on the interferometer is directed approximately due south. Thus the angle $\Gamma$ in equation (17) is zero for our experiments. We can therefore write equation (13) as

$$\beta = -q_{grav} \sin\phi + q_{Sagnac} \cos\phi \tag{27}$$

where

$$q_{grav} = 2\pi m_i m_g \frac{g}{h^2} \lambda_0 A' \tag{28}$$

and

$$q_{Sagnac} = \frac{4\pi m_i \omega A \cos\theta_L}{h} . \tag{29}$$

At Columbia, Mo.,

$$\theta_L = 51.37° \qquad g = 980.0 \text{ cm s}^{-2} . \tag{30}$$

Using these parameters we find for a wavelength $\lambda_0 = 1.430$ Å that

$$q_{grav} = 57.35 \ , \quad q_{Sagnac} = 1.48 . \tag{31}$$

Thus the effect of the rotation of the earth on the neutron phase is about 2.5% of the gravitational effect at this wavelength. This is somewhat surprising since its effect on the neutron trajectory is only of order 1 part in $10^9$ over the distances involved in the experiment, as was pointed out earlier.

Aside from the necessary dependence of the areas $A$ and $A'$ on the neutron wavelength $\lambda_0$ (through the Bragg angle $\theta_B$), the frequency $q_{grav}$ of interference oscillation due to gravity is linearly dependent on $\lambda_0$ while $q_{Sagnac}$ is wavelength independent. We have taken advantage of this difference in wavelength dependence to detect the neutron Sagnac effect.

### 3.4. *Bending*

Unfortunately, as we have pointed out previously (Colella *et al.* 1975) there is an additional effect on the measured phase shift $\beta$ resulting from bending (or warping) of the interferometer under its own weight. This effect is dependent upon the rotation angle $\phi$, since the experiment involves turning the interferometer about an axis which is not an axis of elastic symmetry. We call this effect

$$\beta_{bend} = -q_{bend} \sin\phi \ . \tag{32}$$

We have experimentally measured $q_{bend}$ with X-rays.

The procedure involves using molybdenum $K_\alpha$ X-rays ($\lambda$ = 0.71 Å). We direct a beam of X-rays along the same incident line AB (Fig. 1) and observe the interfering X-ray beams with an X-ray-sensitive proportional gas detector as a function of rotation angle $\phi$. The effect of gravity (gravitational red shift) on the X-rays over the distances involved in the interferometer is negligible. The X-ray Sagnac effect is also negligible here as can be verified from equation (23). Because the X-ray wave field may not remain totally uniform (coherent) over the cross-section of the interfering beams (due to warping of the three interferometer slabs), a plot of intensity *versus* $\phi$ will not necessarily give $q_{bend}$ exactly. In addition, owing to extreme care in the methods we use to mount the interferometer, $q_{bend}$ is fairly small and we do not observe a complete oscillation period for the range of rotation angles $\phi$ which we can safely utilize. We circumvent these problems by measuring the shift in the phase of an interference oscillation pattern resulting from rotating a very flat slab of plastic in the interferometer (see Fig. 6 for the geometry) as a function of the angle $\phi$. Data obtained in this way are shown in Fig. 3 in which molybdenum $K_\alpha$ X-rays were reflected by the (440) planes in the interferometer. We have obtained a similar set of data for the (220) reflection. The results of these experiments are shown in Fig. 4 where we plot the phase shift $\Delta$ of the plastic slab oscillation pattern against sin $\phi$. We observe a linear relation which justifies the form of equation (32). The slope of these plots gives $q_{bend}$. It is reasonably certain that the functional form

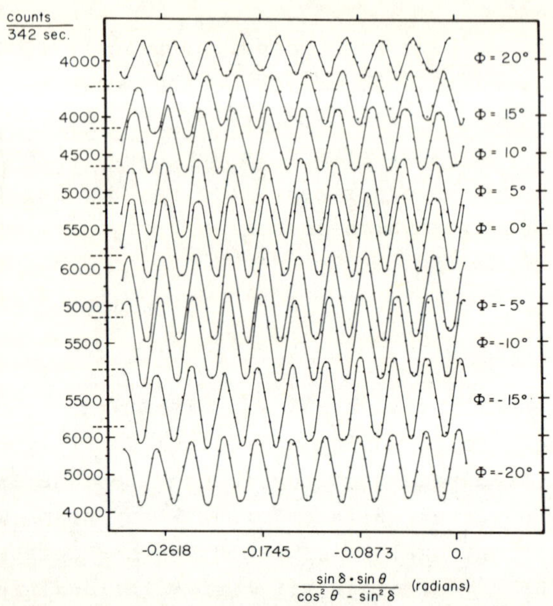


FIG. 3. X-ray (molybdenum $K_\alpha$) intensity as a function of plastic slab orientation angle $\delta$ (see Fig. 6) for various rotation settings $\phi$ of the interferometer. This data is for the (440) reflection in silicon. Only half of the data are shown, i.e. for $\delta < 0$. The shift in phase $\Delta$ of these data as a function of $\phi$ is used to determine $q_{\text{bend}}$.

of the phase shift must be

$$\beta_{\text{bend}} = k\Delta d = \frac{2\pi}{\lambda_0}\,\Delta d \tag{33}$$

where $\Delta d$ is the difference in path length for the trajectory ABD relative to ACD due to bending of the interferometer. From the fact that the slope of the (440) data is very nearly four times the slope of the (220) data, we conclude that

$$\Delta d \propto \sin^2\theta_B\ . \tag{34}$$

Therefore putting together the arguments that have led to equations (32), (33) and (34) we find

$$\beta_{\text{bend}} = -q_{\text{bend}}\sin\phi = -\frac{C}{\lambda_0}\sin^2\theta_B\,\sin\phi \tag{35}$$

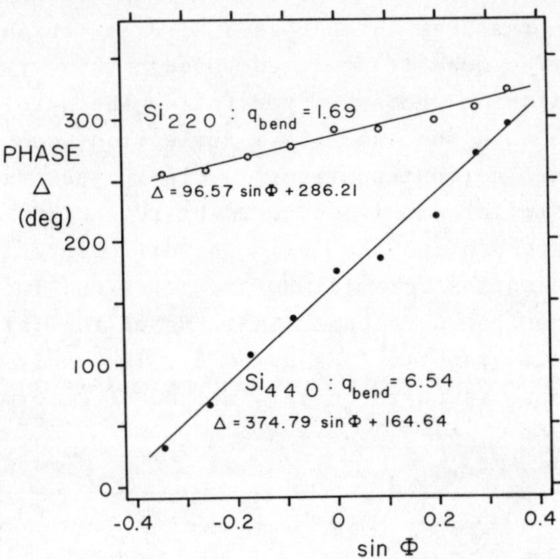


FIG. 4. A plot of the phase shift Δ of the interference oscillations resulting from rotating a plastic slab (Fig. 1) in the interfering X-ray beams for various angles $\phi$. The two sets of data are for reflecting molybdenum $K_\alpha$ X-rays from the (220) and (440) planes in silicon. The slope of these plots gives $q_{bend}$.

the numerical value of the constant is

$$C = 34.57 \ . \tag{36}$$

The reason why the bending effect depends quadratically on $\sin\theta_B$ is not yet understood.

### 3.5. *Neutron wavelength*

The frequency of oscillation of the interference pattern for neutrons obtained by rotating the interferometer about the incident beam direction contains three contributions: gravity, rotation, and bending. The theoretical formulas for the effects of gravity and rotation are given by equations (27) and (28), and $\beta_{bend}$, determined experimentally, is given by equation (35). Each of these terms depends either directly or indirectly on the neutron wavelength $\lambda_0$. It is therefore essential to measure $\lambda_0$ accurately. We have employed a technique which is schematically illustrated in Fig. 5. A pryolytic graphite crystal is placed in the beam which passes directly through the interferometer. The beam transmitted through this crystal is counted with a fission chamber. By rotating the pyrolytic graphite crystal through the same (004) reflection used for monochromation of the incident neutrons, a dip in the transmission is observed. A similar dip is observed by reflecting the beam to the right instead of to the left. The difference in the crystal rotation angles for minimum transmission, i.e. $\theta_2 - \theta_1$, determines the neutron wavelength in terms of the lattice parameter of pyrolytic graphite ($c$ = 6.708 Å). The analysis of the data presented here is based on this method of determining $\lambda_0$.

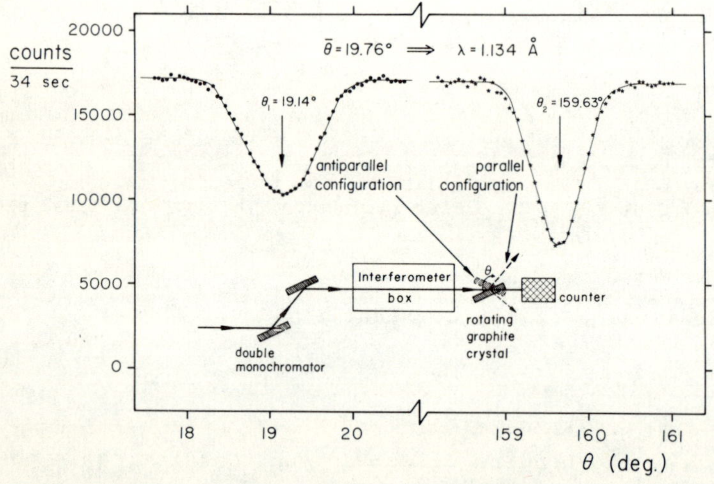


FIG. 5. An example of our method for determining the neutron wavelength by measuring the angular spacing between the transmission dips observed by rotating a pyrolytic graphite crystal in the beam transmitted through the interferometer.

We are in the initial stages of pursuing an alternative and potentially more precise technique which involves measuring the period of oscillation resulting from rotating a slab of silicon in the interferometer. A schematic diagram of the geometry is shown in Fig. 6. This is the standard geometry used by Rauch *et al.* (1976) to measure the coherent scattering lengths of various isotopes at LLL, Grenoble, France. The phase

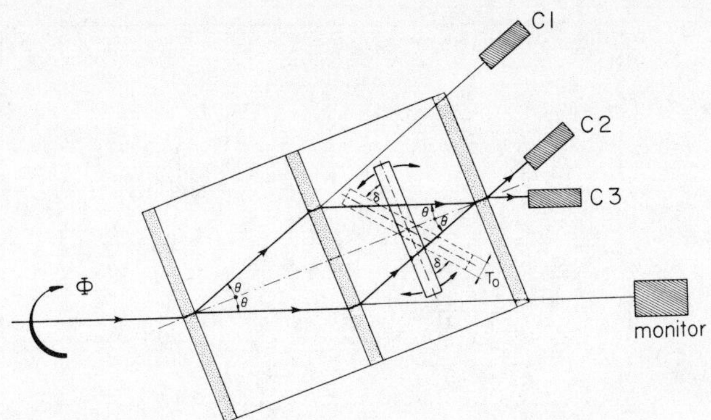


FIG. 6. Geometry of the silicon slab rotation experiments designed to measure the neutron wavelength. The same geometry was used for the X-ray experiments which measure $q_{bend}$. In this case the rotating silicon slab is replaced by a plastic slab.

shift due to the silicon slab, as a function of its orientation δ, is

$$\beta = -\lambda_0 Nb 2T_0 \frac{\sin\delta \sin\theta}{\cos^2\theta - \sin^2\delta} . \tag{37}$$

For silicon, the coherent scattering length $b$ is very accurately known (Shull 1972) and the atom density $N$ is also accurately known from precision lattice parameter measurements (Hart 1975). Therefore we can use this formula to measure $\lambda_0$ precisely. An example of a measurement of this type is shown in Fig. 7 in which a precisely machined single-crystal slab of thickness $T_0 = 0.2931 \pm 0.0001$ cm was used.

### 3.6. *Interference induced by the earth's gravity and rotation*

The total phase shift in these experiments involves three contributions:

$$\begin{aligned} \beta_{tot} &= \beta_{grav} + \beta_{Sagnac} + \beta_{bend} \\ &= -q_{grav} \sin\phi + q_{Sagnac} \cos\phi - q_{bend} \sin\phi . \end{aligned} \tag{38}$$

We see that the Sagnac effect is maximum for $\phi = 0$, while the gravity and bending effects are maximum for $\phi = 90$. We can rewrite equation (38) as

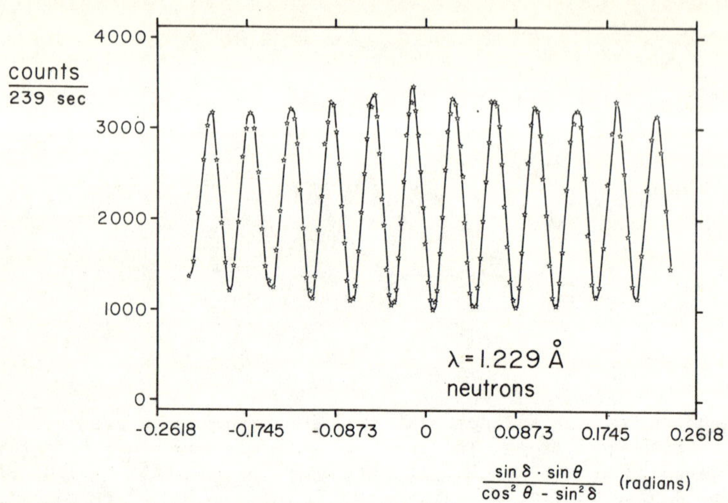


FIG. 7. Example of data obtained by rotating a silicon slab in the interferometer (Fig. 6). The counts in detector $C_3$ are shown.

$$\beta_{tot} = q \sin(\phi - \phi_0) \tag{39}$$

where

$$q^2 = (q_{grav} + q_{bend})^2 + q^2_{Sagnac} \tag{40}$$

and

$$\tan\phi_0 = \frac{q_{Sagnac}}{q_{grav} + q_{bend}} . \tag{41}$$

The fact that the phase shift due to rotation (Sagnac effect) depends upon $\cos\phi$ and not $\sin\phi$ comes about as a result of our selection of due south as the incident beam direction ($\Gamma = 0$ in equation (17)). This leads to an important experimental circumstance: although $q_{Sagnac}$ is of order 2.5% of $q_{grav}$ its contribution to the frequency of oscillation $q$ of the interference pattern as given by equation (40) is very small (of order 3 parts in 10 000). However, it leads directly to a shift $\phi_0$ in the center of the interference pattern. Thus, having measured $q_{bend}$ separately with X-rays, the frequency of oscillation gives us $q_{grav}$ directly.

The counting rates in detectors $C_2$ and $C_3$ are given by (Werner *et al.* 1976):

$$I_2(\phi) = \gamma - \alpha \cos\{q \sin(\phi-\phi_0)+\beta_0\} \tag{42}$$

and

$$I_3(\phi) = \alpha[1 + \cos\{q \sin(\phi-\phi_0)+\beta_0\}] \ . \tag{43}$$

Here $\alpha$ and $\gamma$ are constants dependent upon the incident beam flux and the scattering length and structure of silicon. The phase shift $\beta_0$ exists in any interferometer since the path lengths ABD and ACD are never known to be precisely the same.

It is clear that a measurement of the angle $\phi_0$ requires data to be taken in the range of rotation angles $\phi$ for which $\sin(\phi-\phi_0)$ is non-linear, so that it can be separated from the unknown phase shift $\beta_0$. Since $q_{\text{Sagnac}}$ is independent of wavelength and $q_{\text{grav}}$ depends linearly on $\lambda_0$, we expect $\phi_0$ to decrease with increasing wavelength. The variation is about 1° for $\lambda_0$ between 1 and 2 Å.

The utilization of the double-crystal monochromator assembly is a very important aspect of these experiments. It provides us with a beam of monochromatic neutrons of variable wavelength along a fixed line in space. Thus the interferometer position remains translationally fixed while we are able to perform experiments at a selected neutron wavelength (in practice between 0.9 and 2.1 Å) and also with X-rays.

We show in Figs. 8 and 9 representative data obtained at wavelengths $\lambda_0$ = 1.060 Å and $\lambda_0$ = 1.419 Å respectively. The neutron counting rate in the detector $C_3$ (see Fig. 1) is plotted against the interferometer rotation angle $\phi$. The contrast (maximum/minimum) of this data is seen to be about 3 to 1. Contrasts as high as 4 to 1 have been observed. We are, of course, primarily interested in the frequency $q$ of the fundamental oscillation of this data. To obtain $q$, we numerically Fourier transform the data according to

$$F_q = \sum_{j=1}^{N} I(\sin\phi_j)\exp(iq \sin\phi_j) \tag{44}$$

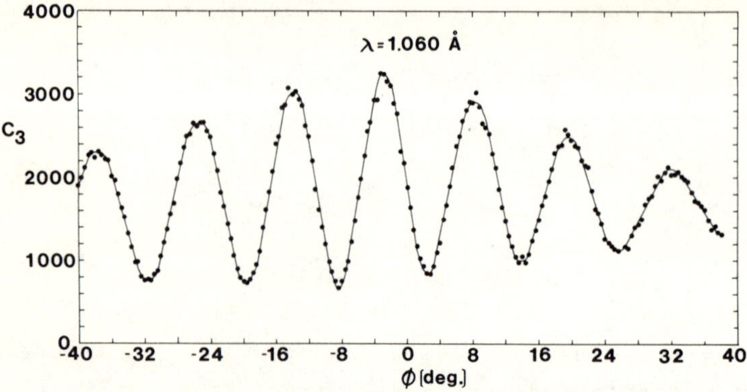


FIG. 8. Gravitationally induced quantum interference experiment at $\lambda$ = 1.060 Å. The counting time was about 5 min per point.

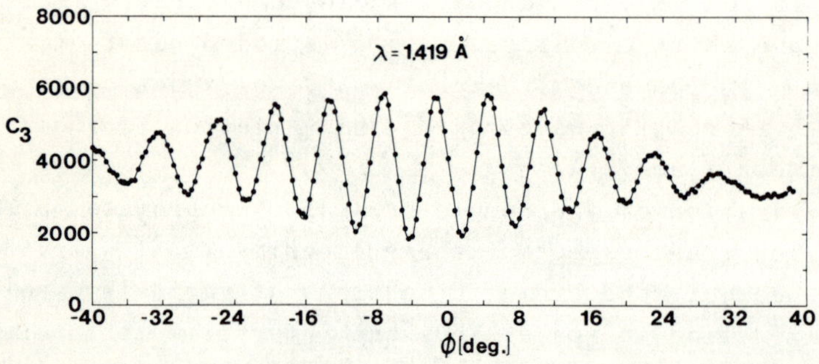


FIG. 9. Gravitationally induced quantum interference experiment at $\lambda$ = 1.419 Å. The counting time was about 7 min per point.

where $I$ is the oscillatory part of the neutron intensity and the index $j$ runs over all the $N$ data points. The Fourier transforms of the data in Figs. 8 and 9 are shown in Figs. 10 and 11.

There is loss of contrast at larger rotation angles $\phi$, which we believe to be due to warping of the interferometer under its own weight as the interferometer is rotated. We are pursuing an explanation of this effect by repeating certain measurements at reduced slit sizes. We have found that the loss of contrast occurs at smaller rotation angles $\phi$ the larger the

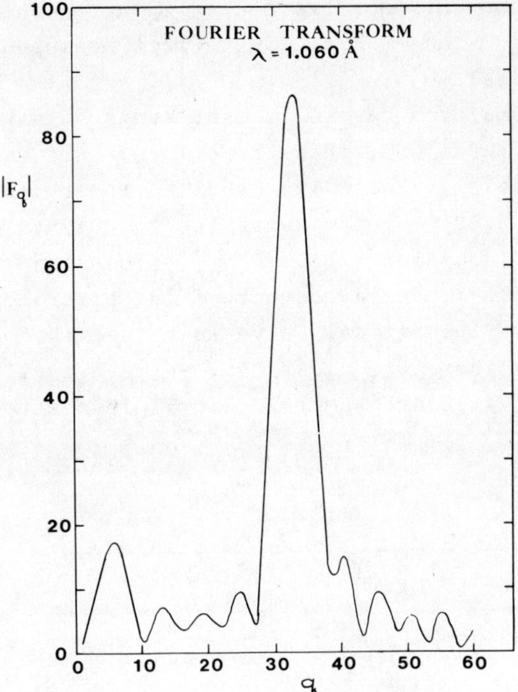


FIG. 10. Fourier transform of the data of Fig. 8.

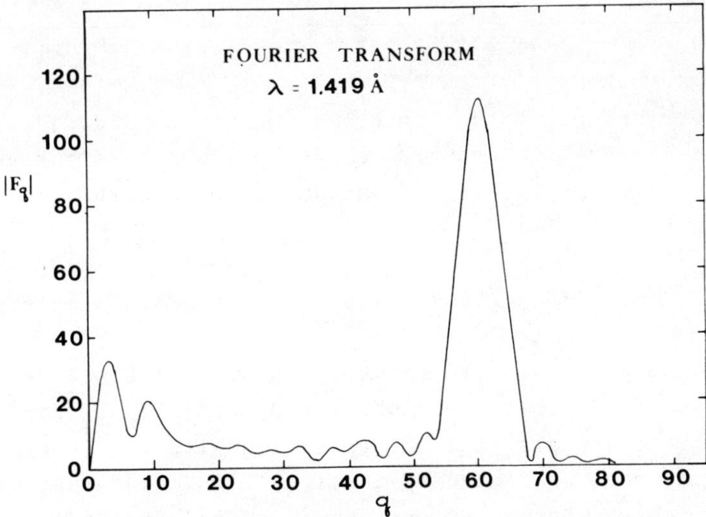


FIG. 11. Fourier transform of the data of Fig. 9.

neutron wavelength. This observation is in agreement with the above explanation, since the bending effect measured with X-rays is proportional to $\sin^2\theta_B (= \lambda_0^2 G^2/4)$.

We have now completed careful measurements at six different wavelengths. The results of these experiments are summarized in Fig. 12 and Table 1. The curve labelled theory is the result of using equation (40) together with the calculated oscillation frequences $q_{grav}$ (equation (28)), $q_{Sagnac}$ (equation (29)), and $q_{bend}$ (equation (35)). The agreement of the experimental measurements with the theoretical curve is better than 1% for each data point. It is therefore clear that we have verified the principle of equivalence to better than 1% in a strongly quantum limit.

TABLE 1

| $\lambda$(Å) | $q_{grav}$ | $q_{Sagnac}$ | $q_{bend}$ | $q$(calc) | $q$(exp) |
|---|---|---|---|---|---|
| 1.060 | 30.50 | 1.06 | 2.48 | 33.00 | 32.98 |
| 1.136 | 35.26 | 1.14 | 2.66 | 37.94 | 37.85 |
| 1.232 | 41.77 | 1.24 | 2.89 | 44.68 | 44.87 |
| 1.419 | 56.41 | 1.46 | 3.32 | 59.75 | 60.30 |
| 1.628 | 75.92 | 2.74 | 3.81 | 79.75 | 80.74 |
| 1.831 | 98.98 | 2.00 | 4.29 | 103.29 | 103.27 |

The determination of the angle $\phi_0$ requires considerable care in data analysis. Recently we have pursued this through direct least squares fitting of the data. We used the functional form

$$I(\phi) = A + B\phi + C\phi^2 + D\cos(Q\sin\phi+\Delta_1)\cos\{q\sin(\phi-\phi_0)+\beta_0\}\ . \tag{45}$$

There are nine parameters to be determined. In Table 2 we show the results of this procedure for determining $\phi_0$ for three wavelengths. It appears that the variation with wavelength is stronger than that predicted by theory. It is clear that we are seeing the neutron Sagnac effect; however, careful analysis of data at a wider selection of wavelengths appears to be necessary.

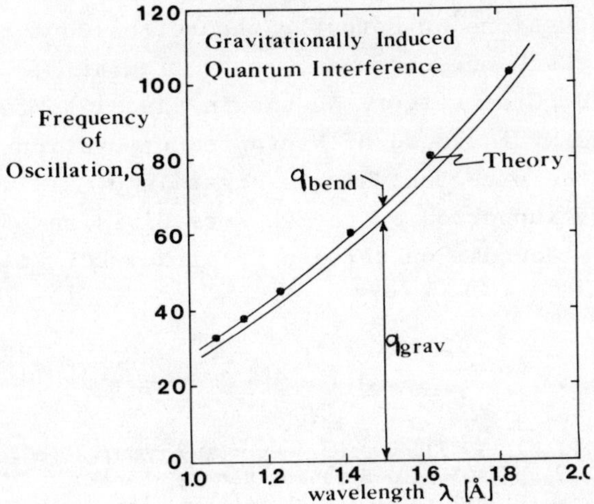


FIG. 12. The frequency of oscillation observed in gravitationally induced quantum interference experiments at various wavelengths. The curve labelled theory is the result of calculating $q_{\text{grav}}$ and adding the measured value of $q_{\text{bend}}$.

TABLE 2

| $\lambda$(Å) | $\phi_0$(calc) (deg) | $\phi_0$(exp) (deg) |
|---|---|---|
| 1.060 | 1.84 | 3.03 |
| 1.136 | 1.72 | |
| 1.232 | 1.59 | |
| 1.419 | 1.40 | 1.35 |
| 1.628 | 1.25 | 1.04 |
| 1.831 | 1.11 | |

## Acknowledgments

The role played by Mr. C. Holmes, Physics Department Machine Shop, University of Missouri, in the success of this experiment was extraordinarily significant. The construction of various mechanical components of the apparatus and the machining of the interferometer in particular required unusual care and precision.

We should like to thank our colleagues for many interesting and helpful discussions on both the theoretical and experimental aspects of these measurements, in particular B. DeFacio, G.W. Ford, and D. Greenberger. We should also like to thank Mr. B. Stone and Mr. J. Burd of Monsanto Corporation for providing us with the excellent single crystals of silicon.

This work was supported by the Physics Division of the National Science Foundation through Grant No. PHY 76 08960.

## References

ANANDAN, J. (1977). *Phys. Rev. D* **15**, 1448.

BONSE, U. and GRAEFF, W. (1977). *Topics in applied physics* (ed. H.-J. Queisser), Vol. 22, pp. 93–140. Springer Verlag, Berlin.

BONSE, U. and HART, M. (1965). *Appl. Phys. Lett.* **6**, 155.

COLELLA, R., OVERHAUSER, A.W., and WERNER, S.A. (1975). *Phys. Rev. Lett.* **34**, 1472.

DOBBS, J.W.T., HARVEY, J.A., PAYA, D., and HORSTMANN, H. (1965). *Phys. Rev. B* **139**, 756.

EÖTVÖS, R. (1890). *Math. Nat. Ber. Ungarn.* **8**, 65.

EÖTVÖS, R., PEKÁR, D., and FEKETE, E. (1922). *Annln. Phys.* **68**, 11.

HART, M. (1975). *Summer School on Electron and Momentum Densities*, Warwick, England.

MICHELSON, A.A., GALE, H.G., and PEARSON, F. (1925). *Astrophys. J.* **61**, 140.

PAGE, L.A. (1975). *Phys. Rev. Lett.* **35**, 543.

RAUCH, H., BADUREK, G., BAUSPIESS, W., BONSE, U., and ZEILINGER, A. (1976). *Proc. Int. Conf. on Interactions of Neutrons with Nuclei, Lowell, Mass.* (ed. E. Sheldon), Vol. II, pp. 1027–1041.

ROLL, P.G., KROTKOV, R., and DICKE, R.H. (1967). *Ann. Phys.* (N.Y.) **26**, 442.

SAGNAC, M.G. (1913). *C.R. Acad. Sci. (Paris)* **157**, 708; **157**, 1410.

SHULL, C.G. (1972). *Phys. Rev. Lett.* **29**, 871.

WERNER, S.A., COLELLA, R., OVERHAUSER, A.W., and EAGEN, C. (1976). *Proc. Conf. on Neutron Scattering, Gatlinburg, Tenn.*, pp. 1060–1072.

# 4. SPIN ROTATION DURING NEUTRON INTERFEROMETRY; OPERATIONALISM AND THE ACTIVE VIEW OF SYMMETRY INVARIANCE

HERBERT J. BERNSTEIN

*Hampshire College, Amherst, Mass. 01002, U.S.A.*

## 1. Introduction

Advances in neutron interferometry have allowed physicists to perform several experiments which show the effects of spin precession on interference patterns. In particular, experiments where the magnetic field is chosen to produce relative spin precessions result in explicit demonstration of the phase factor of -1 associated with fermion rotations through $2\pi$ rad (Rauch *et al.* 1975, Werner *et al.* 1975, Klein and Opat 1976). In effect, these experiments realize some of the *gedanken* and proposed experiments (Aharonov and Susskind 1967, Bernstein 1967, 1969, Klein and Opat 1975) which have elucidated the theoretical considerations of observability of $2\pi$ rotations. These considerations have focused on points of interpretation of the theory, usually assuming that the experiments indeed produce $2\pi$ rotations (Byrne 1975 and 1978, Dowker 1969, Hegerfeldt and Kraus 1968, Mirman 1970, Moore 1970). In this paper we stress the deviations from exact rotations which are present in all experiments, already performed or proposed, and investigate their theoretical implications.

## 2. Existing experiments

The original idea of producing $2\pi$ rotations by magnetic precession perhaps stemmed from consideration of the time-dependent Schroedinger equation. An idealized instantaneously switched-on constant magnetic field imposed during a specific time interval produces an *exact* rotation:

$$\begin{aligned} \mathcal{H} &= -\boldsymbol{\mu}\cdot\mathbf{H} = -\mu_z{}^{\mathrm{op}}H \qquad 0 < t < T \\ \Rightarrow \psi(T) &= \exp\left(\frac{i\mu_z{}^{\mathrm{op}}HT}{\hbar}\right)\psi(0) = \mathcal{R}_z{}^{\mathrm{op}}\left(\frac{|\mu|}{|J_z|}HT\right)\psi(0) \end{aligned} \tag{1}$$

where $\mu$ is the magnetic moment, $J$ the angular momentum, and $\mathcal{H}$ the Hamiltonian. Thus for spin ½ the particle precesses about the magnetic field $\vec{\mathbf{H}}$ at Larmor's frequency $\omega_L = 2\mu H/\hbar$. Note:

$$\mathcal{R}_z{}^{op}(\phi) = \begin{pmatrix} \exp(-i\phi/2) & 0 \\ 0 & \exp(i\phi/2) \end{pmatrix} .$$

In the case of particles flying through static fields, the rotation effect is only approximate for the two spin states effectively have different residence times in the field. Passing through a region of length $L$, a particle with momentum $p$ has a wave function

$$\psi_{\text{field}}(L) = \begin{pmatrix} \exp\left\{\frac{iL(p^2-m\hbar\omega_L)^{\frac{1}{2}}}{\hbar}\right\} & 0 \\ 0 & \exp\left\{\frac{iL(p^2+m\hbar\omega_L)^{\frac{1}{2}}}{\hbar}\right\} \end{pmatrix} \psi(0) \approx \tag{2}$$

$$\approx \exp\left(\frac{ipL}{\hbar}\right)\mathcal{R}_z(\phi)\psi(0)$$

with the approximation

$$(p^2 \pm m\hbar\omega_L)^{\frac{1}{2}} \approx p\left(1 \pm \frac{m\hbar\omega_L}{2p^2}\right) .$$

Thus

$$\psi_{\text{field}}(L) \cong \mathcal{R}_z(\phi)\psi_{\text{zero field}}(L) \tag{3}$$

where the precession angle is the product of the Larmor frequency and the time of passage $mL/p$

$$\phi = \omega_L(mL/p) . \tag{4}$$

The *exact* result is

$$\psi_{\text{field}}(L) = e^{iX} \begin{pmatrix} e^{-i\Upsilon/2} & 0 \\ 0 & e^{i\Upsilon/2} \end{pmatrix} \psi(0) \tag{5}$$

where

$$\Upsilon = \alpha\{(1+\phi/\alpha)^{\frac{1}{2}} - (1-\phi/\alpha)^{\frac{1}{2}}\} \tag{6a}$$

$$X = \frac{\alpha\{(1+\phi/\alpha)^{\frac{1}{2}} + (1-\phi/\alpha)^{\frac{1}{2}}\}}{2} \tag{6b}$$

and $\alpha$ is the propagation phase in the absence of the field

$$\alpha = pL/\hbar \ .$$

In the first instance, therefore, the corrections seem to be an adjustment to the angle of rotation and the propagation phase:

$$\begin{aligned} \Upsilon &= \phi + \frac{\phi}{8}\left(\frac{\phi}{\alpha}\right)^2 + \ldots \\ X &= \alpha - \frac{\alpha}{8}\left(\frac{\phi}{\alpha}\right)^2 + \ldots \ . \end{aligned} \tag{8}$$

These, in practice, are exceedingly small because the ratio $r$ of the precession angle to the total propagation phase is also that of magnetic energy to total energy:

$$r = \frac{\phi}{\alpha} = \frac{\mu H}{p^2/2m} \ .$$

For the experiment of Werner *et al.* (1975), taking the effective field length as 2.7 cm and $\phi \approx 11\pi$, we have $r = 2.9 \times 10^{-8}$. Using an estimated effective length of 3.0 cm for Rauch *et al.* (1975), $r = 4.8 \times 10^{-8}$. Even the experiment of Klein and Opat (1976) with a length of 72 μm gives $r = 3.4 \times 10^{-5}$ for each path. In the first two experiments, with the field present in only one branch of the interferometer, these small corrections can produce a measurable deviation from simple rotation. Indeed, during the course of such experiments, at some point an effect should have been produced which can never be produced by a pure rotation: full polarization of the emerging recombined beam.

A fully polarized beam emerges from an interferometer in the forward direction when illuminated with unpolarized neutrons for certain precession angles. To see this, note that

$$\psi_{\text{combined}} \quad \left\{ e^{i\alpha} + e^{iX} \begin{pmatrix} e^{-iY/2} & 0 \\ 0 & e^{iY/2} \end{pmatrix} \right\} \psi_0 \tag{9}$$

Thus for unpolarized initial neutrons, the emerging net polarization is

$$P_{\text{combined}} = \frac{\mathrm{Tr}(\rho\sigma_z)}{\mathrm{Tr}\rho}$$

$$= \frac{|e^{i\alpha}+e^{i(X-Y/2)}|^2 - |e^{i\alpha}+e^{i(X+Y/2)}|^2}{|e^{i\alpha}+e^{i(X-Y/2)}|^2 + |e^{i\alpha}+e^{i(X+Y/2)}|^2}$$

or

$$P_{\text{combined}} = \frac{\sin(\alpha-X)\sin Y/2}{1 + \cos(\alpha-X)\cos Y/2} \tag{10}$$

$$P_{\text{combined}} = 1 \text{ when } \sin(\alpha-X) = \sin Y/2$$
$$\cos(\alpha-X) = -\cos Y/2$$

$$\text{i.e. } [\alpha-X+Y/2] = \pi(2n+1).$$

Thus when the correction to the propagation phase is the supplementary angle to half the total precession, the emerging forward neutron beam is fully polarized. No polarization effect *at all* can be produced by a pure rotation acting on initially unpolarized neutrons:

$$\psi_{\text{combined}} = \{e^{i\alpha}+e^{i\alpha}\mathcal{R}(\theta)\}\psi_0$$

then

$$P = \frac{|e^{i\alpha}+e^{i\alpha}e^{-i\theta/2}|^2 - |e^{i\alpha}+e^{i\alpha}e^{i\theta/2}|^2}{|e^{i\alpha}+e^{i\alpha}e^{-i\theta/2}|^2 + |e^{i\alpha}+e^{i\alpha}e^{i\theta/2}|^2} = 0 .$$

While the complete polarization must occur at low intensity, it might be possible to observe some polarization with a modification of a previous experiment (Rauch *et al.* 1975). If an effective field of about 10,000 G and an effective length of 12 cm can now be achieved, then picking $\cos(\Upsilon/2) \approx .99$ at a rotation of nearly $3202\pi$ rad will give a polarization of about 3.5% and an intensity of about 1% of the incident beam. This indicates feasibility — more optimum combinations of intensity and polarization are probably achievable.

In all of these static field experiments, the point remains. Small deviations from pure rotation, unmeasurable in some phenomena but significant in others, always occur; they intrinsically prevent the operation of the experiment from being a theoretically perfect embodiment of the symmetry operation, i.e. rotation by $2\pi$.

## 3. Proposed experiments

Rotating magnetic fields — either in time or in space — have been proposed (Aharonov and Susskind 1967, Bernstein 1969, Moore 1970) to 'trap' the precessing spin and turn it in space. This is an improvement over constant fields by showing manifest geometric rotation, and in part answers the complaint that the rotation is otherwise only a calculated quantity. Deviations from strict rotations must occur here also. For simplicity, we treat the time-dependent case.

A strong magnetic field rotates slowly in time perpendicular to the $z$ axis:

$$\mathcal{H}(t)\Psi = \begin{pmatrix} 0 & \mu H e^{i\omega t} \\ \mu H e^{-i\omega t} & 0 \end{pmatrix} \psi \tag{11}$$

$$= i\hbar\frac{d\Psi}{dt} \qquad\qquad \omega \ll \mu H/\hbar \ .$$

The exact solutions are easily found (cf. Landau and Lifschitz 1958, p. 143):

$$\psi_1(t) = \begin{pmatrix} -\dfrac{\alpha_1 B}{\omega_L/2}\exp(i\alpha_2 t) \\ B\exp(i\alpha_1 t) \end{pmatrix} \quad \text{and} \quad \psi_2(t) = \begin{pmatrix} \dfrac{\alpha_2 C}{\omega_L/2}\exp(-i\alpha_1 t) \\ C\exp(-i\alpha_2 t) \end{pmatrix} \tag{12}$$

where

$$\alpha_{1,2} = \mp\frac{\omega}{2} + \frac{1}{2}(\omega_L{}^2 + \omega^2)^{\frac{1}{2}} .$$

Balcar (1979) has shown that solutions to equation (11) can be interpreted as pure rotations of the initial field-free wave functions. The rotation, however, is compounded of the Larmor precession and that of the physically rotating field. Again, as in the homogeneous time-dependent field treated in §2, this approximation ignores the effects of finite turn-on time. Such effects are required by Maxwell's equations.

Indeed, the work of Eder and Zeilinger (1976) gives the adiabatic approximate solution to the related Schroedinger equation for a strong static helical field. Balcar's work shows it is only in this approximation, $\omega_L \gg \omega$, that the helical field can be interpreted as a rotation. We may learn a little of the corrections by examining the expansion of exact time-dependent solutions in the small parameter $\omega/\omega_L$. Here

$$\psi_1(t) \approx \frac{1}{(2-2\omega/\omega_L+\ldots)^{\frac{1}{2}}} \begin{pmatrix} (-1+\omega/\omega_L+\ldots)\exp\{it(\omega_L/2+\omega/2+\omega^2/4\omega_L+\ldots)\} \\ \exp\{it(\omega_L/2-\omega/2+\omega^2/4\omega_L+\ldots)\} \end{pmatrix} \tag{13}$$

and similar approximation for $\psi_2$. Again we see that the wave function is only approximately a rotation (about the $z$ axis with frequency $\omega$) of the field-free case. In fact the largest change is the huge phase of precession $\omega_L t/2$. This phase must be accurately controlled in order to show the much smaller effect of the physically rotating field. The obvious correction does not help; even when the 'twisted' field particle path interferes with one subjected to a constant magnetic field of the same strength, the effect is *not* purely a rotation:

$$\psi_{1\ \text{combined}} \approx e^{it\omega_L/2} \begin{pmatrix} -2^{-\frac{1}{2}} - (2+\ldots)^{-\frac{1}{2}} \exp\left\{it\left(\frac{\omega}{2} + \frac{\omega^2}{4\omega_L} + \ldots\right)\right\} \\ +2^{-\frac{1}{2}} + (2-\ldots)^{-\frac{1}{2}} \exp\left\{it\left(\frac{\omega}{2} + \frac{\omega^2}{4\omega_L} + \ldots\right)\right\} \end{pmatrix}$$

$$\psi_{2\ \text{combined}} \approx e^{-it\omega_L/2} \begin{pmatrix} +2^{-\frac{1}{2}} + (2+\ldots)^{-\frac{1}{2}} \exp\left\{it\left(\frac{\omega}{2} - \frac{\omega^2}{4\omega_L} + \ldots\right)\right\} \\ +2^{-\frac{1}{2}} + (2+\ldots)^{-\frac{1}{2}} \exp\left\{it\left(\frac{\omega}{2} - \frac{\omega^2}{4\omega_L} + \ldots\right)\right\} \end{pmatrix} \tag{14}$$

Once again, the deviations from a pure rotation, including a different rotation angle from $\omega t$, must occur. The experiment may be performed with very high fields and very slow rotation; the rotation effect may therefore be demonstrated to any desired accuracy but never entirely perfectly. Clearly the same sort of effects must be expected for static fields which 'rotate' in space† such as those within a 'polarization twister' or the harmonic helical field given by, for example

$$\mathbf{H} = H_0\left(\cos\frac{2\pi z}{\ell}\cosh\frac{2\pi x}{\ell}, \sin\frac{2\pi z}{\ell}\cosh\frac{2\pi y}{\ell}, \sinh\frac{2\pi y}{\ell}\cos\frac{2\pi z}{\ell} - \sinh\frac{2\pi x}{\ell}\sin\frac{2\pi z}{\ell}\right)$$

$$\mathbf{H} \approx H_0\left(\cos\frac{2\pi z}{\ell}, \sin\frac{2\pi z}{\ell}, 0\right) \qquad \text{for } \frac{x}{\ell}, \frac{y}{\ell} \ll 1 .$$

Thus, even a twisted neutron waveguide with a field perpendicular to the walls (and hence twisting with the guide) will not implement an exact rotation.

## 4. Discussion

In all experiments performed or proposed to date the effect of magnetic fields only approximates that of the perfect symmetry (i.e. covering) operation, rotation by $2\pi$ rad. While it is impossible to analyse all conceivable experiments in advance (and no complete analysis of 'switching-on' effects has been made for time-dependent situations), this basic point is so

†Mezei (1979) has reported the first attempt at such an experiment.

persistent and clear that we may draw some conclusions from the analysis.

(1) All experiments implementing the active symmetry transformation of rotation by using magnetic fields can only approximate the exact symmetry operation.

(2) The approximation can be made as good as desired and the deviations can certainly be made smaller than other experimental errors by judicious choice of the experimental arrangement and parameters.

(3) Finally, we may wish to insist on the necessary exactness of symmetry operations and rule all these experiments invalid as operational representatives of the active viewpoint of symmetry transformations. Such insistence, however, carries general implications for the interpretation of the active view of perfect invariance symmetries.

To elucidate the last conclusion, we note that there is no question about a trivial operationalization of the active view for rotation transformations; one can always do the simple rotation of state-preparing apparatus and call the resulting state the (actively) rotated transform of the original state. However, in order to measure all the potentially observable properties of quantum states one needs not only to prepare those states but to superpose them with other states as well. In particular, phase factors and phase relations generally require such coherent superposition to produce measurable effects. In the remainder of this discussion we shall consider the coherent production of a state and its symmetry transform whenever we speak of preparing the transform state.

It should be noted that it has long been clear that there is no natural way to operationalize exact inversion symmetries such as parity (see Schweber 1962, p. 17 and references therein). Indeed parity violation was discovered precisely by the failure of nature to produce the symmetry-transformed state of the negative—helicity neutrino under circumstances where it otherwise could have been produced.

If we take the strictest interpretation of $2\pi$ rotation symmetry and of conclusions (1) and (2) above, we must seriously question whether the production of a $2\pi$-rotated state can be successfully accomplished by operations of physical experiment.

Unlike neutrino physics, where the experimental impossibility of symmetry-state production meant the violation of a conservation law, the situation here might have a more general significance. Thus we argue that magnetic fields are necessary to 'grasp' and turn the magnetic moment of any fermion. However, as we have seen, such fields invariably produce second-order effects which spoil the exact identification of experimental procedure with producing a rotated state. Such small effects cannot hide the essential intuitively correct *demonstration* of interference effects associated with rotations, but they do make it impossible to give an exact operational meaning to the active rotated state.

Recall that the active view of invariance symmetry $I$ interprets the equation

$$U^{\mathrm{op}}(I)\psi(x) = \psi(I^{-1}x) \tag{15}$$

as giving the values for a wave function of the transformed state $U(I)\psi$ in terms of those for the original state. This contrasts with the passive view, which interprets $U(I)\psi(x)$ as the values of the same state described in a transformed coordinate system. The formalism gives no specific prescription for producing the actively transformed state. In order to test the validity of the active view of symmetries, we must supply an operational definition of the transformed state. As we have seen here the existing and proposed experiments on spin precession do not operationalize an exact complete rotation. We conclude that there may be a general rule at work, evidenced variously in all homogeneous spatial symmetries, i.e. both parity and rotational.

The active symmetry-transformed state cannot be produced operationally. We can never check all the direct experimental consequences of the active view of symmetry; under operationalism only the passive view of exact symmetries can ever be fully verified.

By way of epilogue, we may speculate that the space—time transformations should be the next subgroup scrutinized now that it is clear that active symmetries fail to be operationalized for each of the purely spatial possibilities for covering operations. Furthermore, we may speculate that the implications of this particular restricted sense of not being a $2\pi$ rotation may make the neutron interference experiments ultimately more significant than any argument strenuously maintaining that such experiments do produce measurable $2\pi$ rotations.

## References

AHARONOV, Y. and SUSSKIND, L. (1967). *Phys. Rev.* **115**, 485.
BALCAR, E. (1979). This Conf., paper II.6.
BERNSTEIN, H.J. (1967). *Phys. Rev. Lett.* **18**, 1102.
—— (1969). *Sci. Res. (N.Y.)* **4**, 32.
BYRNE, J. (1975). *Nature (Lond.)* **258**, 385.
—— (1978). *Nature (Lond.)* **275**, 188.
DOWKER, J. (1969). *J. Phys. A* **2**, 267.
EDER, G. and ZEILINGER, A. (1976). *Nuovo Cim. B* **34**, 76.
HEGERFELDT, G.C. and KRAUS, K. (1968). *Phys. Rev.* **170**, 1185.
KLEIN, A.G. and OPAT, G.I. (1975). *Phys. Rev. D* **11**, 523.
—— (1976). *Phys. Rev. Lett.* **37**, 238.
LANDAU, L.D. and LIFSCHITZ, E.M. (1958). *Quantum mechanics*. Addison-Wesley, Reading, Mass.
MEZEI, F. (1979). This Conf., paper II.7.
MIRMAN, R. (1970). *Phys. Rev. D* **1**, 3349.
MOORE, G.T. (1970). *Am. J. Phys.* **38**, 1177.
RAUCH, H., ZEILINGER, A., BADUREK, G., WILFING, A., BAUSPIESS, W., and BONSE, U. (1975). *Phys. Lett. A* **54**, 425.
SCHWEBER, S. (1962). *An introduction to relativistic field theory*. Harper and Row, New York.
WERNER, S.A., COLELLA, R., OVERHAUSER, A.W., and EAGEN, C.F. (1975). *Phys. Rev. Lett.* **35**, 1053.

Note added in proof:

We are not arguing the impossibility of producing equal and opposite relative phase shifts in two orthogonal spin-states. One *can* produce such phase shifts, by using compensating plates or clever cancellation effects in the two arms of an interferometer. But that would not be a satisfactory implementation of a kinematic *exact* rotation; it would simply be a dynamical effect which produced particular phase shifts. For more direct experiments, we have indicated some tell-tale dynamical effects of static fields, helical fields, and time-dependent fields (including spacial inhomogeneities when they are switched on, as required by Maxwell's equations) – effects which challenge the interpretation as a kinematic rotation. Moreover, all of these methods can only implement a *spinor* rotation. This has led us to speculate that nature prevents us from implementing *exact* pure rotations, with the deeper implication that we cannot assign a complete operational meaning to the active view of space symmetries.

# 5. SOME MAGNETIC AND SPIN EFFECTS IN NEUTRON INTERFEROMETRY

A. ZEILINGER†

*Massachusetts Institute of Technology, Cambridge, Mass. 02139, U.S.A.*

## 1. Introduction

A whole field of new experiments in neutron interferometry can be expected as a consequence of the fact that the neutron carries spin and magnetic moment. The first experiments in this field were the experimental demonstration of the $4\pi$ rotational symmetry of spinors (Rauch *et al.* 1975, Werner *et al.* 1975) and of the capability of the neutron interferometer to act as a neutron polarizer (Badurek *et al.* 1976). The first theoretical work was devoted to the study of effects due to ferromagnets and helical magnetic fields (Eder and Zeilinger 1976) and to a general analysis of spin rotations and spin-state superposition (Zeilinger 1976). In this paper we shall first deal with a more detailed discussion of spin-state superposition, secondly we shall present explicit results for ferromagnets as phase shifters, and finally we shall discuss some possible magnetic phase contrast experiments. In most cases we shall present only the results. Details of the calculations can be found in Eder and Zeilinger (1976) and Zeilinger (1976).

As the neutron is a spin-½ particle its wave function can be written as the spinor

$$\psi(\mathbf{r},t) = e^{i\xi}\left[a\begin{pmatrix}1\\0\end{pmatrix} + e^{i\phi}b\begin{pmatrix}0\\1\end{pmatrix}\right] . \tag{1a}$$

where $a$, $b$, $\xi$, and $\phi$ are real quantities which are in general functions of space and time. It is usual to choose the coordinate system in such a way that $\begin{pmatrix}1\\0\end{pmatrix}$ and $\begin{pmatrix}0\\1\end{pmatrix}$ are the eigenstates of $\sigma_z$ with the Pauli spin vector $\boldsymbol{\sigma} = (\sigma_x,\sigma_y,\sigma_z)$ defined as

---

†Permanent address: Atominstitut der Österreichischen Universitäten, Wien, Austria.

$$\sigma_x = \begin{pmatrix} 0 & 1 \\ 1 & 0 \end{pmatrix} \qquad \sigma_y = \begin{pmatrix} 0 & -i \\ i & 0 \end{pmatrix} \qquad \sigma_z = \begin{pmatrix} 1 & 0 \\ 0 & -1 \end{pmatrix} . \tag{2}$$

In equation (1a) $\xi$ denotes the spatial phase and $\phi$ the internal phase of the spinor. The intensity of a spinor as described by this equation is $I = a^2 + b^2$ and its polarization vector is

$$\mathbf{P} = \frac{1}{I}(2ab \cos\phi,\ 2ab \sin\phi,\ a^2-b^2) . \tag{3}$$

Other ways of representing a spinor are for example

$$\begin{aligned} |\psi\rangle &= e^{i\xi}[\alpha|\uparrow_z\rangle + e^{i\phi}b|\downarrow_z\rangle] \\ |\psi\rangle &= e^{i\xi'}[a'|\uparrow_x\rangle + e^{i\phi'}b'|\downarrow_x\rangle] \end{aligned} \tag{1b}$$

and, if the spinor is normalized and we suppress the explicit reference to its space–time variations,

$$|\theta,\phi\rangle = e^{i\xi}[\cos\tfrac{\theta}{2}|\uparrow_z\rangle + e^{i\phi}\sin\tfrac{\theta}{2}|\downarrow_z\rangle] \tag{1c}$$

where $|\uparrow_z\rangle$ is the unit spinor with spin up in the $z$ direction, $\theta$ is the angle of the neutron polarization vector with the $z$ axis, and as above $\phi$ is the angle between its projection onto the $xy$ plane and the $x$ axis. Spin rotations are described by the unitary operator

$$U_R = \exp\left(\frac{-i\boldsymbol{\sigma}\cdot\boldsymbol{\alpha}}{2}\right) = \cos\frac{\alpha}{2} - i\boldsymbol{\sigma}\cdot\tilde{\alpha}\ \sin\frac{\alpha}{2} \tag{4}$$

where $\alpha$ is the angle of rotation and $\tilde{\alpha} = \boldsymbol{\alpha}/\alpha$ is a unit vector parallel to the axis of rotation. It is of particular interest for the interpretation of the experiments discussed in the next section to study the unitary operator describing a 180° rotation

$$U_R(180°) = -\ i\boldsymbol{\sigma}\cdot\tilde{\alpha} . \tag{5}$$

Thus the rotation of a spin pointing initially in the $+z$

direction around an axis normal to it can be written as

$$- i\sigma_x\alpha_x - i\sigma_y\alpha_y|\uparrow_z\rangle = \exp\left\{i\left(\gamma-\frac{\pi}{2}\right)\right\}|\downarrow_z\rangle \qquad (6)$$

where $\gamma$ is the angle between the rotation axis and the $x$ axis. Thus after a spin-flip process a neutron wave contains a phase factor which depends upon the axis around which the rotation was performed.

To avoid an unnecessary complication by a detailed discussion of dynamical diffraction effects and to provide a general analysis applicable to many interferometer types we assume, with one exception, an idealized interferometer experiment using 'half-silvered' mirrors. This idealized interferometer provides two final beams as coherent superpositions of equal amplitudes from both beam paths. These contributions are in phase for the so-called forward beam and out of phase for the deviated beam because of a phase change by $\pi/2$ upon each reflection.

## 2. Spin-state superposition

A neutron interferometer with polarized incident neutrons can be used to demonstrate explicitly some aspects of spin-state superposition. For the experiments proposed in this section we assume that the incoming neutrons are in the spin-up state with respect to the $z$ direction. To obtain a coherent amplitude in the spin-down state we turn the neutron spin in one beam path by 180° around the $y$ direction. The forward and the deviated beam are then both coherent superpositions of spin-up and spin-down states with equal amplitudes. These two states are in phase with each other according to equation (7). Thus the neutron spin in the forward beam is in the positive $x$ direction

$$\begin{aligned} \tfrac{1}{2}|\uparrow_z\rangle + \tfrac{1}{2}\exp\left(\frac{-i\sigma_y\pi}{2}\right)|\uparrow_z\rangle &= \tfrac{1}{2}|\uparrow_z\rangle + \tfrac{1}{2}|\downarrow_z\rangle \\ &= \frac{1}{\sqrt{2}}|\uparrow_x\rangle \ . \end{aligned} \qquad (7)$$

Similarly the spin in the deviated beam is in the negative $x$

direction:

$$\tfrac{1}{2}|\uparrow_z\rangle + \tfrac{1}{2}\exp(i\pi)\exp\left(\frac{-i\sigma_y\pi}{2}\right)|\uparrow_z\rangle = \tfrac{1}{2}|\uparrow_z\rangle - \tfrac{1}{2}|\downarrow_z\rangle$$
$$= \frac{1}{\sqrt{2}}|\downarrow_x\rangle \quad . \tag{8}$$

Presumably the best way to measure these spins is to use d.c. spin turners of the type first introduced by Rekveldt (1971). This experiment would be the realization of a *Gedanken* experiment proposed by Wigner (1963) where he imagines a double Stern—Gerlach arrangement to separate and recombine coherently the $z$-spin components of an $x$-spin particle. This *Gedanken* experiment and its realization as proposed above demonstrate explicitly the coherent spin-superposition law of equations (1a)—(1c). The outcome of these experiments could never be interpreted as an incoherent mixture of the superposed spin states.

The possibilities for direct experimental verifications of equations (1a)—(1c) include additional detailed demonstrations of the spin-superposition law. For example the internal spin phase can be changed directly by shifting the phase of the spin-down beam in Fig. 1 by $\chi$ using a nuclear phase shifter. Thus we obtain for the forward beam

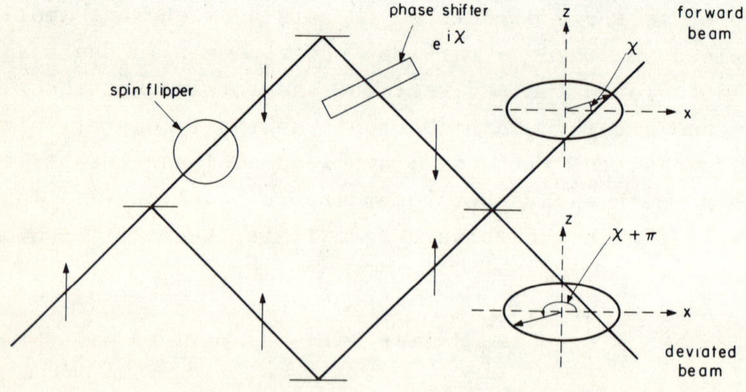


FIG. 1. Experimental test of spin superposition: rotation of the spins in the beams leaving the interferometer by a nuclear phase shift.

$$\begin{aligned} \tfrac{1}{2}|\uparrow_z\rangle + \tfrac{1}{2}e^{i\chi}|\downarrow_z\rangle &= \frac{e^{i\chi/2}}{\sqrt{2}}\left(\cos\frac{\chi}{2}|\uparrow_x\rangle - i\,\sin\frac{\chi}{2}|\downarrow_x\rangle\right) \\ &= \frac{1}{\sqrt{2}}\left|\frac{\pi}{2},\chi\right\rangle . \end{aligned} \tag{9}$$

This means that the spin of this beam can be rotated by a scalar interaction. This seems to be paradoxical but the complete wave function includes the deviated beam and its spin is obtained as

$$\begin{aligned} \tfrac{1}{2}|\uparrow_z\rangle + \tfrac{1}{2}e^{i\pi}e^{i\chi}|\downarrow_z\rangle &= \frac{e^{i\chi/2}}{\sqrt{2}}\left(\cos\frac{\chi}{2}|\downarrow_x\rangle - i\,\sin\frac{\chi}{2}|\uparrow_x\rangle\right) \\ &= \frac{1}{\sqrt{2}}\left|\frac{\pi}{2},\chi+\pi\right\rangle . \end{aligned} \tag{10}$$

Thus the spins of the forward and the deviated beam are always opposite to each other. This implies that the scalar phase shift $\chi$ does not rotate the spin of the complete wave function.

If finally an absorber is brought into one of the beam paths, the final neutron spins can be tipped out of the $xy$ plane because absorption changes the amplitude of one of the states of equations (1a)–(1c). If the intensity of the neutrons passing through the absorber is reduced by $e^{-\alpha}$ we have to reduce the amplitude of its wave function by the factor $e^{-\alpha/2}$ As a particular example we attenuate the spin-down intensity by $e^{-\alpha} = 3 - 2\sqrt{2}$ or its amplitude by $e^{-\alpha/2} = \sqrt{2} - 1$. Thus we obtain for the forward beam

$$\begin{aligned} \tfrac{1}{2}|\uparrow_z\rangle + \tfrac{1}{2}e^{-\alpha/2}|\downarrow_z\rangle &= \tfrac{1}{2}|\uparrow_z\rangle + \frac{\sqrt{2}-1}{2}|\downarrow_z\rangle \\ &= \frac{2-\sqrt{2}}{2}\left(|\uparrow_z\rangle + |\uparrow_x\rangle\right) = \frac{1}{N}\left|\frac{\pi}{4},0\right\rangle \end{aligned} \tag{11}$$

with $N$ being the normalization factor. Thus in this example the spin points in a direction 45° up from the $x$ direction. As this last point deals with amplitude changes its effect could in principle be demonstrated also by a non-interferometric experiment as for example the diffraction experiment proposed by Mezei (1976). This evidently does not hold any more if simultaneously a phase shift is introduced. If performed, the class of experiments proposed in this section would constitute the first direct test of the first principles of the quantum mechanics of spin. Thus these experiments should be performed despite the fact that the principles involved are implicitly tested in existing experiments.

### 3. Magnetic phase shifters

Other applications of the neutron spin character in interferometry can be expected when ferromagnets are used as phase shifters. Their effect on a transmitted neutron beam can be described as a combination of a scalar phase shift and a spin rotation angle. To demonstrate the kind of experimental outcomes we assume that a single-domain magnet, i.e. a magnetically saturated sample, is placed in one beam of the interferometer. In that case the spin rotation angle can be expressed in terms of the forward magnetic scattering length $p_0$ which is a function of the magnetization as

$$\frac{\alpha}{2} = - N\lambda p_0 d \tag{12}$$

where $N$ is the atomic density and $d$ is the thickness of the sample. This is an analogue of the familiar expression of the nuclear phase shift $\chi$ as a function of the nuclear scattering length $b_c$:

$$\chi = - N\lambda b_c d\ . \tag{13}$$

Thus both the nuclear phase shift and the spin rotation angle vary linearly with sample thickness keeping their ratio constant. If we assume that the initial neutron spin points in the $x$ direction and the sample is magnetized in the $z$ direction we obtain for the intensity of the forward beam

$$I = I_0\left\{\frac{1 + \cos\chi\ \cos(\alpha/2)}{2}\right\} \tag{14}$$

where $I_0$ is the intensity without a sample. The polarization of the forward beam is

$$\mathbf{P}_f = \frac{1}{1 + \cos\chi \cos(\alpha/2)} \begin{pmatrix} \cos\frac{\alpha}{2}(\cos\chi + \cos\frac{\alpha}{2}) \\ -\sin\frac{\alpha}{2}(\cos\chi + \cos\frac{\alpha}{2}) \\ -\sin\chi \sin\frac{\alpha}{2} \end{pmatrix} . \quad (15)$$

Figures 2, 3, and 4 show the intensity and the $z$ component of the polarization for iron, cobalt, and nickel as a function of sample thickness. It has to be noted that these particular curves also hold for unpolarized incident neutrons.

The complete behaviour of the polarization vector (equation (15)) is shown in Fig. 5 as the trace of the unit spin vector on a unit sphere again for neutrons initially polarized in the $x$ direction and a single-domain sample magnetized in the $z$ direction. The polarization vector starts at the point where the $x$ axis meets the equator. With increasing sample thickness the polarization vector follows loop trajectories, their particular form being related to the ratio $b_c/p_0$. The connection with absolute thickness values can be obtained by comparison

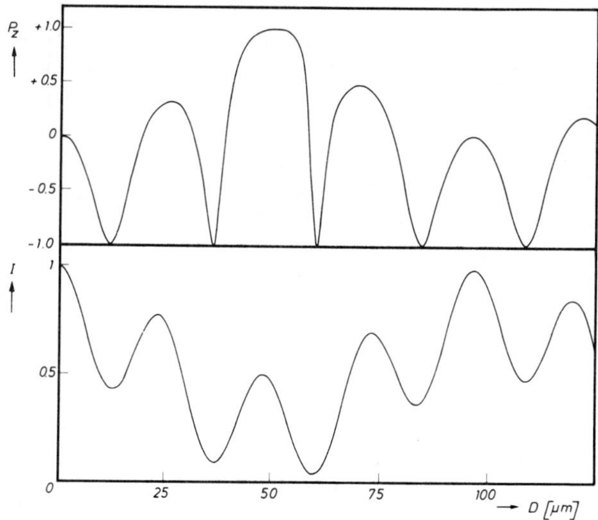


FIG. 2. Polarization (top) and intensity of the forward beam as a function of the thickness of a single-domain (magnetization in the $z$ direction) iron phase shifter in one beam of the interferometer. The incoming neutrons are assumed to be unpolarized or polarized with a polarization direction confined to the $xy$ plane.

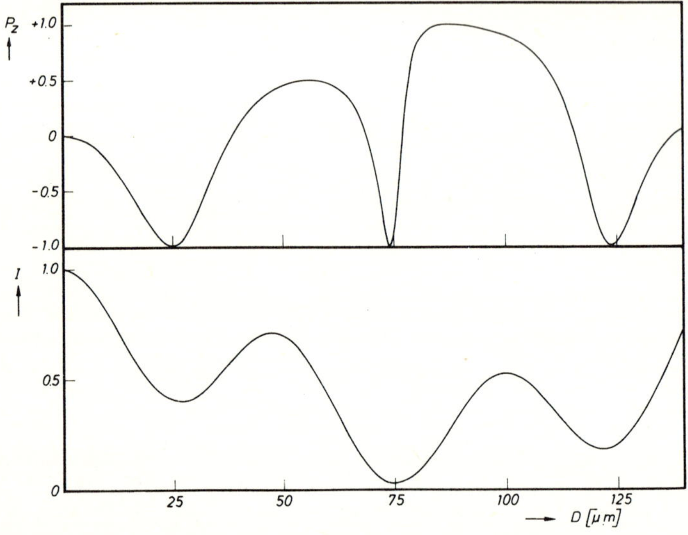


FIG. 3. As Fig. 2 but for cobalt.

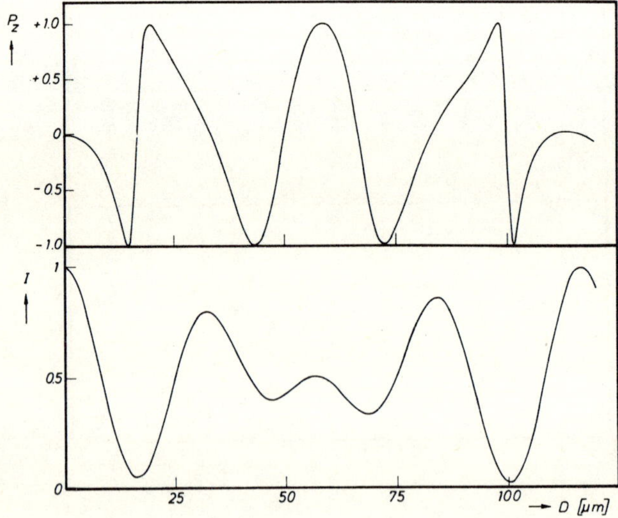


FIG. 4. As Fig. 2 but for nickel.

of the $P_z$ components with those of Figs. 2–4. We point out that, as the motion of the neutron spin in the samples is simply the Larmor precession around the magnetization direction, the tip of the neutron spin would in a single-transmission non-interferometric experiment follow the equator in our representation for all ferromagnets.

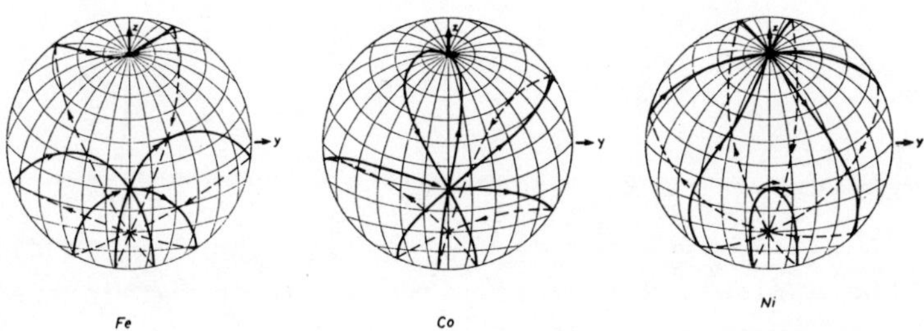


FIG. 5. The behaviour of the polarization vector of the forward beam when a single-domain ferromagnet magnetized in the $z$ direction is used as a phase shifter in one beam of the interferometer. The incoming neutrons are polarized in the $x$ direction.

For magnetic samples with more domains the results can also be derived using the general solutions with the spin rotation angle along one beam path being now a function of the sequence, size, and orientation of the magnetic domains along this beam path. For very small magnetic domains a direct connection to neutron depolarization theory can be found. Here it has to be noted that the first experiments in this field were performed by Korpiun (1966) using the Fresnel biprism type interferometer of Maier-Leibnitz and Springer (1962). He was able to obtain experimental data reflecting correlations between magnetic domains.

## 4. Magnetic phase contrast topography

Recently it was demonstrated that the neutron interferometer can be used to obtain phase contrast images of non-magnetic samples (Schlenker 1978, private communication). One possible approach to phase contrast imaging of magnetic domains would be to use polarized neutrons. Fortunately one can expect to do experiments of this kind even with unpolarized incident neutrons. To discuss this possibility we have to refer to equation (14). Thus a variation of the spin rotation angle $\alpha$ leads to a variation of the intensity of the forward beam. If a sample consists of domains the overall spin rotation angle along one beam path through the sample can be found from the equation

$$\exp\left(\frac{-i\boldsymbol{\sigma}\cdot\boldsymbol{\alpha}}{2}\right) = \prod_i \exp\left(\frac{-i\boldsymbol{\sigma}\cdot\boldsymbol{\phi}_i}{2}\right) \tag{16}$$

where $\boldsymbol{\phi}_i$ is the spin rotation vector in the $i$th domain. Thus a variation of the sequence of the domains can lead to different rotation angles and therefore to variations of intensity.

Another theoretical result of possible relevance to magnetic phase contrast topography is the prediction of the intensity of the forward beam if the spins in both beams in the interferometer are rotated, namely

$$I = \frac{I_i}{2}\left(1 + \cos\frac{\alpha}{2}\cos\frac{\beta}{2} + \tilde{\alpha}\cdot\tilde{\beta}\sin\frac{\alpha}{2}\sin\frac{\beta}{2}\right) \tag{17}$$

where $\boldsymbol{\beta}$ characterizes the spin rotation in the second beam in the same way as $\boldsymbol{\alpha}$ does for the other (equation (4)). One application of this result would be to apply a magnetic field over one beam while the sample is in the other. If the domains in the sample are as thick as the sample itself an intensity variation can be expected which is characterized by the angles between the direction of the magnetic field and the magnetization directions of the individual domains.

A particularly interesting application of the last equation would be the imaging of magnetic domain walls using a two-crystal neutron interferometer. If the wall is in the beam of such an interferometer the neutron spin rotation angles on both sides are different, reflecting different magnetization directions of the domains separated by the wall. Thus if a domain wall is shifted through the beam a variation of the intensity is to be expected.

**Acknowledgments**

I wish to acknowledge useful discussions with D. Atwood, E. Balcar, H. Rauch, C.G. Shull, G.L. Squires, and, in particular, M.A. Horne. For resolving the paradox about the spin rotation by a scalar phase shift, I thank M. Fierz. This work was supported through Wissenschaftsstipendium des Kulturamtes der Stadt Wien.

**References**

BADUREK, G., RAUCH, H., ZEILINGER, A., BAUSPIESS, W., and BONSE, U. (1976). *Phys. Rev. D* **14**, 1177.
EDER, G. and ZEILINGER, A. (1976). *Nuovo Cim. B* **34**, 76.
KORPIUN, P. (1966). *Z. Phys.* **195**, 146.
MAIER-LEIBNITZ, H. and SPRINGER, T. (1962). *Z. Phys.* **167**, 386.
MEZEI, F. (1976). *Commun. Phys.* **1**, 81.
RAUCH, H., ZEILINGER, A., BADUREK, G., WILFING, A., BAUSPIESS, W., and BONSE, U. (1975). *Phys. Lett. A* **54**, 425.
REKVELDT, M.Th. (1971). *J. de Physique* **32**, C579.
WERNER, S.A., COLELLA, R., OVERHAUSER, A.W., and EAGEN, C.F. (1975). *Phys. Rev. Lett.* **35**, 1053.
WIGNER, E.P. (1963). *Am. J. Phys.* **31**, 6.
ZEILINGER, A. (1976). *Z. Phys.* **325**, 97.

# 6. MAGNETIC SUBSTANCES AND FIELDS IN THE NEUTRON INTERFEROMETER

E. BALCAR

*Atominstitut der Österreichischen Universitäten, Schüttelstrasse 115, A-1020 Vienna, Austria*

## 1. Introduction

Rauch *et al.* (1975), Badurek *et al.* (1976) and Werner *et al.* (1975) have made important measurements in neutron interferometry to investigate the effects of magnetic fields acting on part of the spinor wave function of neutrons passing through the interferometer. The corresponding theory has been presented in a comprehensive treatment by Eder and Zeilinger (1976). Although we follow a different approach similar to that of Schwink and Schärpf (1975), we frequently indicate the close connection between both methods and the equivalent results derived for corresponding situations (cf. Bernstein 1979).

In §2 we present the well-known results (see for instance Landau and Lifshitz 1965) describing the spinor part of a neutron wave function under the influence of (a) a magnetic field with varying strength but constant direction and (b) a magnetic field of constant strength with the direction rotating around an axis. In §3 these results are applied to the case of a neutron interferometer. In §§3.1 and 3.2 we give the formulae for the intensities and polarizations respectively for cases (a) and (b) both without and with a nuclear phase shifter. The results for a purely magnetic rotating field are included since it is in principle always possible to cancel the nuclear phase shift through a suitable experimental arrangement. Finally, in §3.3 we list several special cases in which the relevant equations can be considerably simplified by appropriate assumptions.

## 2. Spinor wave functions in magnetic fields

In this section we summarize the known solutions of the time-dependent Schrödinger equation in the presence of magnetic fields. For simplicity we consider a one-dimensional problem with neutrons moving along the $y$ axis:

$$i\hbar\frac{\partial}{\partial t}\Psi(y,t) = \left(-\frac{\hbar^2}{2m}\frac{\partial^2}{\partial y^2} - \boldsymbol{\mu}\cdot\mathbf{B}\right)\Psi(y,t) \quad . \tag{1}$$

The neutron mass is denoted by $m$ and the neutron magnetic moment is given by $\boldsymbol{\mu} = \mu\boldsymbol{\sigma}$, where the $\sigma_i$ are the Pauli matrices and $\mu = -1.91\ \mu_N$ where $\mu_N$ is the nuclear magneton. The magnetic field in equation (1) will affect only the spinor part of the wave function and we therefore separate equation (1) into a space-dependent part $\Phi$ and a spin-dependent part $\psi$

$$\Psi(y,t) = \Phi(y,t)\psi(t) = \exp\left(iky - \frac{\hbar k^2}{2m}t\right)\psi(t) \quad . \tag{2}$$

$\Phi(y,t)$ is given by a simple plane wave in the $y$ direction. Inserting equation (2) into equation (1) gives the equation of motion for the two-component spinor wave function $\psi$:

$$i\hbar\frac{\partial}{\partial t}\psi(t) = -\mu\mathbf{B}\cdot\boldsymbol{\sigma}\psi(t) \quad . \tag{3}$$

An analytic solution of equation (3) is possible for two cases of interest. For a magnetic field with time-dependent field strength $B(t)$ and constant direction $\mathbf{e}_B$ one obtains ($\mathbf{B}(t) = B(t)\mathbf{e}_B$)

$$\psi(\tau) = U(\tau)\psi(0) = \exp\left\{i\frac{\mu}{\hbar}\left(\int_0^\tau B(t)\mathrm{d}t\right)\mathbf{e}_B\cdot\boldsymbol{\sigma}\right\}\psi(0) \quad . \tag{4}$$

This solution is written in a suitable form for later application in the interferometer. We have assumed that the magnetic field only acts along a distance $D$ of the neutron path. The neutron enters this region of space at $t = 0$ and leaves it at $t = \tau$. In equation (4) we have defined a unitary operator $U(\tau)$ which represents the time evolution of the spinor wave function. $U(\tau)$ has the form of a rotation operator around the field direction $\mathbf{e}_B$ and the angle of rotation is

$$\phi(\tau) = \frac{2\mu}{\hbar}\int_0^{\tau} B(t)\mathrm{d}t \ . \tag{5}$$

Another analytic solution of equation (3) is found for the case of a rotating magnetic field of constant strength $\mathbf{B}(t) = B\mathbf{e}(t)$. We assume a field rotating around the $z$ axis in the positive sense

$$\mathbf{B}(t) = B\begin{pmatrix} b_\perp \ \cos\omega t \\ b_\perp \ \sin\omega t \\ b_\parallel \end{pmatrix} = B\mathbf{e}(t) \tag{6}$$

where $b_\perp = B_\perp/B$ and $b_\parallel = B_\parallel/B$. Viewed from a correspondingly rotating co-ordinate system the field appears to be constant and in this case the solution reduces to the simple form given in equation (4) with $\mathbf{e}_B = \mathbf{e}(0)$. We now introduce

$$\psi(t) = \exp\left(-\mathrm{i}\frac{\omega t}{2}\sigma_z\right)\eta(t) \tag{7}$$

and use

$$\mathbf{B}(t)\cdot\boldsymbol{\sigma} = \exp\left(-\mathrm{i}\frac{\omega t}{2}\sigma_z\right)(\mathbf{B}(0)\cdot\boldsymbol{\sigma})\exp\left(\mathrm{i}\frac{\omega t}{2}\sigma_z\right) \tag{8}$$

to obtain from equation (3) the equation of motion for $\eta(t)$:

$$\mathrm{i}\hbar\frac{\partial}{\partial t}\eta(t) = \left\{-\mu\mathbf{B}(0)\cdot\boldsymbol{\sigma} - \frac{\hbar\omega}{2}\sigma_z\right\}\eta(t) \ . \tag{9}$$

The right-hand side corresponds to equation (3) for a time-independent effective field $\mathbf{B}'(0)$ which differs both in size and direction from $\mathbf{B}(0)$:

$$\mathbf{B}'(0) = \begin{pmatrix} B_\perp \\ 0 \\ B_\parallel + \hbar\omega/2\mu \end{pmatrix} = B'\mathbf{e}'(0) \qquad B' = \left\{\left(B_\parallel + \frac{\hbar\omega}{2\mu}\right)^2 + B_\perp{}^2\right\}^{\frac{1}{2}} . \tag{10}$$

Using equations (4) and (7) we obtain the following expression (with $\psi(0) = \eta(0)$):

$$\psi(t) = \exp\left(-i\frac{\omega t}{2}\sigma_z\right)\exp\left\{i\frac{\mu B' t}{\hbar}\mathbf{e}'(0)\cdot\boldsymbol{\sigma}\right\}\psi(0) \ . \tag{11}$$

We see from equation (11) that $\psi(0)$ is first rotated around $\mathbf{e}'(0)$ through an angle $2\mu B' t/\hbar$ in the negative sense. A second rotation around $\mathbf{e}_z$ through an angle $\omega t$ in the positive sense gives the final solution for the rotating magnetic field described by equation (6). With the help of relation (8) the spinor wave function can be rewritten as

$$\psi(t) = \exp\left(i\frac{\mu B' t}{\hbar}\mathbf{e}'(t)\cdot\boldsymbol{\sigma}\right)\exp\left(-i\frac{\omega t}{2}\sigma_z\right)\psi(0) \ . \tag{12}$$

Here $\mathbf{e}'(t)$ appears instead of $\mathbf{e}'(0)$ after rearranging the rotation operators. It is worthwhile demonstrating that the solution for a helical field (Eder and Zeilinger 1976) is also contained in solution (11). If, for the moment, we assume that the neutrons propagate along the $z$ direction, the effect of the rotating field described by equation (6) is similar to a time-independent helical field of extension $D$ along the axis of propagation and with a spatial period characterized by a wave vector $\mathbf{q}$. As mentioned previously, we let the rotating field act during the time interval $0 \leqslant t \leqslant \tau$, where

$$D = \frac{\hbar k}{m}\tau \ . \tag{13}$$

The angle $\omega\tau$ of the rotating field corresponds to the angle $qD$ of the helical field and we have

$$\omega\tau = qD, \qquad \omega = qk\frac{\hbar}{m} \ . \tag{14}$$

With these results we rewrite the solution (11) for the special case $B_{\parallel} = 0$, $B_{\perp} = B$ and obtain

$$\psi(\tau) = \exp\left(-i\frac{qD}{2}\sigma_z\right)\exp\left\{i\frac{\mu B'\tau}{\hbar}(\sin\theta\sigma_x + \cos\theta\sigma_z)\right\}\psi(0) \ . \tag{15}$$

Formula (15) corresponds to the result for a helical field. This is most easily seen from the following expressions ($\mu < 0$ for neutrons) where we use equations (13) and (14):

$$\frac{2\mu B'\tau}{\hbar} = -\left\{(qD)^2+\left(\frac{2Dm\mu B}{\hbar^2 k}\right)^2\right\}^{\frac{1}{2}} \tag{16}$$

$$\cos\theta = -\,qk\left\{(qk)^2+\left(\frac{2m\mu B}{\hbar^2}\right)^2\right\}^{-\frac{1}{2}} . \tag{17}$$

Expressions (15)–(17) contain the rotation angles $\phi_3$, $\phi_2$, and $\phi_1$ defined by Eder and Zeilinger (1976) and this demonstrates the full equivalence of the rotating field solution (11) with the solution for a helical field.

The presence of a mean nuclear scattering potential

$$V_{\mathrm{N}} = \frac{2\pi\hbar^2}{m} b_{\mathrm{c}} N$$

(with coherent scattering length $b_{\mathrm{c}}$ and $N$ nuclei per unit volume) along the distance $D$ of the neutron path leads to an overall phase shift $\chi$ of the wave function. In the usual notation we have

$$\chi = -\frac{V_{\mathrm{N}}\tau}{\hbar} = -\,N\lambda b_{\mathrm{c}} D \tag{18}$$

and the effect of the phase-shifting material is taken into account by a corresponding phase factor

$$\psi(\tau) = \mathrm{e}^{\mathrm{i}\chi}\psi(0) \quad . \tag{19}$$

If both nuclear potential and magnetic fields are present the factor $\mathrm{e}^{\mathrm{i}\chi}$ has to be included in equations (4) and (11).

## 3. Spinor wave functions for an interferometer

### 3.1. *Intensity*

A neutron has two options to propagate through the interferometer which are denoted paths I and II as usual. Quantum mechanics demands that both wave functions $\psi^{\mathrm{I}}$ and $\psi^{\mathrm{II}}$ should add coherently if they are joined again, as is the case in the interferometer. If the neutron emerges into the forward diffracted beam of an ideal focused interferometer, then

$$\psi^{I} = \psi^{II} \tag{20}$$

as shown by Rauch and Suda (1974). To describe the effects of phase-shifting material and/or magnetic fields it is customary to introduce the wave function for the empty interferometer as

$$\psi_i = \psi_i^{I} + \psi_i^{II} = 2\psi_i^{II} \tag{21}$$

and to compare it with the modified wave function

$$\psi_f = \{1+U(\tau)\}\psi_i^{II} \ . \tag{22}$$

To simplify the resulting expressions we assume that only part I of the wave function is modified as expressed by the unitary operator $U$ in equation (22). The extension to the case in which both beams are modified is given by Zeilinger (1976).

With relations (20) and (21) the intensity corresponding to the empty interferometer is

$$I_i = \psi_i^{\dagger}\psi_i = 4 \ . \tag{23}$$

For a purely nuclear phase shifter we use equation (19) and obtain the intensity

$$I_f = \psi_f^{\dagger}\psi_f = \frac{I_i}{2}(1+\cos\chi) \ . \tag{24}$$

Formula (24) shows the well-known modulation with phase angle $\chi$. A similar result for the intensity is derived from equations (4) and (5) for a magnetic field with constant direction

$$I_f = \frac{I_i}{2}\left[1+\cos\left\{\frac{\phi(\tau)}{2}\right\}\right] \ . \tag{25}$$

If both a nuclear potential and a magnetic field with constant direction are present, the intensity is

$$I_f = \frac{I_i}{2}\left\{1+\cos\frac{\phi}{2}\,\cos\chi - \sin\frac{\phi}{2}\,\sin\chi(\mathbf{e}_B\cdot\mathbf{P}_i)\right\} . \tag{26}$$

Here we have introduced the initial polarization $\mathbf{P}_i$ defined as

$$\mathbf{P}_i = \frac{\psi_i^\dagger \boldsymbol{\sigma}\psi}{\psi_i^\dagger \psi_i} . \tag{27}$$

In the special case $\mathbf{e}_B\cdot\mathbf{P}_i = 1$ equation (27) reverts to the simple form

$$I_f = \frac{I_i}{2}\left\{1+\cos\left(\chi+\frac{\phi}{2}\right)\right\} . \tag{28}$$

The intensity for the rotating field derived from formulae (11) and (10) is

$$I_f = \frac{I_i}{2}\left(1+\cos\frac{\phi'}{2}\,\cos\frac{\omega\tau}{2}+e_z'\,\sin\frac{\phi'}{2}\,\sin\frac{\omega\tau}{2}\right) \tag{29}$$

where we have introduced the quantity

$$\phi'(\tau) = \frac{2\mu B'\tau}{\hbar} . \tag{30}$$

$B'$ is defined in equation (10). The corresponding result if, in addition to a rotating field, a nuclear phase shift is to be taken into account has a more complicated appearance. Since $\mathbf{e}'(0)$ lies in the $xz$ plane we set

$$\mathbf{e}'(0) = (\sin\theta,\ 0,\ \cos\theta)$$

and obtain

$$\begin{aligned} I_f = \frac{I_i}{2}\Bigg[1&+\cos\chi\left(\cos\frac{\phi'}{2}\,\cos\frac{\omega\tau}{2}+\sin\frac{\phi'}{2}\,\sin\frac{\omega\tau}{2}\,\cos\theta\right) \\ &-\sin\chi\left\{\sin\frac{\phi'}{2}\,\sin\theta\left(P_{ix}\,\cos\frac{\omega\tau}{2}+P_{iy}\,\sin\frac{\omega\tau}{2}\right)\right. \\ &\left.+\left(\sin\frac{\phi'}{2}\,\cos\frac{\omega\tau}{2}\,\cos\theta-\cos\frac{\phi'}{2}\,\sin\frac{\omega\tau}{2}\right)P_{iz}\right\}\Bigg] . \end{aligned} \tag{31}$$

This is written in such a way that the correspondence with the results for a helical field as given by Eder and Zeilinger (1976) is obvious if we change to their notation by substituting $\theta$ by $\phi_1$, $\phi'$ by $\phi_z$, and $\omega\tau$ by $\phi_3$. The coefficients for the components of $P_i$ appear in a different order since equation (11) is equivalent to a helical field along the $z$ axis as compared with the $y$ axis in the above reference.

We note that in both equations (26) and (31) the presence of a nuclear phase shifter and a magnetic field results in a dependence of the final intensity on the incident polarization. This dependence will be noticeable only for a polarized beam. For an unpolarized beam the terms containing **Pi** will average out.

Unless suitably spaced neutron pulses are used the incident neutrons will enter a rotating field at different phases $\delta$ and consequently $\mathbf{e}(0)$ will have to be phase dependent: $\mathbf{e}(\delta)$. This will result in a zero beam average for the $x$ and $y$ components of $\mathbf{e}$, i.e. all terms containing $\sin\theta$ in equation (31) will vanish

$$I_f = \frac{I_i}{2}\left\{1+\cos\chi\left(\cos\frac{\phi'}{2}\cos\frac{\omega\tau}{2}+\sin\frac{\phi'}{2}\sin\frac{\omega\tau}{2}\cos\theta\right)\right.$$

$$\left.-\sin\chi\left(\sin\frac{\phi'}{2}\cos\frac{\omega\tau}{2}\cos\theta-\cos\frac{\phi'}{2}\sin\frac{\omega\tau}{2}\right)P_{iz}\right\}. \quad (32)$$

In order to describe the effects of a helical magnetic field we have to assume that each neutron enters the field at the same phase ($\delta=0$) and no averaging is necessary.

### 3.2. *Polarization*

The presence of a purely nuclear phase-shifting material in the interferometer leaves the polarization unchanged, i.e. $\mathbf{P}_f = \mathbf{P}_i$, where the definition of $\mathbf{P}_f$ follows by analogy from equation (27). We now consider the case of a magnetic field with constant direction. The spinor wave function (4) inserted into equation (22) leads, together with definition (5), to the following expression for the final polarization:

$$\mathbf{P}_f = \mathbf{e}_B(\mathbf{P}_i\cdot\mathbf{e}_B) + (\mathbf{P}_i\times\mathbf{e}_B)\sin\frac{\phi}{2}$$

$$+ \{\mathbf{e}_B\times(\mathbf{P}_i\times\mathbf{e}_B)\}\cos\frac{\phi}{2}\ . \quad (33)$$

The polarization vector precedes around the direction of the magnetic field. If the effects of a nuclear phase shifter are additionally taken into account, we obtain a more complicated expression for $\mathbf{P}_f$:

$$\mathbf{P}_f = \frac{2}{I_f}\Big((\cos\tfrac{\phi}{2}+\cos\chi)\Big[\mathbf{e}_B(\mathbf{P}_i\cdot\mathbf{e}_B) + (\mathbf{P}_i\times\mathbf{e}_B)\sin\tfrac{\phi}{2}+\{\mathbf{e}_B\times(\mathbf{P}_i\times\mathbf{e}_B)\}\cos\tfrac{\phi}{2}\Big] + \mathbf{e}_B\Big\{(\mathbf{P}_i\cdot\mathbf{e}_B)(1-\cos\chi)(1-\cos\tfrac{\phi}{2})-\sin\chi\,\sin\tfrac{\phi}{2}\Big\}\Big)\,. \tag{34}$$

The corresponding intensity $I_f$ has to be taken from equation (26). Next we consider the polarization for the rotating field case. To simplify the resulting expressions it is of advantage to combine the two unitary operators in equation (11) into one combined rotation:

$$\exp\Big(-i\frac{\omega t}{2}\sigma_z\Big)\exp\Big(i\frac{\mu B' t}{\hbar}\mathbf{e}'(0)\cdot\boldsymbol{\sigma}\Big) = \exp\{i\alpha(t)\mathbf{e}_\alpha(t)\cdot\boldsymbol{\sigma}\}. \tag{35}$$

Defining the rotation angle $\alpha(t)$ by

$$\cos\{\alpha(t)\} = \cos\frac{\phi'(t)}{2}\cos\frac{\omega t}{2} + e_z'(0)\sin\frac{\phi'(t)}{2}\sin\frac{\omega t}{2}\,. \tag{36}$$

we obtain the corresponding axis of rotation as

$$(1-\cos^2\alpha)^{\frac{1}{2}}\mathbf{e}_\alpha(t) = \Big[\Big\{\mathbf{e}'(0)\cos\frac{\omega t}{2} + (\mathbf{e}_z\times\mathbf{e}'(0))\sin\frac{\omega t}{2}\Big\}\sin\frac{\phi'(t)}{2} - \mathbf{e}_z\cos\frac{\phi'(t)}{2}\sin\frac{\omega t}{2}\Big]\,. \tag{37}$$

Setting $\mathbf{e}'(0) = (\sin\theta,\ 0,\ \cos\theta)$ the contents of the square brackets can be rearranged to give

$$(1-\cos^2\alpha)^{\frac{1}{2}}\mathbf{e}_\alpha(t) = \Big\{\mathbf{e}_x\Big(\sin\frac{\phi'}{2}\cos\frac{\omega t}{2}\sin\theta\Big) + \mathbf{e}_y\Big(\sin\frac{\phi'}{2}\sin\frac{\omega t}{2}\sin\theta\Big) + \mathbf{e}_z\Big(\sin\frac{\phi'}{2}\cos\frac{\omega t}{2}\cos\theta-\cos\frac{\phi'}{2}\sin\frac{\omega t}{2}\Big)\Big\}\,. \tag{38}$$

Equation (38) corresponds to $\boldsymbol{\alpha}$ as defined by Eder and Zeilinger (1976). With the help of these definitions we can now use equation (33) to describe the effects of a rotating field on a spinor wave function in the interferometer and the final polarization is given by

$$\mathbf{P}_f = \mathbf{e}_\alpha(\tau)(\mathbf{P}_i\cdot\mathbf{e}_\alpha(\tau)) + (\mathbf{P}_i\times\mathbf{e}_\alpha(\tau))\sin\alpha(\tau)$$

$$+ [\mathbf{e}_\alpha(\tau)\times\{\mathbf{P}_i\times\mathbf{e}_\alpha(\tau)\}]\cos\alpha(\tau) \quad . \qquad (39)$$

Finally we take into account the effects of an additional nuclear phase shift. The expression for $\mathbf{P}_f$ will have the same form as equation (34) but with $\psi/2$ replaced by $\alpha$ and $\mathbf{e}_B$ replaced by $\mathbf{e}_\alpha$:

$$P_f = \frac{2}{I_f}\Big((\cos\alpha+\cos\chi)[\mathbf{e}_\alpha(\mathbf{P}_i\cdot\mathbf{e}_\alpha)+(\mathbf{P}_i\times\mathbf{e}_\alpha)\sin\alpha$$

$$+ \{\mathbf{e}_\alpha\times(\mathbf{P}_i\times\mathbf{e}_\alpha)\}\cos\alpha]$$

$$+ \mathbf{e}_\alpha\{(\mathbf{P}_i\cdot\mathbf{e}_\alpha)(1-\cos\chi)(1-\cos\alpha)-\sin\chi\ \sin\alpha\}\Big) \quad . \qquad (40)$$

For simplicity we have omitted the $\tau$ dependence of $\alpha$ and $\mathbf{e}_\alpha$. The intensity $I_f$ is obtained from equation (31). Expanding the double cross product we collect all terms containing $(\mathbf{P}_i\cdot\mathbf{e}_\alpha)$

$$\mathbf{e}_\alpha(\mathbf{P}_i\cdot\mathbf{e}_\alpha)[(\cos\alpha+\cos\chi)(1-\cos\alpha)+(1-\cos\chi)(1-\cos\alpha)]$$

$$= \mathbf{e}_\alpha(\mathbf{P}_i\cdot\mathbf{e}_\alpha)(1-\cos^2\alpha) \qquad (41)$$

and rewrite equation (40) in the form

$$P_f = \frac{2}{I_f}[\mathbf{e}_\alpha(\mathbf{P}_i\cdot\mathbf{e}_\alpha)(1-\cos^2\alpha)$$

$$+ (\cos\alpha+\cos\chi)\{(\mathbf{P}_i\times\mathbf{e}_\alpha)\sin\alpha+\mathbf{P}_i\ \cos\alpha\}$$

$$- \mathbf{e}_\alpha\sin\chi\ \sin\alpha] \qquad (42)$$

with $I_f$ given by equation (31).

With regard to the phase $\delta$ of the rotating field the same remarks as given at the end of §3.1 also apply to the final polarization.

### 3.3. *Application to special simple cases*

The first measurement using a magnetic field of constant direction (Rauch *et al.* 1975) included an analysis of the intensity oscillations using equation (25). With a neutron wavelength of $\lambda = 2 \times 10^{-8}$ cm ($k \approx 3 \times 10^{8}$ cm$^{-1}$, $v \approx 1.9 \times 10^{5}$ cm s$^{-1}$) a time $\tau \approx 5.3 \times 10^{-6}$ s results if we assume $D = 1$ cm. The magnetic moment is $\mu = 9.66 \times 10^{-24}$ erg G$^{-1}$. If the field (effective along a distance $D$ of path I and with direction parallel to the $z$ axis) is raised from $B = 0$ to $B \approx 130$ G we see from equation (5) that the angle $\phi$ changes from 0 to $-4\pi$ and this is equivalent to one full oscillation according to equation (24). The angle $\chi$ corresponding to an aluminium phase shifter follows from equation (18) for the above wavelength as $\chi \approx -132\pi$ which is again calculated for $D = 1$ cm.

A more interesting situation arises if we consider the interplay of the magnetic and nuclear phase shifts as given in formulae (26) and (34). For an unpolarized neutron beam the average over $\mathbf{P}_i$ will be zero and the expression for the intensity contains an additional modulation due to the nuclear phase shift

$$I_f = \frac{I_i}{2}(1+\cos\frac{\phi}{2}\cos\chi) \ . \tag{43}$$

The corresponding measurements have been reported by Badurek *et al.* (1976). From formula (34) for $\mathbf{P}_f$ we find in this case a final polarization even though $\mathbf{P}_i$ averages to zero:

$$\mathbf{P}_f = -\frac{\sin\chi\ \sin(\phi/2)}{1 + \cos(\phi/2)\cos\chi}\,\mathbf{e}_B \ . \tag{44}$$

If the angles are adjusted such that, for instance, $\phi/2 = \chi + \pi$, equation (44) reduces to $\mathbf{P}_f = \mathbf{e}_B$. We now assume a polarized neutron beam $\mathbf{P}_i = \mathbf{e}_z$ and a magnetic field parallel to the incident polarization $\mathbf{e}_B = \mathbf{e}_z$. The intensity is then given by equation (28) and the polarization remains unchanged. If, on the other hand, $\mathbf{e}_B = \mathbf{e}_x$ and $\chi$ is chosen such that $\cos\chi = 1$, we obtain from equation (34) the result for a purely magnetic field:

$$\mathbf{P}_f = \mathbf{e}_y \sin\frac{\phi}{2} + \mathbf{e}_z \cos\frac{\phi}{2} \quad . \tag{45}$$

In this case the final polarization lies in the $yz$ plane and appears to be rotated about $\mathbf{e}_B$ by the angle $\phi/2$.

Finally, we consider the formulae for a rotating magnetic field with a nuclear phase shifter. A way to simplify the expressions is by adjusting $B_{\parallel}$ such that $B'_z = 0$:

$$B_{\parallel} + \frac{\hbar\omega}{2\mu} = 0 \quad . \tag{46}$$

Then from equation (10) we find $B' = B_{\perp}$ and, furthermore, $\theta = \pi/2$ or $\mathbf{e}' = \mathbf{e}_x$ in formula (31). For an unpolarized neutron beam equation (31) reduces to

$$I_f = \frac{I_i}{2}\left(1+\cos\chi \cos\frac{\phi'}{2} \cos\frac{\omega\tau}{2}\right) \tag{47}$$

and from equations (36) and (37) we obtain

$$\cos\alpha(\tau) = \cos\frac{\phi'}{2} \cos\frac{\omega\tau}{2} , \qquad \phi' = \frac{2\mu B_{\perp}\tau}{\hbar} \tag{48}$$

and

$$\sin\alpha(\tau)\mathbf{e}_{\alpha}(\tau) = \left\{\left(\mathbf{e}_x \cos\frac{\omega\tau}{2}+\mathbf{e}_y \sin\frac{\omega\tau}{2}\right)\sin\frac{\phi'}{2} - \mathbf{e}_z \cos\frac{\phi'}{2} \sin\frac{\omega\tau}{2} \right\} . \tag{49}$$

The unit vector $\mathbf{e}_{\alpha}$ defines the direction of the final polarization for an unpolarized incident beam:

$$\mathbf{P}_f = - \frac{\sin\chi \sin\alpha(\tau)}{1 + \cos\chi \cos\alpha(\tau)} \mathbf{e}_{\alpha}(\tau) \quad . \tag{50}$$

In contrast to equation (44) the direction $\mathbf{e}_{\alpha}(\tau)$ is now a function of $\tau$, but if the neutrons enter the rotating field randomly at all phases $\delta$ the $x$ and $y$ components in equation (40) will average out and $\mathbf{P}_f$ will be parallel to $\mathbf{e}_z$. To avoid the complications introduced by the nuclear phase shift $\chi$ we could assume $\chi$ to be compensated to a multiple of $2\pi$. This would lead us back to formulae (29) and (49) for a rotating field of purely magnetic origin.

## References

BADUREK, G., RAUCH, H., ZEILINGER, A., BAUSPIESS, W., and BONSE, U. (1976). *Phys. Rev. D* **14**(5), 1177–1181.

BERNSTEIN, H.J. (1979). This conf. Paper II.4.

EDER, G. and ZEILINGER, A. (1976). *Nuovo Cim. B* **34**, 76–90.

LANDAU, L.D. and LIFSHITZ, E.M. (1965). *Quantum mechanics*. Pergamon Press, Oxford.

RAUCH, H. and SUDA, M. (1974). *Phys. Status Solidi A* **25**, 495.

RAUCH, H., ZEILINGER, A., BADUREK, G., WILFING, A., BAUSPIESS, W., and BONSE, U. (1975). *Phys. Lett. A* **54**, 425–427.

SCHWINK, CH. and SCHÄRPF, O. (1975). *Z. Phys. B* **21**, 305–311.

WERNER, S.A., COLELLA, R., OVERHAUSER, A.W., and EAGEN, C.F. (1975). *Phys. Rev. Lett.* **35**, 1053.

ZEILINGER, A. (1976). *Z. Phys. B* **25**, 97–100.

# 7. NEUTRON SPIN INTERFERENCE EFFECTS

F. MEZEI

*Institut Laue–Langevin, 156X, 38042 Grenoble cedex, France*

## 1. Introduction

In spite of its familiarity it is still useful to recall the definition of particle interference: if, in a given experimental situation, the quasi-stationary solution $\Psi(r,t)$ of the wave function equation of motion turns out to be approximately the superposition of two (or more) functions $\Phi_i(r,t)$ (which themselves are quasi-stationary solutions of the equation of motion for slightly modified experimental conditions), then we have interference in the measurement of a quantity $A$ whose expectation values contain products of the type $\langle\Phi_i|A|\Phi_j\rangle$, $i \neq j$.

The simplest example of an interference effect with neutrons is the measurement of the $x$ component of the neutron spin in a magnetic field of direction $z$. Indeed, the wave function can be given as a superposition of the two stationary solutions spin-up and spin-down:

$$|\Psi\rangle = \Phi_+|\uparrow\rangle + \Phi_-|\downarrow\rangle \tag{1}$$

where $\Phi_\pm(r,t)$ are spatial wave functions. Then

$$\langle S_x\rangle = \langle\Psi|\tfrac{1}{2}\sigma_x|\Psi\rangle = \tfrac{1}{2}(\langle\Phi_+|\Phi_-\rangle + \langle\Phi_-|\Phi_+\rangle)\,. \tag{2}$$

An interference effect between the spin-up and spin-down states can thus be observed if $\Phi_+$ and $\Phi_-$ are not orthogonal in the same way as spatial interference occurs between non-orthogonal spin states only.

In practice this effect can be observed by measuring Larmor precessions, for example as described by Mezei (1972a). Consider first a neutron beam polarized initially in the direction of the field $B$ which lies along the $Z$ direction. We can produce a wave function of the form (1) by applying a time-dependent homogeneous magnetic field to produce a 90° spin turn. (This is the well-known $\pi/2$ pulse in nuclear magnetic resonance.) This way the spatial part of the wave function is unaffected and we have

$$|\Psi\rangle = \Phi(r,t)\left\{\exp\left(\frac{i}{\hbar}\mu Bt\right)|\uparrow\rangle + \exp\left(-\frac{i}{\hbar}\mu Bt\right)|\downarrow\rangle\right\} \tag{3}$$

which is the solution of the Hamiltonian

$$\hat{H} = \hat{H}_0(r) - \hat{\mu}B$$

where $\hat{H}_0$ is the free-particle term and $\hat{\mu}$ is the neutron magnetic moment operator. From equation (2) it immediately follows that

$$\langle S_x(t)\rangle = \tfrac{1}{2}\|\Phi\|\left(\exp\left(-\frac{i}{\hbar}2\mu Bt\right)+\text{c.c}\right) = \cos\frac{2\mu B}{\hbar}t\ . \tag{4}$$

This behaviour of the observable $S_x$ is called Larmor precession.

## 2. The longitudinal Stern–Gerlach effect

Our main interest here is another conceptually different way of observing Larmor precessions. Wave functions of the form (1) can also be produced by the application of a stationary but spatially inhomogeneous magnetic field. For example assume that a neutron beam initially polarized in the $x$ direction travels into a field $\mathbf{B}\|\mathbf{z}$, necessarily through a field-free region. The Zeeman energy will act as a conservative potential and the solution of the equation of motion leads to de-accelerated spin-up and accelerated spin-down components. This is just the Stern–Gerlach effect with the only difference that here the magnetic accelerations are not perpendicular but parallel to the beam velocity. If the neutron is described by a wave packet with a momentum distribution of width $\Delta k$ sharply peaked around the centre of gravity momentum value $k$ ($\Delta k \ll k$), these (de)acceleration effects are described by the momentum changes $\pm\,\delta k(r) \approx m\mu B(r)/\hbar^2 k$ ($m$ is the neutron mass) with the total energy being conserved. This leads to accumulated phase shifts as compared with the situation without any field present:

$$\beta_{\pm} = \mp \int \delta k(r)\mathrm{d}r = \mp\beta \ . \qquad (5)$$

To proceed we have now to distinguish two limiting cases.

If $\delta k \ll \Delta k$, i.e. the change of the neutron group velocity is negligible compared with the initial velocity spread of the packet, we can write

$$\Phi_{\pm}(r,t) = \Phi(r,t)\exp(\mathrm{i}\beta_{\pm}) \ .$$

If $S_x$ is now measured as a function of the position $r$, e.g. as described by Mezei (1972a), we obtain

$$\begin{aligned}\langle S_x(r)\rangle &= \langle S_x\delta(r)\rangle \\ &= |\Phi|^2\tfrac{1}{2}(\exp\{-\mathrm{i}(\beta_+-\beta_-)\}+\mathrm{c.c.}) \propto \cos(\beta_+-\beta_-) \\ &= \cos 2\beta .\end{aligned}$$

It immediately follows from equation (5) that in a region where $B$ is approximately constant this gives

$$\langle S_x(r)\rangle \propto \cos(2\delta kr+\mathrm{const.}),$$

i.e. a spatial oscillation, which as seen by the moving neutron shows the same Larmor frequency as equation (4) since the packet group velocity is $\hbar k/m$. Thus it is seen that in this case of negligible magnetic acceleration effects the longitudinal Stern–Gerlach effect can still be fully observed as interference 'beating' between the $\Phi_+$ and $\Phi_-$ wave components, which is the Larmor precession.

If, on the other hand, $\delta k > \Delta k$, $\Phi_+$ and $\Phi_-$ will describe two wave packets with significantly different group velocities which sooner or later become separated in real space and time, leading to a classically observable longitudinal Stern–Gerlach effect. Obviously, at the same time the interference between $\Phi_+$ and $\Phi_-$ is destroyed and $S_x(r) \equiv 0$, i.e. no Larmor precession takes place.

Another possible way of observing the longitudinal Stern–

Gerlach effect, which is applicable in both of the above cases, consists of measuring the accumulated phase shifts given by equation (5) via interference with a reference eikonal wave which does not cross the magnetic field, rather than of comparing $\Phi_+$ and $\Phi_-$ with each other in one or other of the above ways. Here the interference periodicity is obviously cos $\beta$ as opposed to cos $2\beta$ in the case of Larmor precession. Such experiments have been performed recently by Werner *et al.* (1975), Rauch *et al.* (1975), and Klein and Opat (1976). It was incorrectly claimed that these experiments provide a direct observation or a new consistency check of the particular behaviour of spinors under space rotation, represented here by the Larmor precessions. In reality, as we have seen, we are only concerned with the well-known 'classical' Stern–Gerlach effect; invariably, whether there are Larmor precessions ($\delta k \ll \Delta k$) or not ($\delta k > \Delta k$) or whether we look at Bosons or Fermions, the interference pattern is actually the Fourier transform of the Stern–Gerlach splitting scheme.

Moreover it is obvious that for spin-½ particles, if $\beta_+ = -\beta_-$ (and $\delta k \ll \Delta k$, which is nearly always true), the simple observation of Larmor precessions which give cos $(\beta_+ - \beta_-)$ = cos $2\beta_+$ is equivalent to the much more complicated and difficult split-beam interference experiments concerned with cos $\beta_+$ = cos $\beta_-$.

## 3. Neutron wave functions in an adiabatically rotating magnetic field

The exact solutions (Landau and Lifschitz 1958, Eder and Zeilinger 1976) of the equation of motion for the spin wave functions in a uniformly rotating magnetic field show that in the adiabatic limit (i.e. when the angular velocity $\Omega$ of the field rotation is negligible compared with the angular velocity $\omega_L$ of the Larmor precession of the particle in the given field) the fermion wave functions show an extra $\pi$ phase shift (sign change) under 360° rotation of the field while this effect is absent for bosons. (Remember that in the adiabatic limit if the particle spin is initially parallel to the magnetic field it will stay parallel to it all the time.) Bernstein (1969) and Moore (1970) suggested that the observation of this effect would give a non-trivial consistency check of the spinor character of fermion spins. Such an experiment was proposed by the author to the ILL (Mezei 1972a) but the proposal was not followed up officially. In what follows an improvised realization is described.

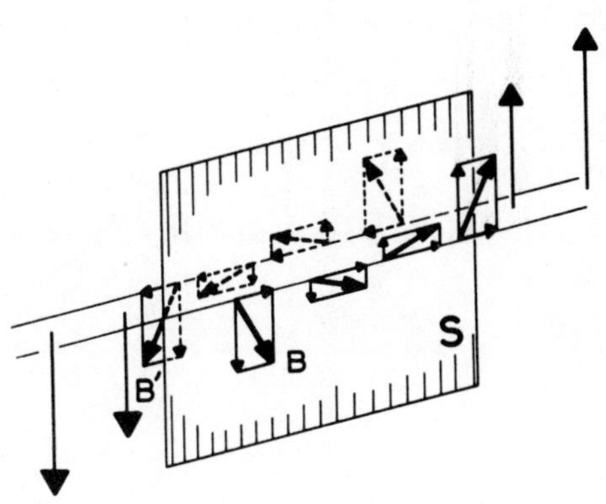


FIG. 1. The rotating magnetic field set-up.

The magnetic set-up is shown in Fig. 1. The layer S consists of thin vertical wires carrying a current. Thus opposite horizontal magnetic fields are created in front of (continuous horizontal arrows) and behind (broken horizontal arrows) this current sheet. The vertical arrows represent the stray field of two oppositely oriented permanent magnets situated on the left and the right respectively of what is shown in the figure. The resulting field pattern in front of and behind the current layer is represented by the solid and broken $B$ and $B'$ vectors respectively. Therefore $B$ and $B'$ rotate through +180° and -180° respectively, which produces the relative effect of a 360° rotation. Experimentally the current sheet was one side of a rectangular coil wound with ∅ = 0.08 mm enamelled copper wire. The field of the return side of the coil was compensated by a larger field bobbin external to the coil. The absolute value of $B$ and $B'$ was between 5 and 6 Oe in the $d$ = 5 cm rotation region; thus for neutrons of wavelength 12 Å travelling horizontally across this pattern $\Omega/\omega_L = 0.2$, which represents a sufficiently good approximation of the adiabatic limit. At the same time the magnetic symmetry assures that the Zeeman energy terms $\exp(-\frac{1}{2}i\omega_L t)$ are sufficiently equal on both sides of the current sheet.

The phase difference between the two wave function components on both sides of the coil have been measured by a Fresnel interferometer which simply consists of two vertical Cd slits of 0.02 mm width separated by a distance of 12.5 m (Fig. 2).

FIG. 2. The Fresnel type neutron interferometer.

The ± 15% monochromatic incoming beam was produced by a helical slot velocity selector on a cold neutron guide which unfortunately was too long (55 m, giving a loss factor of 5–10 above 10 Å). Figure 3 shows the shadow pattern of the current sheet aligned parallel to the slits measured at a neutron wavelength of 6 Å, as indicated in Fig. 2 with the coil being the moving obstacle. The Fresnel diffraction peaks appear clearly and the width of the layer has been found to be 0.115 mm. Thus the periodicity of the interference effects from both sides

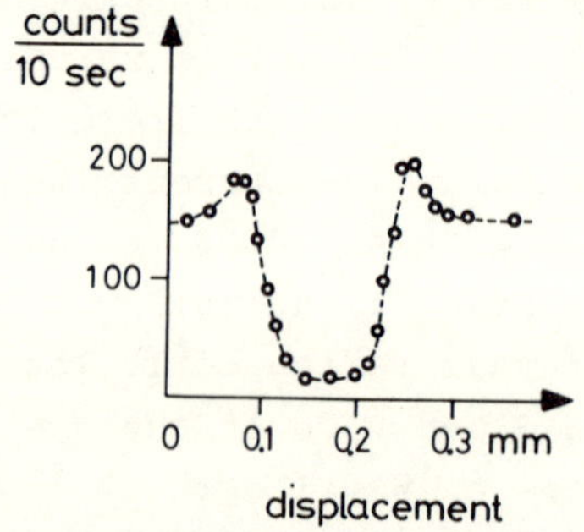


FIG. 3. Fresnel shadow pattern of the layer of wires in Fig. 1.

of the sheet has to be about 0.04 mm at $\lambda$ = 12 Å, i.e. at the limit of the resolution of the system.

Two magnetic field configurations were used.

(1) Longitudinal Stern—Gerlach experiment: a strong external horizontal field is applied and in this way the field rotation is the same in both sides and the current sheet can be used to modulate the absolute value of the field, i.e. to build up a phase shift between the two sides, as seen above. This was used to check the phase and the amplitude of the interference.

(2) Adiabatic field rotation experiment: as described above.

Considering the presence of a fast neutron background, the best chance of observing a two-side interference was at the steep side of the shadow pattern, where only one of the wave amplitudes is small. All measurements were performed once at the 6 Å wavelength (where no interference can occur) to check the absence of geometrical deformation effects and at 12 Å. The first point given in Fig. 4 is the reference with no magnetic phase difference. The next five points show the longitudinal Stern—Gerlach effect with the broken line just representing the calculated period. The last point corresponds to ± 180° adiabatic rotation, a situation clearly showing the sign change we are looking for.

The data represented here consist of one-third of those collected in a 48 h period. The experiment will be repeated with some improvements.

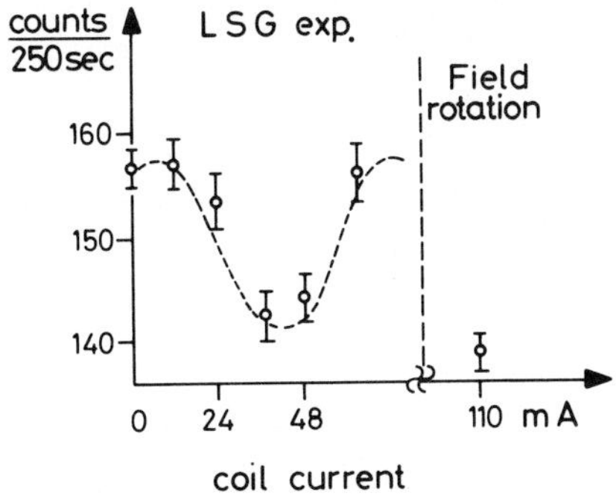


FIG. 4. Measured interference effects in the longitudinal Stern—Gerlach and in the field rotation configurations.

### Acknowledgments

I acknowledge with pleasure the interesting and profitable discussions with many colleagues, especially with Herb Bernstein and Sam Werner, and the technical assistance of Paul Dagleish.

### References

BERNSTEIN, H.J. (1969). *Sci. Res. (N.Y.)* **4**, 32.

EDER, G. and ZEILINGER, A. (1976). *Nuovo Cim.* **34B**, 76.

KLEIN, A.G. and OPAT, G.I. (1976). *Phys. Rev. Lett.* **37**, 238.

LANDAU, L.D. and LIFSCHITZ, E.M. (1958). *Quantum mechanics*. Addison-Wesley, Reading, Mass.

MEZEI, F. (1972a). *Z. Phys.* **255**, 146.

MEZEI, F. (1972b). Research proposal to the Institut Laue—Langevin (unpublished).

MOORE, G.T. (1970). *Am. J. Phys.* **38**, 1177.

RAUCH, H., ZEILINGER, A., BADUREK, G., WILFING, A., BAUSPIESS, W., and BONSE, U. (1975). *Phys. Lett. A* **54**, 425.

WERNER, S.A., COLELLA, R., OVERHAUSER, A.W., and EAGEN, C.F. (1975). *Phys. Rev. Lett.* **35**, 1053.

# 8. THE REPRESENTATION OF NEUTRON POLARIZATION

J. BYRNE

*School of Mathematical and Physical Sciences, University of Sussex, Brighton, United Kingdom*

## 1. Polarization of spin-½ particles

The spin of a particle of finite rest mass can be defined as the angular momentum in its rest frame. For non-relativistic spin-½ particles ($E/mc^2 \rightarrow 1$) the spin operators $\mathbf{S} = \frac{1}{2}\hbar\sigma$ operate in a two-dimensional complex spinor space and the expectation values $\langle\sigma_i\rangle$ form the components of an axial vector called the 'polarization'. For relativistic particles there are a number of alternative and equally valid generalizations of this non-relativistic polarization vector; for example the four-vector polarization (Bargmann, Michel, and Telegdi 1959, Fierz and Telegdi 1970) and the antisymmetric tensor polarization (Rose and Good 1961, Fradkin and Good 1961a,b).

In the present discussion we shall be concerned with the state of spin orientation of beams of particles rather than with the relativistic motion of the spin, and the term polarization will be used in the sense described below. In the Pauli–Dirac representation the spin operators are denoted by $\mathbf{S} = \frac{1}{2}\hbar\sigma'$ where the $\sigma_i'$ matrices operate in a four-dimensional complex spinor space and are the space components of the anti-symmetric tensor $\sigma^{\mu\nu} = (1/2i)[\gamma^{\mu},\gamma^{\nu}]$ (Hamilton 1959). Denoting by $|\pm\frac{1}{2}\rangle$ the four-component wave functions for free particles having spin components $\pm\frac{1}{2}\hbar$ respectively along the momentum (taken as the $l$-axis), the polarization state of a beam of particles may be described as an incoherent superposition of orthogonal pure states

$$\begin{aligned} |a\rangle &= u_1|\tfrac{1}{2}\rangle + u_2|-\tfrac{1}{2}\rangle \\ |b\rangle &= u_2^*|\tfrac{1}{2}\rangle - u_1^*|-\tfrac{1}{2}\rangle \end{aligned} \qquad (1)$$

with probabilities $p_a$ and $p_b$ respectively. The density matrix for the beam can then be written

$$\rho = \sum_{i=a}^{b} |i\rangle p_i \langle i| = \tfrac{1}{2}(P_0\sigma_0 + P\cdot\sigma) \qquad (2)$$

where $P_0$ is the intensity, $\sigma_0$ the unit matrix, and $\sigma_i$ the Pauli spin matrices

$$\sigma_1(\sigma_z) = \begin{pmatrix} 1 & 0 \\ 0 & -1 \end{pmatrix}$$

$$\sigma_2(\sigma_x) = \begin{pmatrix} 0 & 1 \\ 1 & 0 \end{pmatrix} \qquad (3)$$

$$\sigma_3(\sigma_y) = \begin{pmatrix} 0 & -i \\ i & 0 \end{pmatrix}$$

The polarization components for a beam in a mixed state as defined in equation (2) are

$$P_1 = (p_a - p_b)(|u_1|^2 - |u_2|^2) = \frac{P_0 \overline{\langle s_1\rangle}}{\tfrac{1}{2}\hbar}$$

$$P_2 = (p_a - p_b)(u_1u_2^* + u_1^*u_2) = \frac{P_0 \overline{\langle s_2\rangle}(E/mc^2)}{\tfrac{1}{2}\hbar} \qquad (4)$$

$$P_3 = (p_a - p_b)\{\mathrm{i}(u_1u_2^* - u_1^*u_2)\} = \frac{P_0 \overline{\langle s_3\rangle}(E/mc^2)}{\tfrac{1}{2}\hbar}$$

where $\overline{\langle s_i\rangle}$ represents the expectation value of the spin component $s_i$ averaged over the particles in the beam.

It is evident from equation (4) that the quantities $P_i$ reduce to the components of an axial vector in the non-relativistic limit although this is clearly not the case for relativistic particles. However, it is a useful notion to visualize $\mathrm{P}$ as a vector in an abstract polarization space.

Furthermore, since the four quantities $(P_0,\mathbf{P})$ combine in the relationship

$$P_0^2 - \mathbf{P}^2 = P_0^2(1-p^2) \leqslant P_0^2 \tag{5}$$

where

$$p = |\mathbf{P}|/P_0 \tag{6}$$

is a measure of the degree of beam polarization, the quantity $P_0^2 - |\mathbf{P}|^2$ enters in a natural way into any treatment of the interactions of polarized beams. Thus some insight into the behaviour of beam polarization can be gained by characterizing it by a four-vector $(P_0,\mathbf{P})$ in an abstract four-dimensional polarization space of indefinite metric.

The notion of representing the polarization of the beam by a four-vector $(P_0,\mathbf{P})$ derives from the work of Stokes who first applied these ideas to the analyses of polarized light (Stokes 1852). The polarization calculus has been developed by Jones for states of pure polarization and by Mueller for states of partial polarization (Jones 1956). Similar techniques have also been applied to polarization analysis of spin-½ particles (McMaster 1961, Byrne 1971, Byrne and Farago 1971).

## 2. Transfer matrices

We now represent the pure polarization state $|a\rangle$ (equation (1)) in 'spinor' form

$$\mathbf{u} = u_1\begin{pmatrix}1\\0\end{pmatrix} + u_2\begin{pmatrix}0\\1\end{pmatrix} \tag{7}$$

where we have chosen the basis states $\begin{pmatrix}1\\0\end{pmatrix}$ and $\begin{pmatrix}0\\1\end{pmatrix}$ to be states of definite spin component with respect to the momentum. One must emphasize again that $\mathbf{u}$ is a spinor not in space—time but in the abstract polarization space introduced in §1 and $\mathbf{u}$ is in general not a state of definite spin-component in any direction. This is true only in the non-relativistic limit which for present purposes is the most important situation. Furthermore, it is not even necessary to restrict the axis of quantization to be parallel to the momentum although this is very convenient in many cases because then $u_1$ and $u_2$ may be identified with the large components of the Dirac spinor.

Provided we restrict the discussion to processes involving coherent scattering or pure absorption only the transformation of the pure polarization state of a neutron from **u** to **u**′ may be described by the equation

$$\mathbf{u}' = M\mathbf{u} \tag{8}$$

where $M$ is a $2 \times 2$ transfer matrix. Since matrices generated from a given matrix by similarity transformations are equivalent, one is led to consider the invariants of such transformations, namely the trace $T$, the determinant $\Delta$, and of course the matrix rank. In general $2 \times 2$ matrices can have rank 0, 1, or 2. Rank zero matrices are null matrices while rank 1 matrices have vanishing determinant and are of the nature of projection operators. Such matrices are called polarizers since the output beam is fully polarized when either or both of the eigenvalues of $M$ vanish, the sense of the polarization being determined by the non-vanishing eigenvalue.

Within the present context we shall restrict the discussion to matrices of rank 2, i.e. those for which $\Delta$ does not vanish. Since the unit matrix $\sigma_0$ falls in this category and each such matrix possesses an inverse, they constitute a group with respect to multiplication. Further, since $\Delta \neq 0$ we can divide each element of $M$ by $\Delta^{\frac{1}{2}}$, thus reducing the number of real parameters required to specify the matrix from eight to six ($\Delta$ is complex in general). The resultant unimodular matrices can be expressed as elements of the Pauli ring

$$M_2 = \sum_{i=0}^{4} k_i\sigma_i = \begin{pmatrix} k_0+k_1 & k_2-\mathrm{i}k_3 \\ k_2+\mathrm{i}k_3 & k_0-k_1 \end{pmatrix} \tag{9}$$

where the unimodular condition is expressed as a relation between the complex coefficients

$$k_0^2 - (k_1^2+k_2^2+k_3^2) = 1 \ . \tag{10}$$

The advantage of this formulation of the problem is that the unimodular group $C_2$ is homomorphic to the group $L_P$ of proper homogeneous Lorentz transformations in space–time. The properties of this group are of course well known and afford us a geometrical picture of the operations in the abstract polarization space of the various transfer matrices $M_2$ which we encounter. For example every Lorentz transformation is characterized by six real parameters, i.e. one parameter each to specify the degree of 'uniform relative velocity' and the degree of rotation and two parameters each to specify the directions of these operations. For the polarization transfer matrices these two 'directions' coincide in most cases of practical interest and only four real parameters are needed, i.e. two to specify a direction together with the real and imaginary parts of the scattering amplitude. Such matrices are said to be normal and are unitarily similar to diagonal matrices. If the scattering amplitude is real, three real parameters suffice and the transfer matrix is itself unitary.

To find how the components of the Stokes polarization transform under the matrix operation (9) we form the direct product $M_2 \times M_2{}^*$ and then subject it to a similarity transformation generated by the unitary transformation

$$u = \tfrac{1}{2}\begin{pmatrix} 1 & 0 & 0 & 1 \\ 1 & 0 & 0 & -1 \\ 0 & 1 & 1 & 0 \\ 0 & i & -i & 0 \end{pmatrix}. \tag{11}$$

We then find that the matrix

$$M_4 = U(M_2 \times M_2{}^*)U^{-1} \tag{12}$$

is the required transfer matrix. These matrices constitute a $4 \times 4$ representation of $L_P$ (a 'tensor' representation as opposed to the true 'spinor' representation $M_2$); they have an algebraic form which is identical with those describing pure rotations and pure Lorentz transformations of four-vectors in space–time.

As an illustration of the above ideas we can write down

some very general matrices, e.g. the matrix describing the rotation of the polarization through an angle $\phi$ about the axis whose direction cosines are $n_1$, $n_2$, and $n_3$ $(n_1{}^2+n_2{}^2+n_3{}^2 = 1)$ is

$$M_2 = \begin{pmatrix} \cos\phi/2 - \mathrm{i}\, n_1 \sin\phi/2, & -\mathrm{i}(n_2 - \mathrm{i}n_3)\sin\phi/2 \\ -\mathrm{i}(n_2+\mathrm{i}n_3)\sin\phi/2, & \cos\phi/2 + \mathrm{i}\, n_1 \sin\phi/2 \end{pmatrix}. \quad (13)$$

The angle $\phi$ is related to the real part of the scattering amplitude in the process concerned; if there is an absorptive part $\phi$ is complex and $M_2$ is no longer unitary but still normal. The additional parameter describes a pure Lorentz transformation in the polarization space.

In the subsequent discussion we give three applications of these ideas to problems concerned with the motion of the neutron polarization vector in various circumstances. The first two cases concern coherent parity-violating effects (Michel 1964, Stodolski 1974, Kabir, Karl, and Obryk 1974) which give rise to 'weak optical activity' for neutrons in material media and a fictitious parity-violating effect associated with the motion of the neutron magnetic moment in a helical magnetic field analogous to optical activity generated in birefringent media by mechanical torsion (Byrne 1974). The third example concerns the behaviour of the neutron polarization vector in the context of an interference experiment (Rauch *et al.* 1975, Werner *et al.* 1975).

## 3. Coherent parity violation for neutrons

We suppose that the polarization state of the neutron is represented by the spinor $\mathbf{u}(o)\exp(\mathrm{i}kz)$ where $k = \omega/c$ in vacuum and in a medium of refractive index $n(\omega)$

$$k = n(\omega)\frac{\omega}{c} \quad (14)$$

where for $n(\omega) \ll 1$

$$n(\omega) = 1 + 2\pi\left(\frac{c}{\omega}\right)^2 Nf(\omega). \quad (15)$$

$N$ is the number of scattering nuclei per unit volume and

$f(\omega)$ is the coherent forward-scattering amplitude. If there is a weak interaction present then we have, for momentum $p$

$$f(\omega) = f_0(\omega) + \frac{1}{p} \cdot \langle \sigma \cdot p \rangle f_1(\omega) \tag{16}$$

with corresponding indices of refraction

$$n^{\pm}(\omega) = n_0(\omega) \pm n_1(\omega) \tag{17}$$

where the ± signs refer to positive and negative helicity states respectively.

If we assume that the neutron enters the sample at $z = 0$ and leaves at $z = z$ the transmitted neutron is described by the spinor

$$\begin{pmatrix} u_1(z) \\ u_2(z) \end{pmatrix} = \begin{pmatrix} \exp(in^{+}\omega z/c) & 0 \\ 0 & \exp(in^{-}\omega z/c) \end{pmatrix} \begin{pmatrix} u_1(0) \\ u_2(0) \end{pmatrix} . \tag{18}$$

The determinant of the transfer matrix is

$$\Delta = \exp(2in_0\omega z/c) \tag{19}$$

so that the unimodular transfer matrix for the process is

$$M_2 = \begin{pmatrix} \exp(n_1\omega z/c) & 0 \\ 0 & \exp(-in_1\omega z/c) \end{pmatrix} . \tag{20}$$

According to equation (13) this describes a rotation of the polarization vector about the $z$ axis through an angle $\phi$ where

$$\tfrac{1}{2}\phi = -n_1\omega z/c \ . \tag{21}$$

Equation (21) gives the effect known as 'weak optical activity' although in true optical activity the plane ofpolarization rotates through half the angle given by equation (21); this arises from the fact that a rotation of the plane of polarization through an angle $\pi$ reproduces the same polarization state whereas a rotation of $2\pi$ is needed to reproduce the same polarization state for neutrons. This phenomenon is connected with the two-valued nature of spinor representations.

We can now extend the analysis by supposing that $f(\omega)$ has an imaginary part:

$$f(\omega) = \mathcal{R}(\omega) + i\mathcal{I}(\omega) \tag{22}$$

where $\mathcal{I}(\omega)$ is related to the total cross-section $\sigma_t(\omega) = \sigma_s + \sigma_a$ through the optical theorem

$$\sigma_t(\omega) = \frac{4\pi\omega}{c}\mathcal{I}(\omega) = \sigma_0(\omega) \pm \sigma_1(\omega) \; . \tag{23}$$

In this case the determinant of the transfer matrix has the value

$$\Delta = \exp(2in_0\omega z/c - 2N\sigma_0(\omega)z) \tag{24}$$

and the unimodular matrix is

$$M_2 = \begin{pmatrix} \exp(in_1\omega z/c - N\sigma_1(\omega)z) & 0 \\ 0 & \exp(-in_1\omega z/c + N\sigma_1(\omega)z) \end{pmatrix} \tag{25}$$

where $\sigma_1(\omega)$ is the weak contribution to the total cross-section. The factors $\exp(\pm N\sigma_1 z)$ give rise to a differential absorption between positive and negative helicity states characterized by the parameter

$$\beta = 2N\sigma_1(\omega)z \; . \tag{26}$$

By analogy with the 'weak optical activity' this may be termed 'weak circular dichroism'; according to equation (23) the effect is vanishingly small at low energy.

In experiments of course it is the Stokes parameters $(P_0, \mathbf{P})$ that are measured and their transformation matrix $M_4$ is obtained by applying equations (12) and (13) to the unimodular transfer matrix $M_2$ given in equation (15). The result is

$$\begin{pmatrix} P_0 \\ P_x \\ P_y \\ P_z \end{pmatrix}_z = \exp(-N\sigma_0(\omega)z) \begin{pmatrix} \cosh\beta & 0 & 0 & \sinh\beta \\ 0 & \cos\phi & \sin\phi & 0 \\ 0 & -\sin\phi & \cos\phi & 0 \\ \sinh\beta & 0 & 0 & \cosh\beta \end{pmatrix} \begin{pmatrix} P \\ P_x \\ P_y \\ P_z \end{pmatrix}_0 . \quad (27)$$

We note that the transfer matrix reduces to the direct sum of two 2 × 2 unimodular matrices characteristic of pure rotations and pure 'Lorentz transformations' respectively. Writing the result in the form (27) brings out most clearly that not only is an initial transverse polarization ($P_x(0) \neq 0$) rotated but an initial longitudinal polarization ($P_z(0) \neq 0$) is detected through the presence of the element sinh β, and an initially unpolarized beam is polarized through the presence of the diagonally opposite element sinh β.

## 4. Neutron spin rotation in helical magnetic fields

In this example we estimate the rotation of the neutron spin about its momentum, supposedly directed along the $y$ axis, when it interacts non-adiabatically with a helical magnetic field which is twisted about the $y$ axis. Because the magnetic field **B** is a solenoidal vector a helical field cannot be established in free space, but this restriction does not apply in material media which can support internal currents. It turns out that the sense of spin rotation is independent of the sign of the field and in consequence is indistinguishable from a weak parity-violating effect. In this characteristic it contrasts with the action of a residual longitudinal magnetic field which in principle may be eliminated by suitable double-passage techniques.

The interaction of the spin with the magnetic field is described by the Hamiltonian

$$H = -\mu \cdot \mathbf{B} \; ; \quad (28)$$

thus the neutron momentum $p$ in the field is related to the momentum $p_0$ outside the field by

$$\frac{p_0{}^2}{2m} = \frac{p^2}{2m} \mp \tfrac{1}{2}\hbar\omega_L \tag{29}$$

where the $\mp$ signs refer to the spin parallel (antiparallel) to the field and $\omega_L = -geB/2m$ is the Larmor frequency. For $\tfrac{1}{2}\hbar\omega_L \ll p^2/2m$ the plane-wave neutron state in a magnetic field directed along the $z$-axis is represented by $\exp\{i(p_0y/\hbar \mp m\omega_L y/2p_0)\}$ which can be expressed in matrix form (equation (9)) with

$$k_0 = \cos\beta/2, \quad k_y = 0, \quad k_x = 0, \quad k_z = -i\ \sin\beta/2 \tag{30}$$

and

$$\beta = m\omega_L y/p_0 \ll p_0 y/\hbar\ . \tag{31}$$

If the magnetic field is set at an angle $\alpha$ to the $z$ axis equations (30) become

$$k_0 = \cos\beta/2, \quad k_y = 0, \quad k_x = -i\ \sin\alpha\ \sin\beta/2$$
$$k_z = -i\ \cos\alpha\ \sin\beta/2\ . \tag{32}$$

We are, however, interested not in the case where the magnetic field **B** is established at a fixed angle $\alpha$ with respect to some reference direction transverse to the beam momentum but rather in the case that **B** is twisted about the beam and $\alpha$ varies along the beam, i.e.

$$\alpha(y) = \alpha + \eta y\ . \tag{33}$$

In this case equation (32) is no longer correct and a more complex calculation is involved since we encounter difficulties associated with the non-commuting properties of the rotation matrices $\sigma_x$ and $\sigma_y$. The result correct to second order in $\beta$ is

$$k_0 = 1 - \tfrac{1}{2}\left(\frac{\beta}{2}\right)^2 \quad k_x = -i\frac{\beta}{2}(\sin\alpha + \eta y\ \cos\alpha)$$
$$k_y = \frac{i}{6}\left(\frac{\beta}{2}\right)^2(\eta y) \quad k_z = -i\frac{\beta}{2}(\cos\alpha - \eta y\ \sin\alpha)\ . \tag{34}$$

The significant result in equation (34) is the non-vanishing of $k_y$; this corresponds to a rotation of the neutron spin through an angle $\phi$ where

$$\frac{\phi}{2} = \frac{1}{6}\left(\frac{\beta}{2}\right)^2 (\eta y) \ . \qquad (35)$$

The angle of rotation is of first order in the angle of twist $\eta y$ and of second order in the magnetic field; it is therefore independent of the sense of the magnetic field and indistinguishable from the true weak rotation given in equation (21). For thermal neutrons in a sample 1 m long and a magnetic field of $10^{-7}$ Wb m$^{-2}$ there is a rotation of order $10^{-6}$–$10^{-5}$ radians per radian of twist.

## 5. Polarization and interference

The final example we consider is the case of a coherent neutron beam whose polarization state is described by the spinor $\mathbf{u}(0)$ which is divided into two coherent wave trains with amplitudes $A$ and $B$ respectively. One wave train propagates along a path $x$ without altering its polarization state while along the other path the spinor is rotated through an angle $\beta$ (equation (31)) about the $z$ axis. The emergent wave trains are subsequently recombined at the detector where the resultant spinor is

$$\mathbf{u}(x) = \begin{pmatrix} A \exp(\mathrm{i}kx)+B \exp\{\mathrm{i}(kx-\beta/2)\} & 0 \\ 0 & A \exp(\mathrm{i}kx)+B \exp\{\mathrm{i}(kx+\beta/2)\} \end{pmatrix} . \qquad (36)$$

The determinant of the transfer matrix is

$$A = \exp(2\mathrm{i}kx)(A^2+B^2+2AB \cos\beta/2) \ . \qquad (37)$$

The unimodular matrix $M_2$ therefore adopts the form

$$M_2 = \begin{pmatrix} \exp(-\alpha/2) & 0 \\ 0 & \exp(\mathrm{i}\alpha/2) \end{pmatrix} \qquad (38)$$

where

$$\tan\frac{\alpha}{2} = \frac{B \tan\beta/2}{B + A \sec\beta/2} \; . \qquad (39)$$

Equations (38) and (39) show that, whereas neutrons which are constrained to move along either of the two paths through the interferometer suffer spin rotations of zero and $\beta$ respectively, neutrons in the interference pattern, whose path through the interferometer is unknown, undergo a spin rotation of $\alpha$.

The neutron polarization vector in the interference pattern behaves in a markedly different fashion for $A < B$ as compared with $A > B$. When $A < B$ the angle of rotation $\alpha$ reaches the value $\pi$ when $\cos\beta = -A/B$, i.e. close to the interference minimum for $|A-B| < A$, and makes a sharp transition to the full $2\pi$ when $\beta = 2\pi$. When $A > B$ the angle $\alpha$ increases with $\beta$ reaching a maximum value of $2 \tan^{-1}[(B/A)/\{1-(B/A)^2\}^{\frac{1}{2}}]$ when $\cos\beta = -B/A$; when $\beta$ reaches $2\pi$, $\alpha$ drops rapidly back to zero. Thus for $A < B$ the polarization vector carries out a rotation, but for $A > B$ the motion is a nutation. In either case $\alpha \approx \pi$ for $\beta < 2\pi$ and there is a discontinuity in $\alpha$ at $\beta = 2\pi$ for $A/B = 1$. Equation (39) shows that $\alpha = \pi$ at this limit although the intensity vanishes since there is total destructive interference. This rather curious behaviour of the polarization vector near the interference minimum may be summarized in the statement that the 'rotation' of the polarization vector becomes less clearly defined in proportion to the increasing visibility of the interference fringes.

## References

BARGMANN, V., MICHEL, L., and TELEGDI, V.L. (1959). *Phys. Rev. Lett.* **2**, 435.

BYRNE, J. (1971). *J. Phys. B* **4**, 940–53.

—— (1972). *Opt. Commun.* **6**, 67–70.

BYRNE, J. and FARAGO, P.S. (1971). *J. Phys. B* **4**, 954–61.

FIERZ, M. and TELEGDI, V.L. (1970). *Quanta*, pp. 196–209. University of Chicago Press, Chicago.

FRADKIN, D.M. and GOOD, R.H. (1961a). *Nuovo Cim.* **22**, 643–9.

—— (1961b). *Rev. mod. Phys.* **33**, 343–52.

HAMILTON, J.H. (1959). *The theory of elementary particles*, Chap. 8. Clarendon Press, Oxford.

JONES, R.C. (1956). *J. opt. Soc. Am.* **46**, 126–31.
KABIR, P.K., KARL, G. and OBRYK, E. (1974). *Phys. Rev. D* **10**, 1471–4.
McMASTER, W.H. (1961). *Rev. mod. Phys. B* **33**, 8–28.
MICHEL, F.C. (1964). *Phys. Rev. B* **133**, 329.
MUELLER, H. (1943). Report no. 2 of OSRD Project OEMsr-576 cited by Jauch, J.M. and Rohrlich, F. (1955). *The theory of photons and electrons*, p. 42. Addison Wesley, Reading, Mass.
RAUCH, H., ZEILINGER, A., BADUREK, G., WILFING, A., BAUSPIESS, W., and BONSE, U. (1975). *Phys. Lett. A* **54**, 425–7.
ROSE, M.E. and GOOD, R.H. (1961). *Nuovo Cim.* **22**, 565–8.
STOKES, G.G. (1852). *Trans. Camb. phil. Soc.* **9**, 399.
STODOLSKI, L. (1974). *Phys. Lett. B* **50**, 352–6.
WERNER, S.A., COLELLA, R., OVERHAUSER, A.W., and EAGEN, C.F. (1975). *Phys. Rev. Lett.* **35**, 1053.

# 9. INTERFEROMETRY WITH POLARIZED NEUTRONS

G. BADUREK

*Institut Laue-Langevin, Grenoble, France*

## 1. Introduction

About four years ago Rauch, Treimer, and Bonse (1974) and Bauspiess *et al.* (1974) reported a successful attempt to split a monochromatic beam of neutrons into two coherent partial beams with a spatial separation of the order of several centimeters and, after succeeding coherent recombination, to produce interference effects in a way similar to the X-ray interferometer technique developed a decade earlier by Bonse and Hart (1965, 1966).

Neutron interferometry has developed beyond the stage of pure demonstration of its functional principles and is now being used as a novel and, in some respects, unique approach to the study of a number of phenomena of basic, solid state, and nuclear physics. Therefore it seems logical to seek an extension of its possibilities by using polarized rather than unpolarized neutrons and also by implementing spin direction analysis of the beams emerging from the interferometer.

In this paper we therefore outline what might be expected from such an extension, discuss how it could be achieved and review the present state of this special field of interferometry.

## 2. Basic interferometer principles

The basic operation of the triple Laue case (LLL) interferometer which is the most common type of neutron interferometer in use at present is described briefly. Comprehensive theoretical and experimental details of the LLL interferometer are given in the reviews by Bonse and Graeff (1977), and Rauch and Petraschek (1976).

The interferometer consists of an E-shaped arrangement of three plane parallel crystal plates cut from a monolithic block of perfect dislocation-free silicon with high geometric precision. The incident monochromatic neutron wave is split coherently, i.e. with a fixed phase relation, into two partial waves by dynamical Laue diffraction within the first crystal (splitter). Subsequent analogous diffractions at the second (mirror) and third (analyser) crystals allow these two waves to recombine coherently. The wave functions of the two beams emerging from the analyser crystal, i.e. the forward (O) and the deviated (H) diffracted beam, are described by

$$\psi_{0,H} = \psi_{0,H}^{I} + \psi_{0,H}^{II} \tag{1}$$

where the two possible paths within the interferometer are denoted by I and II. For an ideal empty interferometer the contributions in the forward direction are equal, i.e. $\psi_0^{I} = \psi_0^{II}$, whereas the ratio of the deviated contributions shows the so-called Pendellösungs oscillations (Rauch and Suda 1974). If a medium with refractive index $n_\lambda$ for neutrons of wavelength $\lambda$ is inserted between two of the crystals, a phase shift between the partial waves is introduced which for the forward contributions can be expressed by

$$\frac{\psi_0^{I}}{\psi_0^{II}} = \exp\left(2\pi i\frac{1-n_\lambda}{\lambda}\Delta D\right) = \exp\left(2\pi i\frac{\Delta D}{D_\lambda}\right) \tag{2}$$

where $\Delta D$ is the length difference of the two beam paths within this medium and $D_\lambda = \lambda/(1-n_\lambda)$ is its so-called $\lambda$ thickness. For negligible absorption, i.e. real $n_\lambda$, the ratio of the forward intensities with ($I_0'$) and without ($I_0$) phase shifting is easily seen to be

$$\frac{I_0'}{I_0} = \tfrac{1}{2}\left\{1+\cos\frac{2\pi\Delta D}{D_\lambda}\right\} \tag{3}$$

and a similar relation would follow for the H beam because of particle number conservation.

Since the index of refraction is related to the mean potential energy $\overline{V}$ of the neutrons within the corresponding medium, which is just the contribution for zero momentum transfer ($\mathbf{Q}$=0) of the Fourier transformed interaction potential and thus can be expressed in terms of the coherent forward scattering length (per atom) $b_c(0)$

$$n_\lambda = 1 - \frac{\overline{V}}{2E} = 1 - \frac{V(\mathbf{Q}=0)}{2E} = 1 - \frac{\lambda^2 N b_c(0)}{2\pi} \qquad (4)$$

where $E$ is the total energy of the neutrons and $N$ is the number of nuclei per $cm^3$, it is evidently possible to determine $b_c(0)$ with high accuracy by observation of the periodic intensity modulation associated with a variation of the effective thickness or density (for gases) (Bauspiess 1977, Kaiser *et al.* 1977). Furthermore, the extremely strong dependence of the observable interference contrast on neutron optical inhomogeneities of the phase-shifting material suggests an application of the interferometer principle to the study of such phenomena as cluster and precipitation formation in alloys (Rauch, Schindler *et al.* 1975). Possible applications to the special case of magnetic substances are discussed in §4.

## 3. Interaction potential and scattering length

In order to obtain information about the possible advantages of the use of polarized neutrons and of polarization analysis, we have to examine all the contributions to the interaction potential between a neutron and an atom. For a single atom located at the origin of the co-ordinate system the interaction potential is described by

$$V(\mathbf{r}) = -\frac{2\pi\hbar^2}{m}\left(b_{nc} + ib' + \frac{b_i}{\{I(I+1)\}^{\frac{1}{2}}}\mathbf{I}\cdot\boldsymbol{\sigma}\right)\delta(\mathbf{r}) \qquad \text{nuclear interaction}$$

$$+ \mu\boldsymbol{\sigma}\cdot\mathbf{H} \qquad \text{magnetic dipole interaction}$$

| | |
|---|---|
| $-\frac{\hbar}{2mc}\nabla\mathbf{E}\left(\mu+\frac{2mc}{\hbar}\varepsilon_2\right)$ | *Zitterbewegung* of magnetic moment + intrinsic charge distribution (Foldy 1958) |
| $-\frac{\hbar\mu\boldsymbol{\sigma}}{mc}(\mathbf{E}\times\mathbf{k})$ | spin–orbit interaction (Schwinger 1948) |
| $-\frac{1}{2}d_i Z^2e^2(r-R_c)^{-4} - d_p\boldsymbol{\sigma}\cdot\mathbf{E}$ | induced + intrinsic electric dipole moment (Thaler 1959, Shull and Nathans 1967) |

where $\mu = \gamma\mu_n$ is the neutron magnetic moment ($\mu_n$ is the nuclear magneton, and $\gamma = -1.91$), $m$ is the neutron mass, $\mathbf{k}$ is the neutron wave vector, $\mathbf{I}$ is the nuclear spin, $\boldsymbol{\sigma}$ is the Pauli spin operator, $b_{nc}$ is the coherent nuclear scattering length, $b_i$ is the incoherent nuclear scattering length, $b' = \sigma_t/2$ is the absorption contribution to the nuclear scattering length ($\sigma_t$ is the total neutron–nucleus cross-section), $Z$ is the atomic number, $\mathbf{E}$ is the Coulomb field of the atom ($R_c$ is the cut-off radius), $\mathbf{H}$ is the magnetic field of the atom, $\varepsilon_2 = \int r^2\rho(r)d\mathbf{r}$ is the second moment of the radial charge distribution of the neutron, $d_i$ is the induced electric dipole moment of the neutron, and $d_p$ is the permanent electric dipole moment of the neutron.

According to the first Born approximation the coherent elastic scattering amplitude per atom is

$$\hat{b}(\mathbf{Q}) = \frac{m}{2\pi\hbar^2}\sum_t w_t \langle t|V(\mathbf{Q})|t\rangle \tag{6}$$

where $V(\mathbf{Q})$ is the Fourier transform of the interaction potential and $t$, $w_t$ refer to the different states of the scattering target and their relative statistical probability respectively. $\hat{b}(\mathbf{Q})$ still represents an operator in neutron spin space whose eigenvalues are the scattering lengths $b^{\pm}$ for the two spin states. Remembering that according to equation (4) the contributions in the forward direction are relevant only for the interferometer, Fourier transformation of equation (5) for $\mathbf{Q} = 0$ and substitution in equation (6) for a magnetic atom would finally lead to

$$\hat{b}(0) = b_{nc} + \tfrac{1}{2}\gamma r_0 g \langle S\rangle \mathbf{h}\cdot\boldsymbol{\sigma} + \frac{me^2 Z^2}{R_c \hbar^2} d_i + ib' \tag{7}$$

where $r_0 = e^2/m_e c^2$ is the classical electron radius, $2\langle\mathbf{S}\rangle = 2\langle S\rangle\mathbf{h}$ is the polarization of the total atomic spin and $g \approx 2$ is a gyromagnetic factor whose deviation from its spin-only value $g = 2$ takes account of eventually incomplete quenching of the orbital moments in crystal fields in terms of the dipole approximation (Marshall and Lovesy 1971).

As only the magnetic dipole interaction is dependent on the neutron polarization its contribution to the scattering length will be considered in somewhat greater detail. The general expression for coherent magnetic scattering is (Marshall and Lovesy 1971)

$$p(\mathbf{Q})\mathbf{q}\cdot\boldsymbol{\sigma} = \tfrac{1}{2}\gamma r_0 g \langle S\rangle f_m(\mathbf{Q})\{\mathbf{e}(\mathbf{e}\cdot\mathbf{h})-\mathbf{h}\}\boldsymbol{\sigma} \tag{8}$$

where $p(\mathbf{Q})$ is defined as the corresponding magnetic scattering length and $\mathbf{q}$, $f_m(\mathbf{Q})$ and $\mathbf{e}$ are the magnetic interaction vector, the form factor of the magnetic atomic electrons, and a unit vector in the direction of $\mathbf{Q}$ respectively.

Comparison of equations (7) and (8) shows an important fact which has caused some confusion in the literature (Eckstein 1950, Schaerpf 1976): the magnetic scattering contribution can be suppressed by proper arrangement of the scattering geometry except in the forward direction $\mathbf{Q} = 0$. The reason for this surprising difference is that the first Born approximation does not take account of the modification of the forward wave by the scattering process. In deriving equation (7), which would also follow from an exact solution of the Schrödinger equation as in dynamical diffraction theory (Mendiratta and Blume 1976), correctly from equation (8) it has to be considered that in the limit $\mathbf{Q} \to 0$ no definite scattering plane exists and hence $\mathbf{e}$ may point in any direction within a plane perpendicular to the beam direction according to the condition $\mathbf{e}\cdot\mathbf{k} \to 0$. The term $\mathbf{e}(\mathbf{e}\cdot\mathbf{h})$ in equation (8) thus gives no effective contribution to coherent scattering in this case.

Another important feature of the forward magnetic scattering amplitude is its direct response to any constant or non-localized spin density which will not contribute at finite angles of Bragg scattering owing to its rapidly falling form factor and hence can readily be determined by comparing the measured forward amplitude with that obtained by form factor extrapolation of Bragg reflection data to $\mathbf{Q} = 0$ (Schneider and Shull 1971, Stassis and Shull 1972).

By relating the spin polarization to the (ferromagnetic) magnetization **M** and the (paramagnetic) susceptibility $\chi_p$ respectively (in Bohr magnetons $\mu_B$ per atom) the corresponding magnetic coherent scattering amplitudes can be expressed as

$$p_F(\mathbf{Q}) = \frac{\gamma r_0 f_m(\mathbf{Q}) M}{2\mu_B} \qquad \text{collinear ferromagnet with } \mathbf{M} = M\mathbf{h} \tag{9a}$$

$$p_P(\mathbf{Q}) = \frac{\gamma r_0 f_m(\mathbf{Q}) \chi_p H}{2\mu_B} \qquad \text{paramagnet in external field } \mathbf{H} = H\mathbf{h} \tag{9b}$$

The definition of a scattering amplitude per atom for a target system containing in fact many individual scattering centres implies that this target is monoatomic and has a definite direction along which the spins are aligned in order to allow for a separation of purely geometric structure factors. The latter condition, which is trivial for a paramagnet that can scatter coherently only when an external field aligns its spins, is fulfilled for ferromagnets only with collinear single-domain spin structure.

In comparing the amplitudes for different scattering angles proper account has also to be taken of the angular dependence of the Debye–Waller temperature factor $\exp\{-W(\mathbf{Q})\}$ which has to be included in the expressions for the scattering length and is exactly unity for $\mathbf{Q} = 0$.

Whereas nuclear and ferromagnetic scattering amplitudes are of the same order of magnitude (about $10^{-12}$ cm atom$^{-1}$), the paramagnetic amplitudes for typical fields of 20 kG are of the order of $10^{-15}$–$10^{-17}$ cm atom$^{-1}$, which is just the approximate size of diamagnetic scattering from all the atomic electrons. The forward diamagnetic scattering amplitude for a spherical atom can be related to the diamagnetic susceptibility $\chi_D$ by (Stassis 1970)

$$p_D(0) = \frac{-\gamma r_0 \chi_D H}{2\mu_B} . \tag{10}$$

Except perhaps for vanadium (about $10^{-16}$ cm at 20 kG and 20 K) the contributions due to brute force polarization of the nuclear spins within the applied fields are usually negligibly small (about $10^{-19}$ cm).

At this point it should be mentioned, at least in principle, it is also possible to determine scattering amplitudes for $\mathbf{Q} \neq 0$ with the interferometer by using the wave dispersion near the Bragg position in a perfect crystal phase shifter (Graeff *et al.* 1978). In the case of polarized neutron interferometry, however, this method seems to be irrelevant for two reasons: (i) the necessity for perfect magnetic crystals with lattice parameters matched to the interferometer crystal to avoid dispersivity prevents general applications; (ii) the additional polarization-dependent contributions to the scattering length present in non-forward directions owing to the spin–orbit interaction and a hypothetical permanent electric neutron dipole moment are imaginary (no absorption!) and would not enter the real part response of the interferometer (Rauch *et al.* 1976).

## 4. Observable effects of initial neutron polarization on the interference pattern

Neglecting the small imaginary absorption contribution $ib'$ in equation (7), which would be effective as an exponential damping term only, the action of the phase shifter can be described by the unitary operator

$$U = \exp\{i(\chi - \boldsymbol{\alpha}\cdot\boldsymbol{\sigma})\} = (\cos\alpha - i\mathbf{h}\cdot\boldsymbol{\sigma}\ \sin\alpha)\exp(i\chi) \tag{11}$$

where

$$\chi = -2N\Delta D(b_{nc}+d_i me^2 Z^2/R_c\hbar^2) \approx -\lambda N\Delta D b_{nc}$$

and

$$\boldsymbol{\alpha}\cdot\boldsymbol{\sigma} = \alpha\mathbf{h}\cdot\boldsymbol{\sigma} = 2N\Delta D p(0)\mathbf{h}\cdot\boldsymbol{\sigma}$$

represent the phase shifts caused by nuclear and magnetic interaction. It can be readily shown (Eder and Zeilinger 1976, Zeilinger 1976, Badurek *et al.* 1976) that for a completely polarized incident beam, which can be described by a single spinor wave function, the intensity of the forward (0) beam then follows as

$$\begin{aligned} I' &= \psi_0'^{+}\psi_0 = \psi_0^{I+}(1+U^{+})(1+U)\psi_0^{I} \\ &= \frac{I}{2}(1+\cos\chi\ \cos\alpha+\mathbf{h}\cdot\mathbf{P}\ \sin\chi\ \sin\alpha) \end{aligned} \tag{12}$$

where $I$ and $\mathbf{P}$ are the intensity and unit polarization vector respectively of the incident beam. Similarly the unit polarization vector of the 0 beam can be found to be

$$\begin{aligned} \mathbf{P}' &= \frac{\psi_0'^{+}\boldsymbol{\sigma}\psi_0}{\psi_0'^{+}\psi_0} \\ &= \frac{I}{2I'}\{(\cos^2\alpha+\cos\alpha\ \cos\chi)\mathbf{P} + (\mathbf{h}\cdot\mathbf{P}\ \sin^2\alpha+\sin\alpha\ \sin\chi)\mathbf{h} \\ &\quad + (\sin\alpha\ \cos\chi+\sin\alpha\ \cos\alpha)(\mathbf{h}\times\mathbf{P})\}\ . \end{aligned} \tag{13}$$

For an incompletely polarized incident beam with $|\mathbf{P}| < 1$ the corresponding density matrix formalism would be necessary, of course, but since a beam fully polarized perpendicular to the direction of magnetization ($\mathbf{h}\cdot\mathbf{P} = 0$) is equivalent to a completely unpolarized one, the latter situation is also included in equations (12) and (13). An apparent difference between the use of polarized or unpolarized neutrons can be seen immediately by rewriting equation (12) for both cases as

$$I'(\mathbf{P}=\emptyset) = \frac{I}{2}(1+\cos\chi\ \cos\alpha) = \frac{I}{2}\left\{\cos^2\left(\frac{\chi+\alpha}{2}\right)+\cos^2\left(\frac{\chi-\alpha}{2}\right)\right\} \tag{14a}$$

$$I'(\mathbf{P}\cdot\mathbf{h}=\pm 1) = \frac{I}{2}\cos^2\left(\frac{\chi\pm\alpha}{2}\right) \quad . \tag{14b}$$

Whereas for zero polarization the observed intensity pattern will always show a beat effect owing to the linear superposition of two different oscillation frequencies, with polarized neutrons these frequencies can be resolved. This may be of no great advantage as long as the magnetic contribution is comparable to nuclear scattering since then a sufficiently large number of beat nodes occurs as the phase shifter is rotated between its extremal positions to allow for a reasonable fitting procedure. For paramagnetic or diamagnetic samples, however, the observation of a single magnetic beating node would imply a passage of about $10^3$–$10^5$ nuclear oscillations which, apart from geometric limitations, is unlikely to be accomplished because of the finite coherence length of the neutron wave packet. In this case the polarized neutron technique should allow for a determination of the magnetic scattering length by investigating the small change in the oscillation period as the polarization is reversed. As was demonstrated by Rauch *et al.* (1978), rotation of two identical phase shifters in opposite directions may also serve to measure the magnetic scattering separately even with unpolarized neutrons, but it does not seem to be possible to apply this technique to the investigation of weak magnetism in high fields.

The dependence of the observed intensity behind the interferometer on the relative orientation of **P** and **h**, as given by equation (12), immediately suggests its application to the phase contrast radiography of magnetic multi-domain samples. Whereas, using appropriate film registration techniques, this could certainly be accomplished for thin layers with a single-domain arrangement in the transmission direction inserted into one beam path of an interferometer with the small beam spread up (thin crystals) for thick samples only mean values of the domain size, orientation, and magnetization would be derivable by means of a three-dimensional variation of the initial polarization and analysis of the final polarization in a similar manner to the three-dimensional depolarization technique of Rekveldt (1973). In practice, however, enormous complications may be expected from the inevitable small-angle refraction at the domain boundaries to which the interferometer is also extremely sensitive. Nevertheless, Graeff *et al.* (1978) successfully used this method with unpolarized neutrons to visualize the domains of a thin $Fe_3Si$ layer placed in one beam of a thin-mirror interferometer with a suppressed second path.

It follows from equation (13) that by virtue of simultaneous nuclear and magnetic phase shifts the interferometer itself may act as a polarizer, since for an unpolarized incident beam the outgoing beam is polarized according to

$$\mathbf{P}'(\mathbf{P}=0) = \frac{\sin\chi \sin\alpha}{1 + \cos\chi \cos\alpha}\,\mathbf{h} \tag{15}$$

in the direction of the phase shifter magnetization. Zeilinger (1976) has shown in a more general treatment that two orthogonally oriented purely magnetic phase shifters (as magnetic fields) in combination with one nuclear phase shifter would allow the establishment of any desired orientation of the outgoing polarization with respect to the directions defined by the two magnetic phase shifters. However, since this is meaningful only for exactly vanishing fields everywhere outside the phase-shifter fields and the occurrence of any additional magnetic field would cause either strong depolarization or adiabatic rotation of the polarization vector into the field direction such an arrangement is unlikely to be of practical use.

It should be noted at this point that neutron interferometry has been used to demonstrate the $4\pi$ periodicity of spinor wave functions for the first time experimentally (Rauch, Zeilinger *et al.* 1975, Werner *et al.* 1975). By comparing the precession angle of the neutron spin in a homogeneously magnetized material of thickness $\Delta D$,

$$\phi = g_{\mathrm{n}} \int B\mathrm{d}t = \frac{g_{\mathrm{n}}}{v} \int B\mathrm{d}s = \frac{e}{2\pi\hbar c}\gamma\lambda B\Delta D$$

$$= \frac{\gamma r_0}{\mu_{\mathrm{B}}} N\lambda\Delta D = 2p(\emptyset) N\lambda\Delta D \tag{16}$$

where $g_n = 2\mu/\hbar$ is the neutron gyromagnetic ratio and $M = B/4\pi N$ is the magnetization per atom, with the magnetic phase shift $\alpha$ used in equation (11) one finds the important relation $\alpha = \phi/2$. This means that a rotation of the spinor by an angle of $2\pi$ corresponds to a phase shift of $\pi$ only and should therefore give an observable effect on the fringe pattern behind the interferometer. Its actual occurrence is nicely shown in Fig. 1

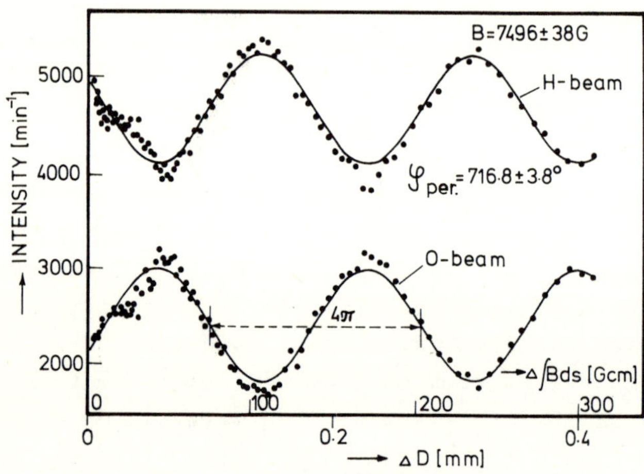


FIG. 1. Observed intensity oscillations behind the interferometer as the spin rotation angle is varied. (After Rauch *et al.* 1978.)

where the result of a refined measurement with two oppositely rotating phase shifters of saturated mu-metal foils is plotted (Rauch *et al.* 1978). The creation of polarization by virtue of simultaneous nuclear phase shift and magnetic spin rotation as shown by equation (15) is demonstrated in Fig. 2 which shows the polarization of the forward beam and the intensity of the deviated beam as a function of the relative nuclear phase shift for three discrete values of spin rotation (Badurek *et al.* 1976). The poor fringe contrast of these measurements is mainly due to field inhomogeneities around the air gap of the electromagnet in which the spin rotation took place. Furthermore, normalizing the observed polarization to the fringe contrast and not to the total intensity, as is done in Fig. 2, shows that the interfering part of the beam may become fully

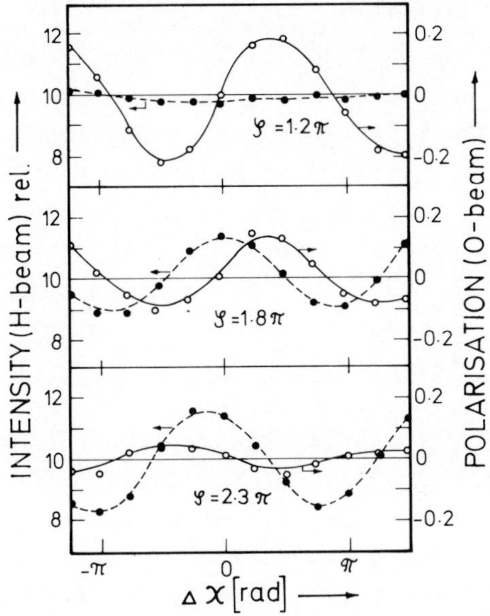


FIG. 2. Intensity and polarization effects behind the interferometer for simultaneous nuclear phase shift and magnetic spin rotation. (After Badurek *et al.* 1976.)

polarized by proper adjustment of nuclear and magnetic phase shift.

Finally, it should be mentioned that polarized neutron interferometry including polarization analysis should allow the demonstration of an important effect concerning the quantum-mechanical superposition principle (Wigner 1963). From equation (13) it follows that if the spin of one of the partial beams within the interferometer is rotated by an angle $\phi = 2d = 180°$ around an axis perpendicular to the initial polarization direction the final polarization in the absence of any nuclear phase shift points in a direction that is exactly orthogonal to the initial direction and the axis of rotation and hence is also orthogonal to the polarization directions of the two partial beams. This implies that the final state has properties that neither of the interfering states had and therefore cannot be interpreted as a classical mixture of the latter.

**5. Experimental realization**

Whereas polarization analysis of the beams leaving the interferometer can be performed using any of the usual techniques (Hayter 1977) and will therefore not be discussed further, the production and application of a polarized neutron beam for the special case of an ideal crystal interferometer is not a trivial goal and is worth some consideration. Since, in general, several instruments have to share the same neutron beam for reasons of efficiency, we may exclude *a priori* any polarizing method that acts across the total available beam cross-sections (polarized targets, polarizing collimators, etc.).

The apparently straightforward replacement of the perfect silicon monochromator by one of the usual polarizing crystals ($Fe_8Co_{92}$, $Cu_2MnAl$, or $Fe_3Si$) is not suitable since, as well as the problem of achieving saturation magnetization within a very limited space (the total available height for the crystal plus the magnet system is about 5 cm for the interferometer at beam guide H25 of the ILL), a large amount of additional shielding would be necessary owing to the much higher integrated reflectivities of these mosaic crystals. However, since this additional intensity has no exact wavelength—angle correlation, as is the case for a perfect-crystal reflection, it is completely useless for the interferometer and leads to a deterioration of the signal/background ratio.

A similar argument shows that it is impossible to use a double-monochromator system consisting of one perfect crystal located within the primary beam and a polarizing mosaic crystal. The mosaic spread of the latter would destroy the wavelength—angle correlation of the reflected neutrons, resulting in a substantial drop in intensity behind the interferometer. Whether or not the replacement of the magnetic Bragg by a magnetic mirror reflection could avoid this decorrelation may only be determined by experiment.

Recently Funahashi (1976) has proposed using the high-energy resolution (about 0.3 μeV) of a back-reflection arrangement of two perfect silicon crystals, one of which is exposed to an intense magnetic field (about 50 kG), to resolve the Zeeman shifting in kinetic energy of the two neutron spin states (about 0.6 μeV).

It is far simpler, however, to incorporate either the interferometer itself or its first crystal into the action of the polarizing device and to make use of the extremely high momentum resolution of perfect crystal Bragg reflections in the non-dispersive (1,-1) position; any angular splitting of the propagation directions of neutrons of different spin orientation that exceeds the rocking width of the interferometer (about 2″ for $\lambda$ = 3 Å) would allow each spin state to be used separately. Spin-sensitive deflection of neutrons can be achieved by transmitting them through a strongly inhomogeneous magnetic field region with triangular boundaries (prism refraction) (Schneider and Shull 1971, Just, Schneider, and Ciszewski 1973).

As in ordinary optics the angular deflection caused by a prism of refraction index $n$ and apex angle $\varepsilon$ for a symmetric passage (optimal with respect to the available beam area) is

$$\delta_{\pm} = 2(n_{\pm}-1)\tan\left(\frac{\varepsilon}{2}\right) \tag{17}$$

where the signs refer to the two neutron spin states. Using equations (4) and (7) their effective angular separation is then

$$\delta = |\delta_{+}-\delta_{-}| = \frac{2\mu B}{E}\tan\left(\frac{\varepsilon}{2}\right) . \tag{18}$$

To proof whether it is possible actually to obtain high degrees of polarization with this technique, we have performed some experimental tests with the arrangement shown in Fig. 3. Using the silicon (111) reflections of the monochromator and the first crystal of the interferometer in the antiparallel (1,-1) position, which for the existing arrangement was the optimal choice with respect to the angular resolution (Badurek *et al.* 1978), the intensity and polarization of the refracted beam behind the second crystal were measured as a function of the rocking angle of this crystal.

The splitting of the neutrons was investigated both for a 'pure' field prism and for a prism cut from a block of mu-metal. The quadratic gap region of the yoke of a small electromagnet

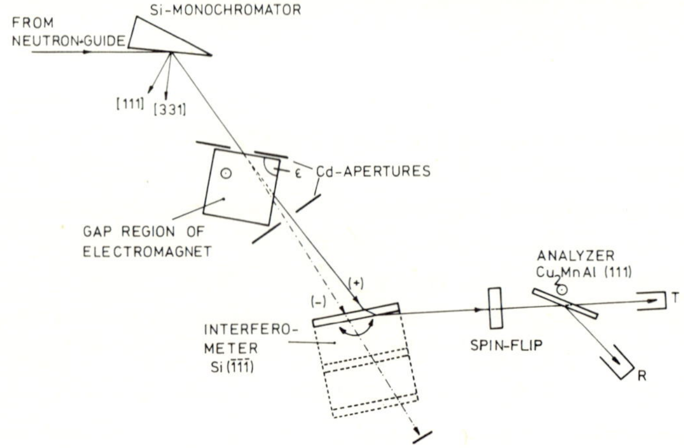


FIG. 3. Experimental set-up to test the applicability of prism refraction bending as a neutron polarizer.

served as the field prism as indicated in Fig. 3. Since the apex angle of this prism was fixed at 90° the height of the magnet air gap had to be as small as 1.5 mm in order to obtain a reasonable induction of about 15 kG. The beam cross-section was therefore restricted to 1.5 mm × 15 mm. The mu-metal prism had an apex angle of 150°, a thickness of 5 mm, and a height of 20 mm. The average transmission over a beam area of 13 mm × 3.5 mm was determined to be 0.24 ± 0.02, corresponding to a linear attenuation coefficient $\Sigma_a = 1.1 \pm 0.15\ \text{cm}^{-1}$. The results of the measurements using both these prisms are shown in Figs. 4 and 5 respectively. The rocking curves were measured with the analyser crystal removed, the intensity being normalized to a cross-sectional area of 1 $\text{mm}^2$. The effective degree of polarization which includes both the polarization efficiency $P_2$ of the analyser and the depolarization coefficient $D$ of the set-up was determined by means of a single-coil d.c. flipper whose efficiency was calibrated by the iron shim method. The observed splitting of the two spin components was 3.9″ for the field prism which agrees well with the theoretical value of 4.1″ obtained using equation (18). The measured splitting of 8.1″ for the metal prism was used to determine the magnetic induction (8.2 kG).

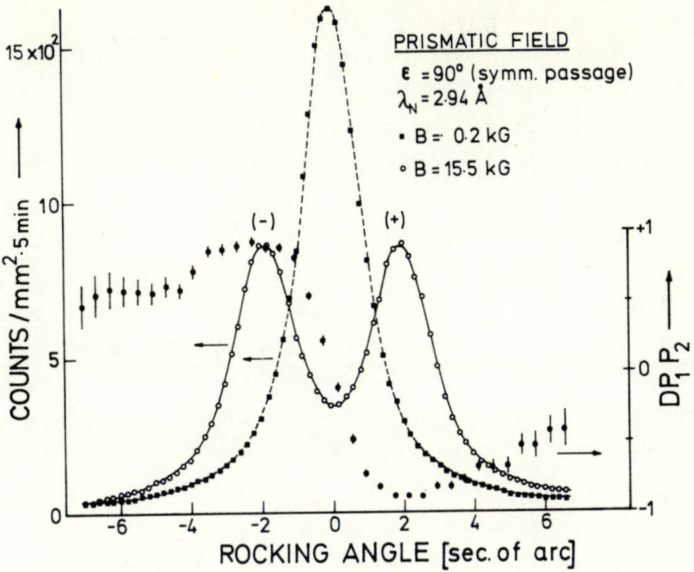


FIG. 4. Normalized intensity and polarization of the diffracted beam *versus* the rocking angle of the second crystal for the pure field prism.

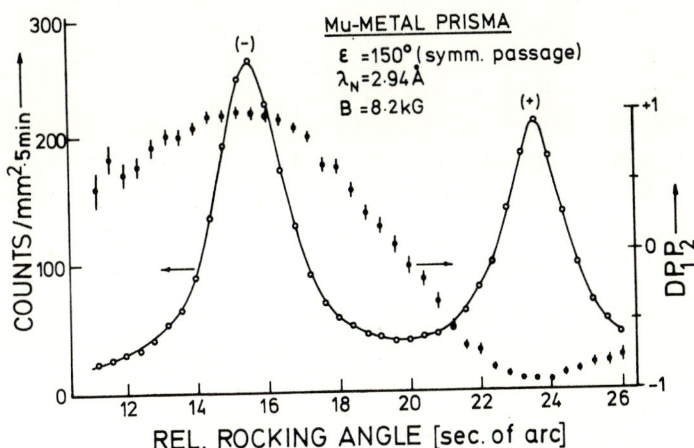


FIG. 5. As Fig. 4 but for a mu-metal prism.

The results clearly show that degrees of polarization close to unity can be achieved. Because of the large and extended stray field of the analyser magnet no extra guide field was necessary to avoid marked depolarization. The poor average transmission of the metal prism could, in principle, be improved by using a closely packed configuration of several smaller prisms with equal apex angles (Schneider and Shull 1971), but the disadvantage of inhomogeneous illumination of the beam cross-section would still be present and would seriously disturb any film-registration technique. With an optimally designed electromagnet or permanent magnet system, on the other hand, it should certainly be no problem to establish a pure field splitter with a beam cross-section of at least 1 cm × 3 cm.

Since it is important for the implementation of polarization turners, it finally should be mentioned that insertion of the anodized aluminium wires (Ø = 0.35 mm) of a spin-flip coil between the monochromator and the interferometer crystal caused a 25% reduction of the height but no corresponding broadening of the width of the rocking curves. For a complete understanding of this behaviour, which is certainly due to small-angle diffraction effects, further investigations are necessary.

## 6. Conclusion

It has been shown that at least for interferometric investigations of magnetic materials polarized neutrons are of considerable advantage and that the change in the experimental set-up which is necessary for their application is essentially small and easy to accomplish.

## References

BADUREK, G., RAUCH, H., WILFING, A., BONSE, U., and GRAEFF, W. (1978). *J. appl. Phys.* **12**, 186.

BADUREK, G., RAUCH, H., ZEILINGER, A., BAUSPIESS, W., and BONSE, U. (1976). *Phys. Rev. D* **14**, 1177.

BAUSPIESS, W. (1977). Thesis, University of Dortmund.

BAUSPIESS, W., BONSE, U., RAUCH, H., and TREIMER, W. (1974). *Z. Phys.* **271**, 177.

BONSE, U. and GRAEFF, W. (1977). *Topics appl. Phys.* **22**, 97.

BONSE, U. and HART, M. (1965). *Appl. Phys. Lett.* **6**, 155.
—— (1966). *Z. Phys.* **194**, 1.
ECKSTEIN, H. (1950). *Phys. Rev.* **78**, 731.
EDER, G. and ZEILINGER, A. (1976). *Nuovo Cim. B* **34**, 76.
FOLDY, L.L. (1958). *Rev. mod. Phys.* **30**, 471.
FUNAHASHI, S. (1976). *Nucl. Instrum. Methods* **137**, 99.
GRAEFF, W., BAUSPIESS, W., BONSE, U., and RAUCH, H. (1978). 11th Int. Congr. of Crystallography, Warsaw, *Acta Cryst.* A **34**, 54, 238.
GRAEFF, W., BAUSPIESS, W., BONSE, U., SCHLENKER, M. and RAUCH, H. (1978). 11th Int. Congr. of Crystallography, Warsaw, *Acta Cryst.* A **34**, S 239.
HAYTER, J. (1978). In *Neutron diffraction* (ed. H. Dachs), Chap. 2, Springer Verlag, Berlin, Heidelberg, New York.
JUST, W., SCHNEIDER, C.S., CISZEWSKI, R., and SHULL, C.G. (1973). *Phys. Rev. B* **7**, 4142.
KAISER, H., RAUCH, H., BAUSPIESS, W., and BONSE, U. (1977). *Phys. Lett. B.* **71**, 321.
MARSHALL, W. and LOVESY, S.W. (1971). *Theory of thermal neutron scattering*, Clarendon Press, Oxford.
MENDIRATTA, S.K. and BLUME, M. (1976). *Phys. Rev. B* **14**, 144.
RAUCH, H., BADUREK, G., BAUSPIESS, W., BONSE, U., and ZEILINGER, A. (1976). *Proc. Int. Conf. on Interaction of Neutrons with Nuclei* (ed. E. Sheldon), Lowell, Mass., p. 1027, National Technical Information Service, Springfield, Virginia 1976.
RAUCH, H. and PETRASCHEK, D. (1976). *Theorie des Interferometers*, Rep. AIAU 76 401, Atominstitut, Vienna (unpublished).
RAUCH, H., SCHINDLER, W., BAUSPIESS, W., and BONSE, U. (1975). Meeting of the Austrian and German Physical Societies, Munich, Lecture D 16-2 (unpublished).
RAUCH, H. and SUDA, M. (1974). *Phys. Status Solidi A* **25**, 495.
RAUCH, H., TREIMER, W., and BONSE, U. (1974). *Phys. Lett. A* **47**, 369.
RAUCH, H., WILFING, A., BAUSPIESS, W., and BONSE, U. (1978). *Z. Phys. B* **29**, 281.
RAUCH, H., ZEILINGER, A., BADUREK, G., WILFING, A., BAUSPIESS, W., and BONSE, U. (1975). *Phys. Lett. A* **54**, 425.
REKVELDT, M.TH. (1973). *Z. Phys.* **259**, 391.
SCHAERPF, O. (1976). Habilitationsschrift, Tech. Univ. Braunschweig.
SCHNEIDER, C.S. and SHULL, C.G. (1971). *Phys. Rev. B* **3**, 830.
SCHWINGER, J. (1948). *Phys. Rev.* **73**, 407.
SHULL, C.G. and NATHANS, R. (1967). *Phys. Rev. Lett.* **19**, 384.
STASSIS, C. (1970). *Phys. Rev. Lett.* **24**, 1415.
STASSIS, C. and SHULL, C.G. (1972). *Phys. Rev. B* **5**, 1040.
THALER, R.M. (1959). *Phys. Rev.* **114**, 827.
WERNER, S.A., COLELLA, R., OVERHAUSER, A.W., and EAGEN, C.F. (1975). *Phys. Rev. Lett.* **35**, 1053.
WIGNER, E.P. (1963). *Am. J. Phys.* **31**, 6.
ZEILINGER, A. (1976). *Z. Phys. B* **25**, 97.

# 10. SOME PRINCIPLES OF THE BEHAVIOUR OF NEUTRONS IN MAGNETIC STRUCTURES RESULTING FROM THE DYNAMIC THEORY OF DIFFRACTION

O. SCHÄRPF

*Institut A für Physik der Technischen Universität, Mendelssohnstrasse 1, D-33 Braunschweig, Federal Republic of Germany*

Investigations of the internal magnetic structures in thick iron and nickel single crystals by small-angle neutron scattering (Schärpf, Vehoff, and Schwink 1973, Schärpf and Berkefeld 1974, Schärpf 1975, Schärpf and Seifert 1975, Schärpf, Seifert, and Schwink 1976, Schärpf 1976a, b, Schärpf and Strothmann 1976, Schärpf and Brandt 1976, Schärpf 1978a, b, c) have answered some questions related to the dynamical theory in magnetic structures. These problems and the corresponding answers are presented here.

In nearly all the literature on magnetic neutron scattering in solid state physics the undiscussed fundamental starting point are the relations

$$Q_{\perp} = \hat{\boldsymbol{\kappa}} \times (\hat{\mathbf{m}} \times \hat{\boldsymbol{\kappa}}) = \sum_{\beta} (\delta_{\alpha\beta} - \hat{\kappa}_{\alpha}\hat{\kappa}_{\beta})\mathbf{m}_{\beta}$$

$$|\mathbf{Q}_{\perp}| = \sin(\hat{\mathbf{m}}\hat{\boldsymbol{\kappa}})$$

where $\hat{\boldsymbol{\kappa}}$ is a unit vector in the direction of the scattering vector $\boldsymbol{\kappa}$ and $\hat{\mathbf{m}}$ is a unit vector in the direction of the manetization. This expression goes back to Halpern and Johnson (1939) and has been used by Blume (1963), Marshall and Lovesey (1971), Bacon (1975), Rauch and Petrascheck (1974, 1977), Mendiretta and Blume (1976), and Stassis and Oberteufer (1974). It states that only the magnetization component $\mathbf{m}_{\perp}$ normal to the scattering vector $\boldsymbol{\kappa}$ is effective for neutron scattering. Only a few authors (Ekstein 1949, 1950, DeBenedetti 1964, Rauch and Petrascheck 1977, Schmidt and Deimel 1976, Schärpf and Strothmann 1976, Schärpf 1976a, b, 1978a, b) have considered the problem of what occurs when the scattering vector $\boldsymbol{\kappa}$ goes to zero, i.e. for pure forward scattering. Even in this group of authors there is no agreement at all about the result, which is to be expected. The difference is not significant as long as one assumes that the magnetization direction is suitably arranged in the experiment (i.e. normal to the scattering vector) when all theories give the same result, but the formalism in the literature is more comprehensive and is not justified by its agreement with experiment only in this special case. However, what is the correct theory? In particular, what is the direction normal to the scattering vector if the scattering vector is zero?

In the dynamical theory of diffraction of neutrons in magnetic materials it is essential to answer this question because one always has to take the direct beam (0,0,0) into account. The method of procedure is also twofold. Most authors first determine the scattering by single atoms and then sum over the crystal, arriving in this manner at an expression for the interaction potential as a function of the scattering vector $\boldsymbol{\kappa}$. In an ideal crystal $\boldsymbol{\kappa}$ is identified with the reciprocal lattice vector $\mathbf{G}$ and thus directly gives the Fourier series needed for use in the dynamical theory. In this case the Halpern magnetic scattering vector $\mathbf{Q}_{\perp}(\mathbf{G}) = \hat{\mathbf{G}} \times (\hat{\mathbf{m}}\times\hat{\mathbf{G}})$ is a consequence of the Born approximation, for in this approximation the scattering amplitude is given by the Fourier transform of the interaction potential. Marshall and Lovesey (1971), Rauch and Petrascheck (1974) and Mendiretta and Blume (1976) proceeded in this manner but did not draw any conclusions related to a magnetization direction that was not perpendicular to the necessary two or more scattering vectors and without answering the question of whether the condition that the magnetization be normal to all the scattering vectors simultaneously is possible.

Rauch and Petrascheck (1977) also drew conclusions related to a magnetization direction that was not perpendicular to the scattering vector. They assumed that when the magnetization direction was not perpendicular to $\mathbf{G}$ for $|\mathbf{G}| \to 0$, i.e. for the forward direction, $\hat{\mathbf{G}}$ has to be substituted by the unit vector normal to the surface $\hat{\mathbf{n}}$. The consequence of this will be discussed below in connection with the construction of the refraction diagram with the aid of the dispersion surface.

Apparently Schneider and Shull (1971) also used this method of procedure because they were surprised that the forward scattering amplitude differed considerably from the value which would be expected by extrapolation of the measured magnetic scattering amplitudes at Bragg reflections using theoretical wave functions for 3d electrons. They suggest that this lends support to the existence of an anomalous magnetic form factor in iron in the small-angle region, a consequence which need not be drawn for the reasons discussed below.

A second group of authors (Ekstein 1949, 1950, Schärpf 1976b, 1978a, b, Schmidt and Deimel 1976) do not proceed from the single scatterers, i.e. the atoms, but treat the crystal as a continuous three-dimensionally periodic distribution of magnetically scattering matter irrespective of whether this magnetization density originates from the spin or the orbit. This magnetization density can be represented by means of a suitable triple Fourier series. This corresponds to von Laue's development of the dynamical theory in terms of a continuous charge distribution for X-rays. The influence of the different terms in the Fourier series upon the neutrons is then determined according to whether one has a single-beam, double-beam, etc. case. In general a different dependence on the magnetization direction is obtained for the forward-scattered beam, which is always an excited wave, than for the Bragg-reflected waves, namely the forward-scattering amplitude does *not* depend on $Q_\perp$ and is always the same irrespective of the direction of the magnetization. It depends only on $|\mathbf{B}|$. This is shown in the following manner.

The Fourier coefficients of the magnetization $\mathbf{M}(\mathbf{r})$, field $\mathbf{H}(\mathbf{r})$, magnetic induction $\mathbf{B}(\mathbf{r})$, and nuclear potential $V(\mathbf{r})$ are given by

$$\mathbf{m}_n(\mathbf{G}_n) = \frac{1}{\tau}\int_\tau \mathbf{M}(\mathbf{r})\exp(2\pi i\mathbf{G}_n\cdot\mathbf{r})\mathrm{d}\mathbf{r}$$

where $\mathbf{G}_n$ is the reciprocal lattice vector, $\tau$ is the unit cell, and analogous expressions hold for $v_n$, $\mathbf{h}_n$, and $\mathbf{b}_n$. Since for $\mathbf{H}$ and $\mathbf{B}$ the Maxwell equations must be satisfied in the form rot $\mathbf{H} = 0$ and div $\mathbf{B} = 0$, one obtains $\mathbf{G}_n \times \mathbf{h}_n = 0$ and $\mathbf{G}_n \cdot (\mathbf{h}_n + 4\pi\mathbf{m}_n) = 0$; using the relation

$$\mathbf{G}_n \times (\mathbf{G}_n \times \mathbf{h}_n) = \mathbf{G}_n(\mathbf{G}_n \cdot \mathbf{h}_n) - \mathbf{h}_n \mathbf{G}_n^2 = 0$$

this gives for $\mathbf{G}_n \neq 0$

$$\mathbf{h}_n = -\frac{4\pi \mathbf{G}_n(\mathbf{G}_n \cdot \mathbf{m}_n)}{\mathbf{G}_n^2}$$

and

$$\mathbf{B}_n^{\perp}(\mathbf{G}_n) = 4\pi\left\{\frac{\mathbf{G}_n(\mathbf{G}_n \cdot \mathbf{m}_n)}{\mathbf{G}_n^2} - \mathbf{m}_n\right\}$$

but for $\mathbf{G}_0 = (0,0,0)$ one obtains

$$\mathbf{B}_0(\mathbf{G}_0) = \frac{1}{\tau}\int_\tau \mathbf{B}(\mathbf{r})\,d\mathbf{r} = \mathbf{B}_{Av} ,$$

i.e. the macroscopic average **B** and not just the normal component of its Fourier component.

This simple result that the Fourier transform for $\mathbf{B}_0(\mathbf{G}_0)$ yields the average induction $\mathbf{B}_{Av}$ irrespective of the boundary direction and the direction of the magnetization and that for $\mathbf{G}_n$ with $n \neq 0$ there results only $\mathbf{B}_n^{\perp}(\mathbf{G}_n)$, the component of $\mathrm{B}_n^{\perp}$ normal to the scattering vector $\mathbf{G}_n$ (which here is a reciprocal lattice vector), leads to essentially different dispersion surfaces for the forward-scattered and the Bragg-scattered waves.

Here we consider in particular the single-beam case because we have constructed an experiment for this case. In the Ekstein method for the single-beam case one obtains simple spheres as dispersion surfaces with radii $\sqrt{(k^2 \pm k_B^2)}$ with $k_B^2 = 2m\mu|\mathrm{B}_{Av}|/\hbar^2$ and the total $B_{Av} = \sqrt{(B_x^2 + B_y^2 + B_z^2)}$ irrespective of the direction of the scattering vector and the magnetization and also irrespective of the direction of the boundary (Fig. 1(a)).

In the method of Rauch and Petrascheck (1977) the radius of the dispersion surface depends on the direction of the scattering vector (according to Rauch and Petrascheck the normal to the refracting boundary) and on the direction of the magnetization. If the angle between **B** and **n** is β then the radius of the dispersion surfaces for the single-beam case is $\sqrt{(k^2 \pm k_{Bi}^2 \sin^2\alpha)}$ (= $k'^+$ or $k'^-$ respectively) with $k_{Bi}^2 = 2m\mu|4\pi M_{Av}|/\hbar^2$. In the

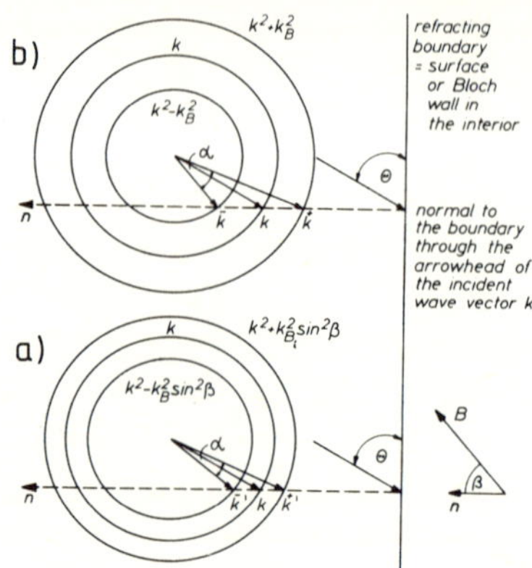


FIG. 1. Dispersion surfaces and refraction for the single-beam case according to (a) Ekstein, Schärpf, and Schmidt and Deimel and (b) Rauch and Petraschek ($B_i = 4\pi M_{Av}$).

single-beam case these are also simple spheres (Fig. 1(b)). Surfaces of this type are the usual ones around the reciprocal lattice point $\mathbf{G}_m$ for beams not in the forward direction and not in the direction of a Laue reflection or near a Laue reflection. However, $k_{Bi}^2$ has then to be multiplied by the form factor $F(\mathbf{G}_n)$ to obtain $\mathbf{m}_n = \mathbf{M}_{Av}F(\mathbf{G}_n)$, and $k_{Bi}^2 = 2m\mu|4\pi m_n|/\hbar^2$. This corresponds to the fact, discussed by Mendiretta and Blume (1976) in connection with their equation (12), that the component of the spin along the direction of the scattering vector does not contribute to magnetic scattering. However, there remains the question of whether this is also the case for forward scattering, i.e. in the single-beam case. This behaviour has not been discussed in the literature, although it is indispensible for the use of all the very complicated formalisms of, for example, Mendiretta and Blume, which are useless without the solution of this problem. Ekstein (1949) and Schmidt and Deimel (1976) considered this but came to no conclusions.

We have performed experiments which decide between these two methods in favour of that of Ekstein (1949) and Schmidt and Deimel (1976). We used large Fe—4 at. % Si single crystals (15 mm in diameter and 10 cm long) with well-defined magnetic domain structures for our measurements: the domains are disc-shaped and lie parallel to each other and normal to the ⟨110⟩ crystal axis and to the surface. The walls between the domains are 90° Bloch walls. The measurements were performed using a double-crystal spectrometer with silicon crystals with a resolution of 5″ which was first used by Schneider and Shull (1971) and was described further by Schärpf and Strothmann (1976) who also gave the measured curves. Figure 2 shows these results. The angle of deflection $\alpha$ is given as a function of the angle $\theta$ of grazing incidence at the wall (see Fig. 1); $\alpha$ is the angle between $\mathbf{k}^+$ and $\mathbf{k}^-$ in Fig. 1. We measured the refraction at the boundary between two magnetic domains where the magnetizations on each side of the boundary make an angle of $2\vartheta_0 = 90°$ with each other, i.e. an angle of $\beta = 45°$ between $\mathbf{B}$ and the normal to the refracting boundary. This angle can be changed further by an applied field. The refraction can easily be shown quantitatively with the aid of the dispersion surfaces described above. To obtain the angle of deflection one has to draw the normal to the refracting boundary from the endpoint of the incident wave vector $\mathbf{k}$. The deflected wave vector then lies in the direction of the wave vectors $\mathbf{k}^+$ or $\mathbf{k}^-$ as is shown in Fig. 1. In our case the incident wave vector $\mathbf{k}$

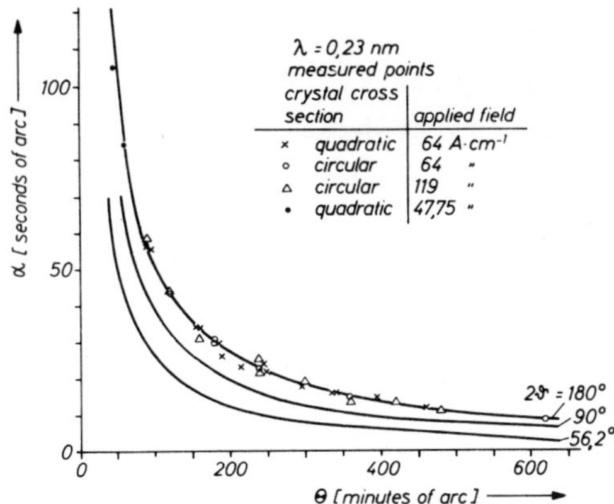


FIG. 2. Angle of deflection $\alpha$ (the angle between $\mathbf{k}^+$ and $\mathbf{k}^-$ in Fig. 1) as a function of the angle of grazing incidence $\theta$. The curves for $2\vartheta = 90°$, $562°$ were obtained using the theory of Rauch and Petraschek; the points were measured and lie on the curve calculated according to the method of Ekstein, Schärpf, and Schmidt and Deimel.

at the crystal surface is not deflected because this surface is nearly normal to $\mathbf{k}$. In the interior of the crystal there are only the wave vectors $\mathbf{k}^+$ or $\mathbf{k}^-$. Therefore at the Bloch wall, as at the refracting boundary in the interior of the crystal, the incident wave vector is either $\mathbf{k}^+$ or $\mathbf{k}^-$ and the deflected wave vector is then $\mathbf{k}^-$ (from $\mathbf{k}^+$ before deflection) or $\mathbf{k}^+$ (from $\mathbf{k}^-$ before deflection). Here the wave vectors $\mathbf{k}^+$ or $\mathbf{k}^-$ are of nearly grazing incidence at the Bloch walls with angles of grazing incidence $\theta$ in the range 45′–600′ and show measurable angles of deflection $\alpha$ in the range of 10″–105″ (Fig. 2). The distribution of intensities between the deflected and undeflected waves is determined by the resolution of the spin direction into one component parallel to the magnetization and one component antiparallel to the magnetization. This is treated further by Schärpf (1976b, 1978a) together with the change of this distribution by the helical structure of the boundary (Schärpf 1978b).

From the above construction one can immediately see that in the first case (Ekstein 1949, Schmidt and Deimel 1976) the angle of deflection depends only on $\theta$ and not on $\beta$, i.e. there is no dependence of the angle of deflection on the direction of magnetization, because the dispersion surfaces are unchanged. In the second case one would obtain spheres with smaller distances between the dispersion surfaces and therefore smaller angles of deflection if the angle $2\vartheta$ between the directions of magnetization is changed. This behaviour is given in Fig. 2 for $2\vartheta = 180°$, $90°$, and $56.2°$ together with our measured results for 90° Bloch walls, i.e. with an angle of $2\vartheta = 90°$ (without an applied field), and with different applied fields. The curve for 56.2° would be for an applied field of 200 A $\mathrm{cm}^{-1}$. These measurements show that irrespective of the angle of magnetization one always has the same angle of deflection and that this angle is the maximum possible angle of deflection, i.e. not just a component of $\mathbf{B}$ contributes to the radius of the dispersion surface.

The consequence of this is that neither the Born approximation for the forward scattering in the dynamical theory nor the relation $\hat{\kappa} \times (\hat{\mathbf{m}} \times \hat{\kappa})$ where the limiting process for $\boldsymbol{\kappa} \to 0$ is not unique are allowed. The method of Rauch and Petrascheck (1977) is therefore wrong. Also the method proposed by Badurek (1979), namely to average in $\hat{\kappa} \times (\hat{\mathbf{m}}\times\hat{\kappa}) = \hat{\mathbf{m}} - \hat{\kappa}(\hat{\mathbf{m}}\cdot\hat{\kappa})$ over $\hat{\kappa}$ so that $\hat{\kappa}(\hat{\mathbf{m}}\cdot\hat{\kappa})$ gives no effective contribution, in general is not correct because it does not take into account $H_{Av}$ but the contribution of $\mathbf{M}_{Av}$. There is no question that refraction also takes place for $\mathbf{H}_{Av}$ alone (see the experiments described by Badurek).

The above results are treated further and more quantitatively by Schärpf (1976b, 1978a). They are also extended to helix structures comprising a long helix of about 1000 atoms (Bloch walls). These are treated extensively by Schärpf (1976b, 1978b). Comparisons with measurements are given by Strothmann and Schärpf (1978) and Schärpf (1978c).

## Acknowledgments

The author wishes to thank Prof. Dr. Ch. Schwink and the reactor group of the Physikalisch Technische Bundesanstalt in Braunschweig especially Prof. Dr. V. Siegel, director of the PTB reactor, and Dr. H. Strothmann.

## References

BACON, G.E. (1975). *Neutron diffraction* (3rd edn.). Clarendon Press, Oxford.
BADUREK, G. (1979). This Conf., Paper II.9.
BLUME, M. (1963). *Phys. Rev.* **130**, 1670–76.
DeBENEDETTI, S. (1964). *Nuclear interactions*. John Wiley, New York.
EKSTEIN, H. (1949). *Phys. Rev.* **76**, 1328–31.
—— (1950). *Phys. Rev.* **78**, 731–2.
HALPERN, O. and JOHNSON, M.H. (1939). *Phys. Rev.* **55**, 898–923.
MARSHALL, W. and LOVESEY, S.W. (1971). *Theory of thermal neutron scattering*. Clarendon Press, Oxford.
MENDIRETTA, S.K. and BLUME, M. (1976). *Phys. Rev. B* **14**, 144–154.
RAUCH, H. and PETRASCHECK, D. (1974). Grundlagen für ein Laue-Neutronen-interferometer: Teil I: Dynamische Beugung 3. Auflage, AIAU 74405b, Atominstitut der Österreichischen Hochschulen, Vienna.
—— (1977). In *Topics in current physics* (ed. H. Dachs), vol. 6, *Neutron diffraction*. Preprint; corrected after this workshop in the final version (Springer, Berlin 1978), citing Bacon 1975, who is also not correct (see Schärpf 1978a).
SCHÄRPF, O. (1975). *Physica B* **80**, 289–300.
—— (1976a). *Proc. Conf. on Neutron Scattering, Gatlinburg, Tennessee, USA*. Oak Ridge National Lab. Rep. ORNL USERDA UC-34, CONF-760601-P2, pp. 706–712.
—— (1976b). *Habilitationsschrift*. Technische Universität, Braunschweig.
—— (1978a). Determination of the scattering angle. *J. appl. Crystallogr.* (Int. Conf. Small-Angle Scattering, Gatlinburg, 1977). *J. appl. Cryst.*, 11, 626-630.

—— (1978b). Behaviour of the intensity after transition of neutrons through a helical magnetic structure. *J. appl. Crystallogr.* (Int. Conf. on Small-Angle Scattering, Gatlinburg, 1977). **11**, 631-636.

—— (1978c). *J. Magn. magn. Mater.* (Arbeitsgemeinschaft Magnetismus Conf.) **9**, 249-251.

SCHÄRPF, O. and BERKEFELD, J. (1974). *Proc. Int. Conf. on Magnetism ICM-73*, vol. 4, pp. 192–6. Nauka, Moscow.

SCHÄRPF, O. and BRANDT, K. (1976). *AIP Conf. Proc. 34* (ed. J.J. Becker and G.H. Lander), pp. 116–118. American Institute of Physics, New York.

SCHÄRPF, O. and SEIFERT, R. (1975). *Proc. Neutron Diffraction Conf.*, pp. 90–100. RCN-Rep. 234, Reactor Centrum Nederland, Petten.

SCHÄRPF, O., SEIFERT, R., and SCHWINK, CH. (1976). *J. Magn. magn. Mater.* **2**, 93–98.

SCHÄRPF, O. and STROTHMANN, H. (1976). *Proc. Conf. on Neutron Scattering, Gatlinburg, Tennessee,* Oak Ridge National Lab. Rep. ORNL USERDA UC-34, CONF-760601-P2, pp. 713–719.

SCHÄRPF, O., VEHOFF, H., and SCHWINK, CH. (1973). *Int. J. Magn.* **5**, 223–29.

SCHMIDT, H.H. and DEIMEL, P. (1976). *Phys. Status Solidi B* **73**, 87–93.

SCHNEIDER, C.S. and SHULL, C.G. (1971). *Phys. Rev. B* **3**, 830–835.

SCHWINK, CH. and SCHÄRPF, O. (1975). *Z. Phys. B* **21**, 305–311.

STASSIS, C. and OBERTEUFFER, J.A. (1974). *Phys. Rev. B* **10**, 5192–202.

STROTHMANN, H. and SCHÄRPF, O. (1978). *J. Magn. magn. Mater.* **9**, 257-260.

# 11. MATTER AND LIGHT WAVE INTERFEROMETRY IN GRAVITATIONAL FIELDS

L. STODOLSKY

*Max-Planck-Institut für Physik und Astrophysik, München, Federal Republic of Germany*

## 1. Introduction

An elegant experiment has demonstrated quantum-mechanical interference fringes for particles (neutrons) traversing different paths in the earth's gravitational field (Overhauser and Colella 1974, Colella, Overhauser, and Werner 1975). Since the effect observed can be calculated from a Newtonian potential inserted in the Schrödinger equation, the experiment contains no real surprises, amusing though it is. Nevertheless, it does by implication suggest an interesting problem: what is the description of such quantum interference effects when the gravitational field is not simply Newtonian, i.e. when the full tensor character of gravitation comes into play?

I shall attempt to answer this question in the simplest case, i.e. in the 'semi-classical limit' where the particle is taken to travel along its classical path. This means that the gravitational fields vary slowly in a distance of the order of the de Broglie wavelength of the particle. It also means that the fields must vary slowly in a distance comparable to the intrinsic size of the particle itself otherwise the internal structure of the particle will come into play. The semi-classical motion also implies that the position and momenta are roughly defined at the same time. This means that important variables should not change significantly over the band of frequencies or wavelengths necessary to define a localized wave packet. On the other hand our wave trains must be long enough for the packets traversing different paths to overlap upon reaching the interference point.

## 2. The quantum phase

Since we assume the semi-classical limit, the quantity of

interest is simply the quantum-mechanical phase accumulated by the particle as it travels along its classical path. This phase may then be brought into play when different classical paths are made to interfere, as in an interferometer.

Now what can this phase be? In the rest frame of a free particle with mass $m$ the quantum factor is $e^{imt}$. If we wish the phase to be invariant, the only plausible generalization appears to be to let $t \to \int ds$. The proper time $ds$ is the 'distance' element of general relativity: $ds^2 = g_{\mu\nu} dx^\mu dx^\nu$. Thus a path in space–time from point A to point B has a phase factor $e^{i\Phi}$ associated with it (in units of $\hbar = c = 1$) with

$$\Phi = \int_A^B m \, ds \; . \tag{1}$$

This answer also appears reasonable from the Feynman path integral approach to quantum mechanics where the phase is the classical action since $m\int ds$ can be taken as the action in relativity (Landau and Lifshitz 1958). If we assume that (1) is also valid for paths adjoining the classical path then we find in the usual way that the actual classical path is the one for which $\delta\Phi = 0$, i.e. a geodesic.

Equation (1) may also be given the aspect of a typical quantum action integral if we divide the defining equation for $ds^2$ by $ds$ and define $P^\mu \equiv m dx^\mu/ds$; then

$$\Phi = \int_A^B g_{\mu\nu} P^\mu dx^\nu = \int_A^B P_\nu dx^\nu \; . \tag{2}$$

This latter form, involving the canonical momentum, suggests the possibility of further generalization but we shall not pursue this here. With no gravitational fields present $\Phi$ becomes $\Phi^0 = p_\mu (X^A - X^B)^\mu$ as expected. For weak fields we call the small deviation from the Minkowski metric $h_{\mu\nu}$ so that $g_{\mu\nu} = g_{\mu\nu}^{\;0} + h_{\mu\nu}$. The total phase can now be split into $\Phi^0$ and the extra 'gravitationally induced' phase $\phi$:

$$\Phi = \Phi^0 + \phi \; . \tag{3}$$

By using

$$ds = \{(ds^0)^2 + h_{\mu\nu}dx^\mu dx^\nu\}^{\frac{1}{2}} \approx ds^0 + \tfrac{1}{2}h_{\mu\nu}\frac{dx^\mu}{ds}dx^\nu$$

where $(ds^0)^2 = g_{\mu\nu}{}^0 dx^\mu dx^\nu$ we find

$$\phi = \tfrac{1}{2}\int_A^B h_{\mu\nu}p^\mu dx^\nu \qquad (4)$$

where $p^\mu$ is now the usual four-momentum of special relativity: $p^\mu \equiv m\, dx^\mu/ds_0$.

Equation (4) will be our basic formula in most of the following and should be adequate for practically all imaginable experiments, *gedanken* or actual. It can be thought of as the gravitational analogue to the formula for a charged particle in a vector potential, $e\int A_\mu dx_\mu$, with the role of 'charge' now played by the four-vector $p^\mu$. We verify that in the Newtonian limit it gives the expected result. In that limit we ignore all components of $h_{\mu\nu}$ except $h_{00}$, which is twice the Newtonian potential $V$ of non-relativistic physics; then equation (4) yields

$$\phi = \int VP_0 dt$$

$$= VMt \quad \text{for low velocities.} \qquad (5)$$

Equation (5) is of course just what we would have anticipated from the idea that the gravitational potential couples to the energy of the particle. Before turning to some more novel applications of the formulae, we consider an interesting theoretical point, that of 'gauge invariance'.

## 3. Gauge invariance

The idea of gauge invariance is suggested by analogy with the equivalent problem in electrodynamics where a charge particle moving from A to B in a potential $A_\mu$ (the electric and magnetic fields are given by $F_{\mu\nu} = \partial_\mu A_\nu - \partial_\nu A_\mu$) has the phase

$$\phi_{\text{e.m.}} = e \int_{x_A}^{x_B} A_\mu \mathrm{d}x_\mu \quad . \tag{6}$$

Now this phase is gauge dependent; its value depends on the actual value of $A_\mu$ which is not observable. We are 'rescued' from being able to observe $A_\mu$ by the fact that $\phi$ is also unobservable; only phase *differences* can be observed. Therefore if we consider a second path from A to B, the difference

$$e \int_{\text{path 1}} A_\mu \mathrm{d}x_\mu - e \int_{\text{path 2}} A_\mu \mathrm{d}x_\mu = e \oint A_\mu \mathrm{d}x_\mu = e \int F_{\mu\nu} \mathrm{d}A_{\mu\nu}$$

(by Stokes theorem) is an observable and is also gauge invariant. It might be thought that something analogous occurs in the gravitational case, with $g_{\mu\nu}$ playing the role of $A_\mu$, but actually the situation is quite different.

First, what transformations of $g_{\mu\nu}$ might we entertain as possible 'gauge transformations'? If $g_{\mu\nu}$ is to be replaced by another metric tensor which really describes a different geometry, this cannot be thought of as a gauge transformation because a different geometry implies a physical, observable, difference. Thus the remaining choice for the 'gauge transformation' of $g_{\mu\nu}$ is a metric tensor describing the same geometry but in another co-ordinate system:

$$g_{\mu\nu} \to \frac{\partial X'_\mu}{\partial X_\alpha} \frac{\partial X'_\nu}{\partial X_\beta} g_{\alpha\beta} \; .$$

Such a transformation would be a reasonable candidate for the 'gauge transformation'. However, in this case we find that the phase itself, $m\int_A^B \mathrm{d}s$, is (by definition) invariant. In the electromagnetic problem the phase itself was, however, gauge *variant*. Thus there appears to be no simple analogue to the electromagnetic gauge invariance, if by that we mean a transformation of the $g_{\mu\nu}$ which changes the phases but not the physical situation.

We argue that the phase $m\int \mathrm{d}s$ is not gauge variant; indeed

$\int ds$ is in principle a measurable quantity in relativity theory. Is it then somehow observable in the phase? In fact it is, again owing to a striking difference between the familiar electromagnetic problem and the gravitational problem. Although two different charges cannot interfere, two different masses can. Thus in gravity we can have a new kind of interference between two different things.

Let us imagine a system which is not in a mass eigenstate but in a coherent superposition of mass eigenstates instead. A famous example of this is the $K^0$ meson complex (Feynman 1965). When the two components are made to interfere, the relative phase $(m_1-m_2)\int ds$ can be observed. This amounts to a 'clock' in the rest frame of the system running proportionally to the proper time. In the $K^0$ system the 'clock' can be 'read' by following the oscillations in the strangeness of the system. This clock will be affected by the value of $g_{00}$. This is the 'red shift'.

Incidentally, we see that it is a good thing that different charges do not interfere, otherwise by the same argument we could make a clock whose rate would indicate the absolute level of the electromagnetic potential $A_0$. The $K^0$ clock, on the other hand, slows down when we put it in a region of smaller $g_{00}$, but then so do all other clocks — the red shift is universal. We are rescued from observing $A_0$ locally by the super-selection rule which forbids interference between different charges; we are rescued from observing the value of $g_{00}$ locally by a general co-ordinate transformation which compensates for the equal slowing down of all clocks.

## 4. Interferometry

We can consider various examples where interference between two paths for the same particles is observed. An incoming beam is split, sent over the two paths, and coherently recombined at a detector. If $A_1$ and $A_2$ with $A = |A|e^{i\Phi}$ are the amplitudes for the paths, then the intensity at the detector is proportional to

$$|A_1|^2 + |A_2|^2 + 2|A_1A_2|\cos(\Phi_1-\Phi_2) \ . \qquad (7)$$

If something is now to change $\Phi_1 - \Phi_2$, it affects the interference term and can be observed as a change in the intensity. In the following we shall assume that the gravitational changes introduced affect $\Phi$ only and not $|A|$; in extreme cases this assumption might have to be re-examined. The paths along which $\Phi_1$ and $\Phi_2$ are to be calculated start from a common point in space–time, travel along the trajectories taken by the respective wave packets, and recombine at another common point in space–time. Under our conditions the result is independent of the precise path used in the 'world channel' swept out by the wave packet.

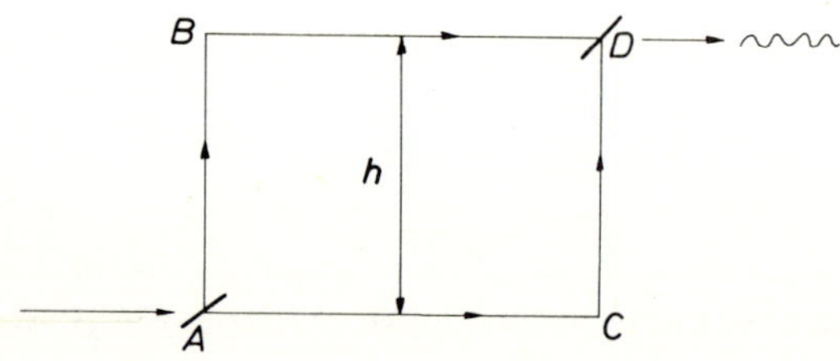


FIG. 1. Schematic representation of an interferometer of height $h$.

Let us first consider the experiment of Overhauser and Colella (1974). Schematically, the experiment may be thought of as a rectangular interferometer (Fig. 1) where the paths ABD and ACD interfere. When the apparatus is vertical the ABD path spends more time at a higher value of the earth's potential; when the apparatus is horizontal the two paths are equivalent. Thus as the apparatus is rotated from vertical to horizontal a certain number of fringe shifts appear. The calculation of Overhauser and Colella (1974), using the Newtonian potential $mgh$ in the Schrödinger equation, gives for the number of fringes ($\hbar=1$)

$$l_{BD}M^2gh\frac{\lambda}{2\pi}\ .$$

On the other hand, from our equation (5) we expect that

$$MVt = Mght$$

where $t$ is the time spent at the higher potential in the vertical arrangement. Since $\lambda = 2\pi/p$, $t = l_{BD}M/P$, we see that the two answers are equivalent. Deviations from equation (5) are to be expected to order $(v/c)^2$, owing to both the variation of $E$ with velocity and the presence of $h_{ij}$ terms in the metric. For thermal neutrons this is very small: $(v/c)^2 = 10^{-10}$.

## 5. Gravity waves

More novel effects appear if we consider non-Newtonian problems such as a gravity *wave* incident on an interferometer. If we take the rectangle of Fig. 1 to be in the (1,2) plane, then an incident gravity wave travelling in the 3 direction can be described by $h_{11} = -h_{22} = h\cos\omega t$; other components of $h_{\mu\nu}$ are zero. (If the rectangle is taken to be oriented along the 1 and 2 axes, the other polarization state of a gravity wave, $h_{12} = h\cos\omega t$, has no effect.) According to equation (4) we can calculate the phase shift for a particle entering the apparatus at time $t$ and moving along the path ABD:

$$\phi_{ABD} = \int_{ABD} h_{ij}p^i dx^j = \int_{ABD} h_{ij}p^i p^j \frac{dt}{E}$$

$$= h\frac{p^2}{E}\left(\int_t^{t+\tau} dt\cos\omega t - \int_{t+\tau}^{t+2\tau} dt\cos\omega t\right).$$

We have for simplicity taken the rectangle to be a square and $\tau$ is the flight time of the particle along one side of the square. In this case $\phi_{ACD} = -\phi_{ABD}$ and

$$\phi_{ABD} - \phi_{ACD} = \frac{hp^2}{\omega E} 2\{2\sin\omega(t+\tau) - \sin\omega t - \sin\omega(t+2\tau)\}. \quad (8)$$

If the flight time is short compared with the gravity wave ($\omega\tau \ll 1$) the vertical and horizontal legs tend to cancel out. However, at $\omega\tau = \pi$ (or $n\pi$) there is a maximum effect

$$\phi_{ABD} - \phi_{ACD} = \frac{-hp^2}{\omega E} \, 8 \, \sin\omega t$$

or (9)

$$= \frac{-hpl}{\pi} \, 8 \, \sin\omega t$$

when $\omega = (P/E)(\pi/l)$, where $l$ is the length of a side. If we set the instrument at the midpoint of the interference pattern, $\Phi_1 - \Phi_2 \approx \pi/2$ in equation (7), then the gravitational phase affects the intensity linearly:

$$|A_1 A_2| \frac{hpl}{\pi} \, 8 \, \sin\omega t \; . \tag{10}$$

The gravity wave will thus produce a time-varying signal at the detector at frequencies near $\omega = 1/\tau$. Equation (10) indicates that the sensitivity of the 'device' is governed by $l/\lambda$, where $\lambda = 1/P$ is the de Broglie wavelength. Interferometry with thermal neutrons where $\lambda \sim 10^{-8}$ cm is now possible. Therefore $pl \sim 10^8$ for a 1 cm interferometer. Since $h$ for astronomically produced gravity waves is at least as small as $10^{-17}$ (and probably very much less than $10^{-20}$) (Kafka 1977a, b, c) our 'instrument' would have to be very sensitive and is probably only of conceptual interest.

In this discussion we have assumed that the parts of the apparatus stay at fixed co-ordinate values. This can be achieved if these parts are free masses since a metric of the gravity-wave type (with no 0 components) does not affect the positions of non-moving bodies (Landau and Lifshitz 1958, Vol. 3).

Other variations on the basic idea may be considered. Instead of a square interferometer, we might have an L-shaped 'Michaelson interferometer'. Instead of being in the plane perpendicular to the direction of propagation of the gravity wave, the interferometer might have one side parallel to the propagation direction.

## 6. Rotation

Another interesting class of non-Newtonian situations is related to rotational effects. The simplest case obtains when we have no 'true' gravitational fields but are at rest in a rotating system. In this case the metric tensor components which differ from unity are $g_{00} = 1 - (\Omega r)^2$ ('centrifugal potential') and $g_{0i} = \mathbf{g}_i$ ('Coriolis potential') where the spatial three-vector $\mathbf{g}$ is tangential to the circle of rotation and of magnitude $g = \Omega r$, where $r$ is the radial distance from the axis and $\Omega$ the angular frequency of rotation. If $\Omega r \ll 1$ we have small deviations from $g_{\mu\nu}{}^0$ and so equation (4) applies, giving

$$\phi = -\tfrac{1}{2}\int E(\Omega r)^2 \mathrm{d}t + \int E\mathbf{g}\cdot\mathrm{d}\mathbf{x} \qquad (11)$$

It can easily be seen that the second term of equation (11) leads to a description of a Michelson—Gale—Pearson type of effect for matter waves (Michelson 1925, Michelson, Gale, and Pearson 1925, Page 1975). The first term which comes from the centrifugal force may not be negligible, even if $\Omega r \ll 1$, since for slow massive particles $\mathrm{d}t \gg \mathrm{d}x$ (recall our $c = 1$ units). There is a difference with the same problem involving photons, since for light $\mathrm{d}t = \mathrm{d}x$. In interferometry with neutrons in rotating systems the centrifugal term should not, in general, be forgotten.

More interesting, perhaps, is the case where the field $g_{0i} = \mathbf{g}_i$ is produced in the space around a rotating massive body. This field, yet to be detected experimentally, is one of the more interesting predictions of General Relativity. This means that in the presence of rotating bodies, in addition to whatever other fields may be present, there is a contribution to the phase of

$$\phi = \int E\mathbf{g}\cdot\mathrm{d}\mathbf{x} \ . \qquad (12)$$

For particles of approximately constant energy (such as slow particles with $E \approx m$) this is like a charged particle in a magnetic vector potential with 'charge' $E$ and 'vector potential' $\mathbf{g}$.

The phase given by equation (12) may be observed by interference of two paths enclosing a finite area. Using Stokes' theorem we see that the effect is proportional to the enclosed 'flux':

$$\phi_1 - \phi_2 = E \int (\nabla\times\mathbf{g})\cdot d\mathbf{A} \tag{13}$$

where dA is an element of area and we have assumed $E$ constant and $E$ and **g** time independent. As in the Michelson–Gale experiment the effect might be shown by comparing the fringes between paths traversed in opposite directions for a large and a small enclosed area. Other possibilities could include varying $E$ or **g**. Landau and Lifshitz (1958) give a formula for **g** when the fields are not too strong:

$$\mathbf{g} = \frac{2G}{r^2}\,\hat{\mathbf{n}} \times \mathbf{L} \tag{14}$$

where $L$ is the angular momentum of the rotating body and $\hat{\mathbf{n}}$ the unit vector at the point of observation.

If we consider a labratory experiment the effect seems hopelessly small. For a circuit that just encloses the rotating body once with a slow neutron ($E \approx m$) we have

$$\phi_1 - \phi_2 = 4\pi GM\frac{L}{R}\ . \tag{15}$$

Since the gravitational constant is only $G = 2.6 \times 10^{-65}$ cm$^2$ or for neutrons $GM = 1.3 \times 10^{-52}$ cm, a body of mass $10^3$ kg and radius 1 m rotating at $10^3$ rev s$^{-1}$ gives us $\phi_1 - \phi_2 \approx 10^{-14}$. We might imagine the neutron going around the rotating body many times. Then in the lifetime ($10^3$ s) of a thermal neutron ($v = 10^5$ cm s$^{-1}$) it can travel $10^6$ m, bringing the effect to the $10^{-8}$ level.

The **g** field is much stronger if we use the earth as the rotating body. The angular momentum in equation (14) is now ($\hbar$ = 1 units) $I_\oplus \approx 10^{67}$. We cannot (at least with neutrons) easily consider encircling the earth, so we use equation (13) and obtain

$$\phi_1 - \phi_2 \approx (\text{area}) \times 2\frac{GML}{R^3}\ . \tag{16}$$

At or near the surface of the earth ($R = 6 \times 10^3$ km) we find, if (area) = 1 km$^2$, an effect of order 0.1 fringe. There seems in this case no intrinsic objection to this proposal but there is at present no known technique for such large-area neutron interferometry. Here the idea of multiple enclosures of a smaller area might be promising, since a distance of the order of only 1 km or less is necessary. For an experiment performed on the surface of the earth the usual Coriolis Michelson–Gale–Pearson effect would of course have to be carefully separated from the general relativistic effect.

We might also consider the possibility that equation (16) could be used to measure the angular momentum of the Galaxy. If we take the Galaxy to have a mass of $2 \times 10^{11}$ solar masses and a radius of $3 \times 10^4$ light years and to be rotating with a period of $2 \times 10^8$ year, then $L_{\mathrm{gal}} \approx MR^2/\mathrm{period} \approx 10^{100}$. The factor $GM(L/R^3)$ in equation (16) is then about $10^{-9}\ \mathrm{km}^{-2}$. Thus the area to be enclosed (by means of single or multiple traversals) must be large but again not, in principle, impossible during the lifetime of a thermal neutron.

## 7. Optical interferometry

We have preferred to use the classical phase formulation as our starting point, since it suggests, via the Feynman approach, how we can go beyond the semi-classical limit. For the semi-classical limit, however, we could also have used the eikonal equation in general relativity (Landau and Lifshitz 1958):

$$g^{\mu\nu} \frac{\partial\Phi}{\partial x^{\nu}} \frac{\partial\Phi}{\partial x^{\nu}} - m^2 = 0 \ . \tag{17}$$

The eikonal equation for light (Landau and Lifshitz 1958) is simple (equation (17) with $m = 0$). Even though the physical interpretation of the waves in question is entirely different, this then permits a parallel treatment of matter waves and optical interferometry. In particular we can derive our basic formula (equation (4)). Putting $g^{\mu\nu} = g^{0\mu\nu} - h^{\mu\nu}$ in equation (17), and working to first order in $h$, we obtain for the 'local momentum'

$$K_{\mu}(X) = \frac{\partial\Phi}{\partial x^{\mu}} = K_{\mu}{}^{0} + \tfrac{1}{2}h_{\mu\sigma}K^{0\sigma} \tag{18}$$

where $K_\mu^{\ 0}$ is a constant four-vector with $K_\mu^{(0)}K^{\mu(0)} = m^2$ or 0, for matter or light respectively, with the assumption that $K_\mu(x)$ goes smoothly over to $K^0$ as $h^{\mu\nu} \to 0$. Since $\Phi = \int K_\mu \mathrm{d}x^\mu$ we again obtain equation (3) and

$$\phi = \tfrac{1}{2} \int h_{\mu\nu} K^{0\nu} \mathrm{d}x^\mu \tag{19}$$

which is equivalent to equation (4). This derivation is actually superior to that involving the expansion of the square root which precedes equation (4) since the eikonal is the distance along the actual geodesic. Thus the curvature of the path is taken into account. The formula then also applies when $\mathrm{d}s_0$ is small, as with light.

Although the formulae for matter and light appear to be the same the difference in the momentum vector $K_\mu^{\ 0}$ may be important. For effects where only the three-vector part of $K_\mu$ enters, as in the discussion of the 'gravity-wave detector', what matters in comparing $\phi$ for matter and light is the comparison of the de Broglie wavelength and the optical wavelength. For thermal neutrons ($\lambda \approx 1$ Å) compared with laser light ($\lambda \approx 10^3$ Å) the neutrons are a factor of $10^3$ more effective. On the other hand X-rays with $\lambda \approx 1$ Å are equivalent to neutrons. The situation is different when the *time* component of $K_\mu$ enters, as in equation (5) or as in the effects discussed in §6. There what matters is the *mass* of the particle compared with the energy of the photon, which for neutrons compared with optical photons is GeV/eV $\approx 10^9$. Thus, although it would be very interesting to repeat the experiment of Overhauser and Colella (1974) to test the relativistic description of light, we see that the effect is a factor of $10^9$ smaller due to the mass and $10^5$ smaller due to the shorter time, yielding $10^{-13}$ over one cm. This is perhaps within the range of laser techniques, but the apparatus must be mechanically stable to this accuracy (Winkler 1977). An interesting point to note here is that the Newtonian and non-Newtonian metric components enter with equal importance in the optical case.

Although optical fringe shifts will thus always be much smaller than those for neutrons, the much greater sensitivity of laser (or perhaps microwave) techniques at the present time makes it worthwhile to reconsider some of the effects described in §6.

In particular, with light or microwaves it is perhaps conceivable that the rotational **g** field could be established by a Michelson—Gale—Pearson type experiment between space stations. Briefly, in such an experiment two beams are sent around an identical closed path in opposite directions and made to interfere. The effect is then manifested as a shift between the fringes for a large enclosed area and a small enclosed area. (With a laboratory rotating body the effect could also be manifested by changing the sense of rotation of the body.) If we imagine that the space stations are so arranged that the beams encircle the earth, then we can use equation (15) with, as just explained, the neutron mass replaced by the photon energy ω:

$$\phi_1 - \phi_2 = 4\pi\frac{G\omega L}{R} . \tag{20}$$

For $\omega = 1$ eV, $L_{\oplus} = 10^{67}$, and $R = 10^4$ km we obtain the rather substantial effect $\phi_1 - \phi_2 \approx 10^{-2}$. With microwaves the effect is smaller in proportion to ω, but perhaps the technology is better adapted to the purpose. If we do not wish to encircle the earth we may imagine an area in space enclosed by a system of mirrors or antennas and we can use equation (16):

$$\phi_1 - \phi_2 = \text{area} \times 2\frac{G\omega L}{R^3} . \tag{21}$$

With the same values for ω, $L_{\oplus}$, and $R$ we find $\phi_1 - \phi_2/\text{area} = 10^{-11}\ \text{km}^{-2}$. Since a resolution approaching $10^{-8}$ fringe can now be achieved using laser techniques (under laboratory conditions of course) we need only consider a region of linear dimensions of some tens or hundreds of kilometres, which is perhaps not unthinkably large.

If we wish to use equation (21) to measure the rotation of the *Galaxy*, then with $\omega = 1$ eV, $L = 10^{100}$, and $R = 3 \times 10^4$ light years, we find $\phi_1 - \phi_2 = 10^{-18}\ \text{km}^{-2}$. If we take $10^{-8}$ to be the limit of sensitivity then the linear dimensions of the system need 'only' be of the order of $10^5$ km, i.e. about the size of the earth. It is amusing to consider that if all the many practical problems could be overcome and the systematic errors understood, it would be possible to measure the angular momentum of the Galaxy with an experiment confined well within the solar system. As emphasized earlier it makes no difference from the purely formal point of view if the necessary area is enclosed by going once around a large area or many times around a small area.

We might even speculate that some day such experiments, enclosing sufficiently large areas, could search for a 'cosmological' **g** field produced by rotation at very large distances (Collins and Hawking 1973) – a kind of latter-day Newton's ice pail experiment!

Many interesting generalizations and extensions of these problems suggest themselves which I hope to treat in a later publication.

## Acknowledgment

I am thankful to J. Ehlers for helpful proposals concerning the manuscript and many discussions concerning relativity.

## References

COLELLA, R., OVERHAUSER, A.W., and WERNER, S.A. (1975). *Phys. Rev. Lett.* **33**, 1237.

COLLINS, C.B. and HAWKING, S.W. (1973). *Mon. Not. r. astron. Soc.* **162**, 367.

FEYNMAN, R. (1965). *Lectures on physics*, 11–12, Vol. 3, Addison-Wesley, Reading, Mass.

KAFKA, P. (1977a). In *Proc. Int. School of Cosmology and Gravitation, Erice, 1975* (ed. J. Weber). Plenum Press, New York.

—— (1977b). *Some remarks on gravitational wave experiments*. Preprint Max-Planck-Institut, Munich.

—— (1977c). Private communication.

LANDAU, L. and LIFSHITZ, E. (1958). *The classical theory of fields*. Addison Wesley, Reading, Mass.

MICHELSON, A.A. (1925). *Astrophys. J.* **61**, 137.

MICHELSON, A.A., GALE, H.G., and PEARSON, F. (1925). *Astrophys. J.* **61**, 140.

OVERHAUSER, A.W. and COLELLA, R. (1974). *Phys. Rev. Lett.* **33**, 1237.

PAGE, L.A. (1975). *Phys. Rev. Lett.* **35**, 543.

WINKLER, W. (1977). Private communication.

# 12. PHASE CONSIDERATIONS IN A ROTATING SYSTEM

LORNE A. PAGE

*Department of Physics and Astronomy, University of Pittsburgh, Pittsburgh, Pa. 15260, U.S.A.*

## 1. Introduction

The sensitivity and stability achieved in recent years with X-ray and neutron interferometers (Bonse and Hart 1965, 1966, Rauch, Treimer, and Bonse 1974, Colella, Overhauser, and Werner 1975, Werner *et al.* 1975) has been impressive; over path lengths of a few centimetres some $10^8$ wavelengths typically, it is now possible to sustain a lateral separation between two coherent beams which again amounts to some centimetres. In particular the sensitivity is such that in 1975 it was possible to demonstrate convincingly (Colella, Overhauser, and Werner 1975) that the wavelength for a neutron is shortened according as the neutron travels in a deeper gravitational potential, resulting in a gravitational phase shift. As was anticipated at that time, continuing work along the lines of that pioneering gravitational experiment is now succeeding in isolating and identifying the ancillary effect, namely the phase shift due to absolute rotation of the earth plus the interferometer. In terms of fringe shift this latter effect would typically be a factor 30 or so smaller than the gravitational effect, each shift being directly proportional with minor corrections to the 'area' of the interferometer. A detailed report has been made at this conference on such ongoing experiments at Missouri (Werner *et al.* 1979).

We recall that with optical photons there have been two classic experiments closely analogous to the ongoing neutron work in so far as the rotational phase shift is concerned. Sagnac (1913) rotated an optical interferometer encompassing an area of about $10^3$ $cm^2$ at several revolutions per second and demonstrated a discernible phase shift between its two beams, one being directed progressively and the other retrogressively; a previous experiment by Sagnac using an interferometer which was somewhat larger and fixed to the earth had not succeeded principally for lack of sufficient area. Michelson, Gale, and Pearson (1925) employed as their main interferometer a grand optical system embracing some $10^9$ cm$^2$ which was fixed to the earth and lay flat, and with light from an arc source demonstrated with some difficulty the earth's rotation with respect to the fixed stars. A subsidiary interferometer of much smaller area served as their null reference.

In contrasting a neutron interferometer in rotation with the more standard photon interferometer in rotation it does not seem amiss, even at this late date, to enquire as to what basic physics is the same between the two seemingly different experimental situations. To this end we consider a paradigmatic interferometer in the following sections and wherever possible make no distinction between massive particles and photons. The presentation proceeds at a low level of sophistication, and is essentially a brief review of some ways in which one may picture what is happening in a rotating interferometer to first order in the angular velocity. None of this can be called original thinking. One approach (§4) makes manifest the inherent and therefore inescapable time shift (see e.g. Landau and Lifshitz 1975) around a closed loop which is in constant absolute rotation; this approach appears not to have received much notice in the present context, at least where a massive particle is involved. Interestingly, it is just such an approach which affords a rather direct unification between the behaviour of photons, neutrons or electrons, assuming a grossly simplified interferometer.

No attempt is made here to go into the *dynamics* of the splitter, the mirrors, etc., important though the dynamics must surely be in a practical instrument which has to deliver a usable intensity to the analyser. In the paradigm used here (§2) the 'interferometer' consists almost wholly of empty space. Another simplifying strategy is to have the instrument rotating about its centre and not about an axis some thousands of kilometres distant.

In §5 we get rid of the interferometer entirely and review some simple hand-waving arguments regarding phase when we picture an essentially plane wave from the vantage point of a system in slow rotation and require that the wave 'veer' according to the classical trajectory. We agree not to consider photons in relation to a true gravitational field; no anomalies present themselves.

## 2. The paradigm

Arranged rigidly in a square for simplicity of calculation we have a sender of waves (the splitter) at position 1, the ultimate receiver (the analyser) at 4, and two identical transceivers (loosely the mirrors) at locations 2 and 3 (see Fig. 1).

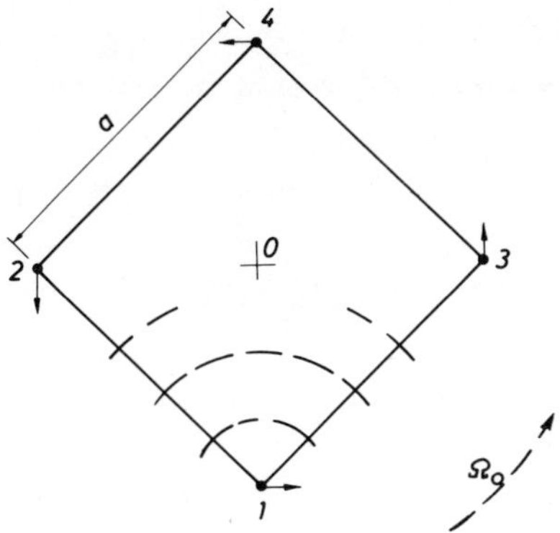


FIG. 1. The four elements of the rudimentary interferometer arranged in a square of side $a$. The array rotates at $\Omega_0$ rad $s^{-1}$ about its centre 0 which is fixed in the Mach Lab. The instantaneous velocity vectors are indicated.

The discussion is to be kinematic only; the dynamic details of how the sender produces the wave and by what mechanism the re-emitters 2 and 3 function are not specified. The four active elements are assumed to be small relative to dimension $a$; also they are assumed to have no 'directionalizing' properties implying smallness with respect to wavelength, which is stringent indeed.

To obviate time dilation, Lorentz contraction, and the like we require the relevant speed parameter $a\Omega_0/c$ to be negligible in second order. We take the centre 0 of our imaginary interferometer to be anchored to the 'Mach Lab' — that truly inertial system with gravity excluded.

## 3. Review of the photon case

Without resorting to a further diagram we may view the course of a photon in the Mach Lab. We may contemplate the present and suitable future positions of the four elements of our interferometer. If we keep only the first order in $a\Omega_0/c$ then the path from 1 to 3 in terms of distance (or in terms of time since we are in a vacuum) is increased from $a$ to $a(1+a\Omega_0/2c)$ and similarly the path from 3 to 4. The correction term for the path 1 to 2 to 4 has the opposite sign. One then obtains the standard result (see Michelson, Gale, and Pearson 1925) for the phase shift: path 3 minus path 2, i.e. $\Delta\Phi_{32} = 2\omega a^2\Omega_0/c^2$ for light of frequency $\omega$. With each beam we go half-way around the instrument instead of the whole way round as in the work of Michelson, Gale, and Pearson (1925).

This standard result can be recast formally as

$$\Delta\Phi_{32} = \frac{2}{\hbar}\left(\frac{\hbar\omega}{c^2}\right)a^2\Omega_0 \ . \qquad (1)$$

We notice that the same result would necessarily apply for an extremely fast particle of total energy $E$, and thus for a particle of relativistic mass $M = E/c^2$, for we have been concerned only with the kinematics in our supposed inertial frame in a vacuum. For what it may be worth then

$$\Delta\Phi = \frac{2}{\hbar}Ma^2\Omega_0 \qquad (2)$$

at least for an extremely relativistic particle. In §4 we shall find out that expression (2) has a more general validity; in fact it works out that the factor $M$ is simply the total energy divided by $c^2$ for any degree of 'relativisticness' of the particle at hand.

## 4. The case of a wave of general phase velocity

At the instant in the Mach Lab (call it time zero) when sender 1 is moving just along $x$ (see Figs. 1 and 2) we make a Lorentz transformation along $x$ at $\beta = a\Omega_0/\sqrt{2}c$, which makes the speed of 1, but not its acceleration, zero. We call this Lorentz

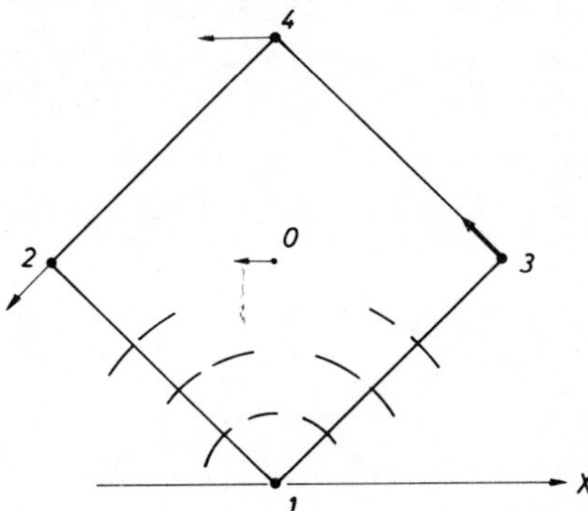


FIG. 2. Interferometer at $t = 0$ in the inertial system CM-1 which 'co-moves' with the sender. In system CM-1 a given wavefront arrives with great precision simultaneously at transceivers 2 and 3. The ultimate receiver is at 4.

system CM-1 since it 'co-moves' with point 1. The utility of this truly inertial system CM-1 is that, barring effects of order $(a\Omega_0/\tilde{v})^2$ where $\tilde{v}$ stands for the actual mechanical speed of our particle or the nominal speed of the centre of mass of a wave packet yet to be constructed, it can be shown that in the time of the CM-1 system each wavefront released from sender 1 in the 'reasonable neighbourhood' of time zero will sweep through transceiver 3 at the *same instant* as through transceiver 2. This quite plausible intermediate result will be explained in the following paragraph. We have pictured as in Fig. 2 a sequence of spherical wavefronts of frequency $\omega$ and phase velocity $v_\phi$ emanating from sender 1. Actually the sphericity is only exact, the $\omega$ and the $v_\phi$ are only precisely isotropic as it were, if sender 1 remains exactly at rest in CM-1 – which it does not of course – but the fractional error in the overall calculated phase shift will turn out to be no worse than a multiple of $(a\Omega_0/\tilde{v})^2$. If we accept that we deal with simultaneous reception at 2 and 3, thus $\Delta t = 0$ in CM-1, and note that this instantaneity has associated with it a non-zero $\Delta x = \sqrt{2}a$ along the direction of transformation, we have in the Mach Lab the corresponding non-zero $\Delta t_{\text{Mach}} = a^2\Omega_0/c^2$ which is crucial to the further argument.

If we are concerned about the non-sphericity of the originally emitted wavefront in CM-1 it can be worked out that to order $\Omega_0^{\ 3}$ there is a minuscule negative correction to $\Delta t_{\text{Mach}}$ which fractionally is only something like unity times $a^2\Omega_0^{\ 2}/\tilde{v}^2$ and stems directly from an exceedingly mild Doppler effect, which is apparent in CM-1 but not apparent to the mirrors and is due to the extremely small $x$ motion in CM-1 of the sender (to the left in Fig. 2) if we take a sufficiently broad segment of time (say several $a/\tilde{v}$ s) containing $t = 0$; this is necessary to obviate troubles with the Causality Principle. This Doppler shift has the effect of increasing $v_\phi$ in the general direction of receiver 3 and conversely for receiver 2. One may transform according to the minor motion of sender 1 either the superluminal $v_\phi$ vector or the subluminal $\tilde{v}$ since we are only calculating kinematics. One uses of course $v_\phi \tilde{v} = c^2$. *If* one were to send thermal neutrons, for example, half way around the grand circuit of Michelson on the rotating earth then the fractional correction to $\Delta t_{\text{Mach}}$ would be only about $10^{-12}$ at most.

We now imagine the re-radiated wavefronts from 3 and 2 proceeding on their way to the general future vicinity of receiver 4; from the point of view of the 'co-moving' system CM-4 there would be zero phase shift at 4 in our approximation if both senders were synchronized in that system. For in steady-state rotation the frequencies received by 4 via the two different paths are rigorously equal (beats in time are not allowed in steady-state rotation since wavefronts are countable); the optical path-lengths in question are equal from symmetry and because fractional effects of higher order than $a\Omega_0/\tilde{v}$ are judged to be negligible. There is, however, a non-trivial time offset between the two re-radiation events from the point of view of CM-4; what was time separation zero in CM-1 is for the same events a time separation in CM-4 amounting to $2a^2\Omega_0/c^2$ s. Finally, then, we may invoke $\Delta\Phi = E\Delta t/\hbar$ yielding $\Delta\Phi = 2Ma^2\Omega_0/\hbar$, which is just expression (2) again.

However, the expression can now be read as saying that each component (of a wave packet) of wave number $k$ will suffer the first-order phase shift given by expression (2) where the relativistic mass $M$ is generally a function of $k^2$. In the non-relativistic domain, as in a normal neutron experiment, the mass factor in the numerator is of course exceedingly sharp.

The time offset just derived, and upon which the final result rests, is just the one discussed in textbooks (e.g. Landau and Lifshitz 1975); without its existence we should now have proved that no rotational phase shift of order $\Omega_0$ could exist, even in the limit of a non-relativistic beam particle.

In the paradigm just worked through a square interferometer was used for reasons of transparency. As one might surmise, expression (2) can be shown to hold for any polygon, regular or otherwise, provided we replace $a^2$ by the projected area. The formula restricting to non-relativistic speeds has been given earlier (Page 1975) on the basis of simple hand-waving. A quite formal treatment (Anandan 1977) which included gravitational as well as rotational effects has yielded a relativistic first correction to the rest mass $M_0$ as an additive $p^2/2M_0c^2$ in so far as the rotational effect in first order in $\Omega_0$ is concerned; expression (2) and this result are therefore compatible.

## 5. A free wave packet in a slowly rotating system

Dispensing now with the transceivers or mirrors and not enquiring into where the waves originate, we picture an unconstrained and essentially plane wave packet travelling in a quasi-inertial system which rotates steadily and very slowly with respect to the Mach system. At the outset we take the fixed axis of rotation to be in the general vicinity of the region of space shown in Fig. 3. We make the sensible requirement that if we were to follow the course of the packet for a time while it travels through some distance $s$ shown in Fig. 3 the packet has to show us a slanting angle $\delta\theta$ which must evolve with time as $\delta\theta = 2\Omega_0 t$. We relate the nominal distance $s$ travelled to the elapsed time $t$ by $t \sim s/\tilde{v}$ which in turn is about $s\tilde{M}/\hbar\tilde{k}$. The group velocity is $\tilde{v}$. The nominal relativistic mass is $\tilde{M}$ and the nominal wave number is $\tilde{k}$. We attribute the required 'veering' of a given component of the group (a component having wave number $k$) to the fact that the phase difference, station 4 minus station 3, namely $\Phi_{43}$, exceeds the phase difference $\Phi_{21}$. Thus

$$\Phi_{43} - \Phi_{21} \equiv \Delta\Phi_k \sim kw\delta\theta \sim 2\frac{\tilde{M}}{\hbar}ws\Omega_0. \qquad (3)$$

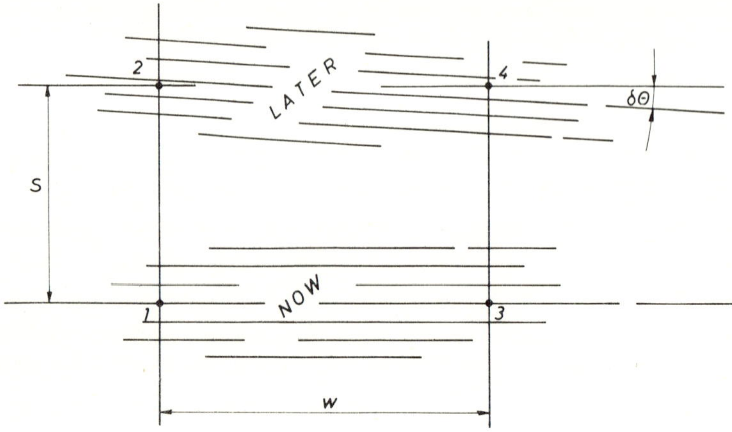


FIG. 3. A 'plane' wave packet seen veering slightly in a Cartesian system which is not strictly inertial. The veering is attributed to an apparent excess phase at point 4 in relation to point 2, thus proportional to both the span $w$ and the nominal distance travelled $s$.

Since the product $ws$ is the area of the rectangle 1243, equation (3) is seen to be in a sense a replay of the standard rotational phase shift (2). We might now let the axis of rotation be quite far to the left in Fig. 2 and thus generate some centrifugal acceleration to the right in the figure, or we might suppose that the instantaneous axis of rotation is being accelerated to the left; in either case we shall have a phenomenological acceleration **g** in our quasi-inertial system. The factor $2\Omega_0$ in expression (3) would then go to $2\Omega_0 + g/\tilde{v}$ and of course we shall have synthesized by means of the new term the standard gravitational phase shift of Collela, Overhauser, and Werner (1975). One could mention the profound but not especially startling result that the kinetic energy of the particle in this example 'gravitates' just as the rest energy does.

Returning to Fig. 3 we might set $\Omega_0$ to zero, let the particle be charged, and let there be a uniform pure magnetic field normal to the plane of the figure. In a backhand manner the Larmor frequency $\Omega_L = qB/2Mc$ will now play the role of the $\Omega_0$ used throughout the previous discussion. The manifest veering which takes place in the inertial Mach Lab can, as we know, be removed to first order in $\Omega_L$ by going to Larmor's rotating system. The correspondence between the previously considered mechanical rotation and now Larmor's rotation (waiving algebraic signs) are

$$\mathbf{\Omega}_L = \tfrac{1}{2}\frac{q}{Mc}\nabla \times \mathbf{A} \leftrightarrow \Omega_0 = \tfrac{1}{2}\nabla \times (\mathbf{\Omega}_0 \times \mathbf{r})$$

$$\hbar\mathbf{k} = M(\tilde{\mathbf{v}}+\frac{q}{Mc}\mathbf{A}) \leftrightarrow \hbar\mathbf{k} = M(\tilde{\mathbf{v}}+\Omega_0\times\mathbf{r})$$

and for phase shift around a loop

$$\Delta\Phi = \frac{q}{\hbar c}\oint \mathbf{A}\cdot d\mathbf{s} \leftrightarrow \Delta\Phi = \frac{M}{\hbar}\oint(\mathbf{\Omega}_0\times\mathbf{r}_0)\cdot d\mathbf{s}\,,$$

which yields $2(M/\hbar)$ (area) $\Omega_0$.

This kind of correspondence can be seen in the work of Stodolsky (1979) and appears for example in the non-formal portion of Anandan (1977). To summarize, the rotational phase shift can be said to arise from a velocity-dependent potential energy in the Langrangian and is therefore akin to the well-known Bohm—Aharanov phase shift arising similarly from a velocity-dependent potential. A reason for the present digression into the case of a charged particle is that one might contemplate for electrons a simultaneous superposition of a Bohm—Aharanov shift and a shift due to mechnical rotation in a single experimental set-up. Zeilinger (1978) has remarked upon this possibility. The two shifts would be equal in magnitude when the mechanical $\Omega_0$ equals a pseudo-Larmor frequency $\Omega_L' = (q/2Mc)B$, where $B$ is the average over the area of the interferometer of $\mathbf{B}$ normal to the area. Should we worry about whether to have the flux of $\mathbf{B}$, presumably in the Bohm—Aharanov type whisker or coil located in the apparatus, carried around with the interferometer or sitting still while the interferometer is in motion, it turns out that the correction to $\Omega_L'$ is fractionally no worse than the first power of $(a\Omega_0/\tilde{v})$ in the notation of §4.

## 6. Size of the rotational phase shift

We have seen that the rotational phase shift is quite small in all cases one might reasonably envision. We note that it is proportional to the product of the relativistic mass, the area, and the angular velocity with respect to the local Mach system, and we have noted that we can substitute photon mass in the same expression applicable to particles to the extent that true gravitational fields at least are ruled out of consideration in our simple view of the effect.

One can compare roughly the effective product $MA\Omega_0$ for several experimental situations. Sagnac (S) used 3 eV/$c^2$ photons sent around an area of about $10^3$ cm$^2$ and with a reversible angular velocity of about 10 rad s$^{-1}$ obtained a nominal 0.1 fringe shift in his interferometer. Michelson (MGP) went to almost $10^6$ times the area of S with similar photons and using the irreversible angular velocity of the earth, down by a factor of $10^5$ from S, achieved about 0.2 fringe shift. These two experiments were compatible with each other, with known wavelengths of light and known velocity of light, to within about 2 per cent. Both results could be explained on the basis of a circulating 'aether wind' if one so chose, and Sagnac did in fact view his result this way; Michelson, however, noted that this kind of experiment could not distinguish between the Special Relativity approach and the more naïve approach.

Turning to neutron experiments carried out thus far, the mass of $10^9$ eV/$c^2$ is certainly favourable compared with the classic experiments of S and MGP; in fact despite the relatively small area of present-day neutron interferometers, the earth's rotation, if reversed with respect to the apparatus for argument, could give in principle something like one full fringe shift. Rotation of a neutron interferometer system substantially faster than the earth may well be technically extremely difficult; it appears that there is an extremely long way to go in terms of $\Omega_0$ before the appearance of any effect more subtle or more fundamental than the basic lowest-order result we have discussed in this paper, and this lowest-order effect has already been demonstrated (Werner *et al.* 1979).

As for X-ray interferometers (see the review by Hart and Siddons (1979)) a 10 keV/$c^2$ photon is more advantageous than the photons Sagnac used — 3000 times more so. It becomes a question of how large an effective area one might be able to rotate smoothly; one recalls the massive stone mounting used by Michelson and Morley which floated on mercury. In electron interferometry (see the review by Möllenstedt and Lichte (1979)) the mass factor would improve some $10^5$ over an optical photon yet it might be important to strive for somewhat larger effective areas than are currently in use. In the electron case one would have perhaps quite a variety of options for resetting to essentially zero phase shift, including for example electrostatic means or the approach using a hidden magnetic flux which was mentioned at the end of §5.

## 7. General speculations

To close on a truly speculative note, we know that various schemes are proposed to test for a 'dragging around' of the local Mach Lab in the near vicinity of many kilograms of swiftly rotating matter; see for example a recent review by Braginsky, Caves, and Thorne (1977). In virtually all the schemes the detectors are high $Q$ and classical; except for a proposal involving microwaves in an exquisitely perfect cavity they will rely upon integrating a tiny effect over many periods of a macroscopic system. In contrast an interferometric device would be, as it were, *on line* at all times and have essentially no inertia. The relatively instantaneous response to any change in signal would no doubt be heavily buried in noise; integration would then presumably be performed, say by computer processing or even by mechanical integration in a low-loss pendulum. The contrast between the ponderomotive approach and the interferometric approach can be made clear in a simple-minded example. Suppose the question were upon how fast, right at this moment, is the earth rotating; on the one hand we should think of the quite slow response of the ponderomotive devices such as Foucault's pendulum or a mechanical gyroscope, and on the other hand we should think of the quick response of a laser gyroscope.

Now whether interferometry will one day be in a position to compete with the high-$Q$ macroscopic responders as standardly envisioned is naturally less than clear; however, X-ray or neutron interferometers using perfect crystals as paradigms represent in a sense extremely high-$Q$ devices themselves, although they are essentially achromatic. Although their great sensitivity to certain effects appears to reside in their large $\oint \mathbf{k}\cdot d\mathbf{s}$ it can be said to be practically enabled by the high directivity supplied by the anomalous transmission (Borrmann) effect, which is itself a high-$Q$ phenomenon.

The most attractive point to be made concerning the *direct* use of interferometry to detect extremely small effects such as gravity-wave effects (we do not refer to the indirect use where one observes with great precision a displacement of some macroscopic object which constitutes the primary responder) seems to be the following. The primary responder in an interferometer is the individual beam particle, as opposed to a large collection of some Loschmidt numbers of atoms. With an interferometer then one is not bound or confined to a certain bandwidth at certain frequencies but can in principle select frequencies and bandwidths not irrevocably as in the painstaking fabrication of the hardware but at will in the data analysis process; the flexibility thus provided should be welcome.

## References


ANANDAN, J. (1977). *Phys. Rev. D* **15**, 1448–57.
BONSE, U. and HART, M. (1965). *Appl. Phys. Lett.* **6**, 155–56.
—— (1966). *Z. Phys.* **194**, 1–17.
BRAGINSKY, V.B., CAVES, C.M., and THORNE, K.S. (1977). *Phys. Rev. D* **15**, 2047–68.
COLELLA, R., OVERHAUSER, A.W. and WERNER, S.A. (1975). *Phys. Rev. Lett.* **34**, 1472–74.
HART, M. and SIDDONS, D.P. (1979). This Conf., paper III.2.
LANDAU, L.D. and LITSHITZ, E.M. (1975). *The classical theory of fields* (4th edn.), Chap. 10. Pergamon Press, Oxford.
MICHELSON, A.A., GALE, H.G., and PEARSON, F. (1925). *Astrophys. J.* **61**, 140–45.
MÖLLENSTEDT, G. and LICHTE, H. (1979). This Conf., paper II.1.
PAGE, L.A. (1975). *Phys. Rev. Lett.* **35**, 543.
RAUCH, H., TREIMER, W., and BONSE, U. (1974). *Phys. Lett. A* **47**, 369–71.
SAGNAC, G. (1913). *C.R. Acad. Sci. Paris*, **157**, 1410–13.
STODOLSKY, L. (1979). This Conf., paper II.11.
WERNER, S.A., COLELLA, R., OVERHAUSER, A.W., and EAGON, C.F. (1975). *Phys. Rev. Lett.* **35**, 1053–55.
WERNER, S.A., STAUDENMANN, J.L., COLELLA, R., and OVERHAUSER, A.W. (1979). This Conf., paper II.3.
ZEILINGER, A. (1978). Unpublished work.

# 13. NEUTRON VERSUS ELECTRON DIFFRACTION: A PARADOX RESOLVED

DANIEL M. GREENBERGER
*City College of New York, New York, N.Y. 10031, U.S.A.*

A. W. OVERHAUSER
*Purdue University, West Lafayette, Ind. 47907, U.S.A.*

The first successful electron diffraction experiment, which spatially separated and subsequently coherently recombined an electron beam, was performed by Marton in 1952 (Marton 1952, Marton, Simpson, and Suddeth 1953, 1954). This classic experiment was important not only in its own right but also because when the Aharonov–Bohm (AB) effect was discovered (Aharonov and Bohm 1959) the interpretation of the Marton experiment played a key role in the understanding of the AB effect. At first it was thought that the Marton experiment contradicted the AB effect because of the presence of stray magnetic fields in the laboratory (Mendlowitz 1960). However, it was pointed out by Werner and Brill (1960) that the opposite was true: the AB effect was *necessary* in order to interpret the Marton experiment. This beautiful and simple result has become the standard interpretation of the Marton experiment. The problem is that their argument, which claims that the AB effect caused the stability of Marton's fringes and that a stray magnetic field would not affect the pattern, can also be used to show that the gravity experiment of Colella, Overhauser, and Werner (1975) (COW), or any similar experiment, cannot work.

In these experiments it is the 'stray' gravity field which cannot affect the pattern. Thus there is a direct contradiction between the accepted interpretation of the electron experiment of Marton and the result of the COW neutron-gravity experiment. This is the 'paradox'.

Marton's experiment consisted of setting up a system of three successive diffraction gratings (very thin films of copper) in a modified electron microscope, and by producing first order diffraction at each of the three films he was able to separate and recombine the electron beam (Fig. 1). The angle of diffraction $\theta$ was about 1°. Besides the difficulty of making the films, which had a short lifetime and took 3 h to adjust by hand, it took over 1200 exposures of 6 min each before he located his first fringe pattern. These patterns never lasted longer than six exposures, were almost invisible when they occurred, and had to be measured by a microdensitometer. Despite his extreme care in eliminating systematic errors highly visible fringes were never seen, a fact which we consider very important. Subsequent experimental work using improved techniques has yielded very beautiful fringe patterns (Chambers 1960, Möllenstedt and Bayh 1962a,b, Bayh 1962) and has totally confirmed the AB effect.

The AB effect can be explained by pointing out that a magnetic field can be incorporated into the Hamiltonian of a problem, either classically or quantum mechanically, by changing the momentum $p_\mu$ to $p_\mu - (e/c)A_\mu$, where $A_\mu$ is the vector potential. Then in the Schrödinger equation

$$p_\mu\psi \rightarrow \{p_\mu - (e/c)A_\mu\}\psi \ .$$

Now if one writes

$$\psi = \psi_0 \exp \frac{ie}{\hbar c}\int^x A_\mu dl_\mu) = \psi_0 \exp\{i\phi(x)\}$$

where $\psi_0$ satisfies the field-free Schrödinger equation, then

$$(p_\mu - \frac{e}{c}A_\mu)\psi_0 \exp(i\phi) = (p_\mu\psi_0)\exp(i\phi)$$

and it looks as if the whole electromagnetic effect factors out into a multiplicative phase term, leaving the free-particle equation as before. So where is the reality of the electromagnetic field?

The answer is that the function $\phi(x)$ exists only if the defining line integral is single valued. However, this will be true only if the curl of $A_\mu$ vanishes

$$\partial_\mu A_\nu - \partial_\nu A_\mu = F_{\mu\nu} = 0$$

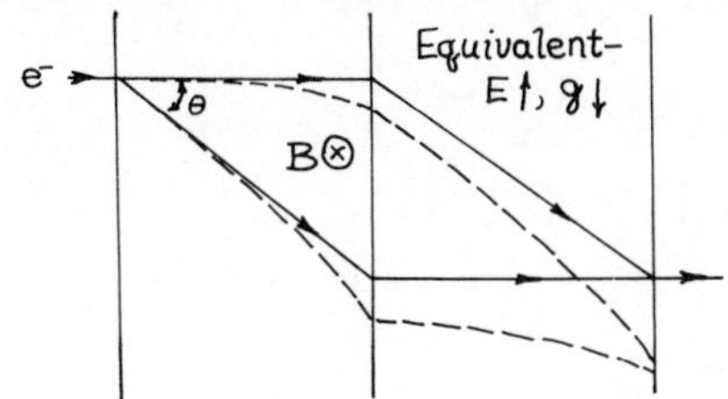


FIG. 1. The Marton experiment. Three thin copper films were used to separate and recombine an electron beam (solid line) by first order diffraction at the successive films. A stray magnetic field (into the paper) would deflect the beam trajectories into the arcs of circles (dotted lines). The effects of this were to change the effective path lengths of the separated beams, shift the beams along the third film by different amounts, and change the relative phase of the beams by an amount proportional to the flux through the area bounded by the beams. Equivalently, an electric or gravitational field as shown would produce similar results (even though the trajectories are now parabolas).

or if there is in fact no magnetic field. Otherwise $\phi = \phi(x,P)$, where $P$ is the path of integration, and the integral is path dependent. Now AB pointed out that if an electron beam is coherently split and taken around opposite sides of a solenoid, and then recombined (Fig. 2), although $B$ is non-zero only in the solenoid, $A$ is necessarily non-zero outside the solenoid. For the combined beam

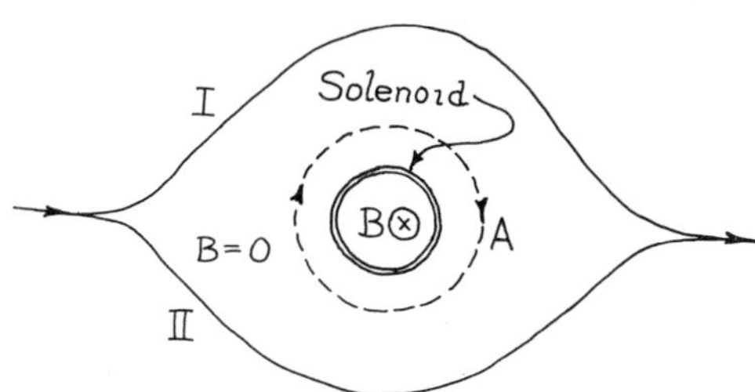


FIG. 2. The Aharonov—Bohm Effect. A beam is split and coherently recombined. The separated beams pass on opposite sides of a solenoid whose axis is perpendicular to the paper. The magnetic field $B$ is confined to the region inside the solenoid so the particle beams never pass within it nor feel any force from it. Yet the vector potential $A$ is non-zero outside the solenoid and produces a phase shift between the beams which is gauge invariant and proportional to the flux of $B$ in the area between the beams.

$$\psi = \psi_{\mathrm{I}} + \psi_{\mathrm{II}} = \psi_0 \exp(i\phi_{\mathrm{I}}) + \psi_0 \exp(i\phi_{\mathrm{II}}) = \psi_{\mathrm{I}}\left[1+\exp\{i(\phi_{\mathrm{II}}-\phi_{\mathrm{I}})\}\right]$$

$$= \psi_{\mathrm{I}}\left\{1+\exp\left(\frac{ie}{\hbar c}\oint A\,dl\right)\right\}$$

$$= \psi_{\mathrm{I}}\left\{1+\exp\left(\frac{ie}{\hbar c}\int B\,dS\right)\right\}.$$

Thus the phase of the recombined beam depends on the flux of $B$ through the solenoid even though *neither* beam ever passes through a magnetic field.

When this astounding effect was first proposed, it was pointed out that there had been stray magnetic fields in the laboratory during Marton's experiment and that therefore his beams had a flux passing between them which, using the parameters of his apparatus, should have moved his fringe pattern back and forth 60 times per second with an amplitude of about 100 fringes. Therefore his patterns should have been washed out and he should have seen nothing. It was thought that this disproved the AB effect.

However, it was further observed by several people, and explicitly calculated by Werner and Brill (1960), that there is another effect, namely that the beams in this experiment are in a real magnetic field unlike the original AB case. This deforms them slightly, into arcs of circles (the dotted paths in Fig. 1), and Werner and Brill showed that the extra path-length difference between the beams due to this effect exactly cancels the AB effect to the lowest order. This lovely result showed the necessity for the AB effect in explaining Marton's experiment, as it is needed to cancel the bending effect, and has seemed to be the final word on the subject.

However, one can extend their results. In the case of a pure electric field (as shown in Fig. 1), the phase factor $(e/\hbar c)\oint A\,dl$ becomes $-(e/\hbar)\int V\,dt$. Furthermore, non-relativistically an electric potential is equivalent to any other type of potential, and so for any other force with potential $U$ the phase factor becomes $-(1/\hbar)\int U\,dt$ (which is equivalent to $(1/\hbar)\int \delta p\,dr$, in all three cases). Thus the Werner–Brill argument goes through for the gravitational force (in spite of the fact that the paths are parabolas rather than circles) and so the presence of a gravitational force should not produce a phase shift because it too will be cancelled by the bending beam.

Therefore the Werner–Brill argument, in spite of its historical importance (at the time they had to convince the sceptics of the reality of the AB effect), contradicts experiment. It also seems to indicate that there is an intrinsic difference between a diffraction grating, which Marton used, and a deflected free-particle beam as used in later experiments.

In order to resolve the paradox one can perform an alternative calculation of the beam propagation from first principles by keeping track of the wave fronts rather than following individual rays (Greenberger and Overhauser 1979). If one wants to know the effect of a small perturbation like a magnetic or gravitational field acting on an originally free particle passing through a grating, one can proceed as follows. First one calculates the change in phase at a fixed point on the receiving screen (like point C in Fig. 3(a)) rather than following the bending beam (to point C′). One then realizes that the bending of the beam produces no change in phase at a fixed

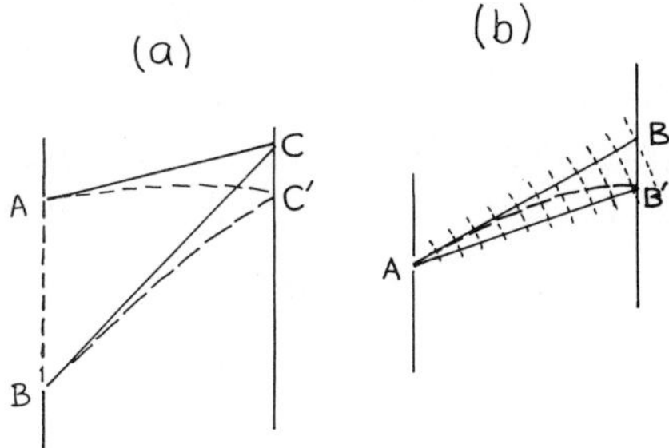


FIG. 3. Effect of a perturbation on the beam. (a) One calculates the phase of the wave function on the screen past the grating at a fixed point C. When the beam is shifted to C′ by a perturbing force, one still calculates the phase at C rather than follow the deflected ray. The result is the phase shift $\exp\{-(i/\hbar)\int U dt\}$ calculated along the unperturbed beam. (b) There is no extra phase shift due purely to the bending of the beam. While the deflected beam along the dotted path AB′ has a different phase than AB, the phase is the same as along the undeflected beam, the solid line AB′. Thus though a different ray strikes B′ than in the undeflected case, the phase is unaffected by the bending and the total effect is the phase shift given above.

point. For example in Fig. 3(b), if the beam bends from AB to the dotted path AB′, the phase striking B′ will be different from that striking B but it will be the same, to first order, as that arriving along the unperturbed straight line path AB′. Thus there is no change in phase induced at the fixed point B′ owing to the fact that the beam is bending. The only effect along such a path, say the straight line AB or AB′, is due to the fact that the wavelength may be changing. This effect is accounted for by the extra phase factor

$$\psi = \psi_0 \exp\left(-\frac{i}{\hbar}\int U \mathrm{d}t\right)$$

integrated along the unperturbed path (Landau and Lifschitz 1958).

This result produces two effects which alter the pattern observed on the screen: first the accumulated phase shifts have the effect of lowering the beam envelope by the amount of the classical motion (to lowest order); second there is an overall phase factor on the wave function at the pattern. This would have the effect that if two beams have been separated and recombined this phase factor will shift the fringes of the recombined beams by exactly the same amount (to lowest order) as the shifting of the beam envelope. Thus the entire fringe pattern, fringes plus envelope, shifts as a unit. This fringe shift $\exp\{-(i/\hbar)\int \Delta U \mathrm{d}t\}$, where $\Delta U$ is the potential difference between the separated beams, is exactly that measured in the COW experiment. This result is true not only for a grating but also for a three-dimensional crystal or a deflected free-particle beam, and agrees with the results of Chambers and the other AB experiments.

However, it also shows, when applied to the Marton experiment, that his fringes did indeed oscillate back and forth owing to this same factor. How then did he see any effect? The answer is that an oscillating fringe pattern actually spends most of its time at its endpoints and so does not wash out completely. If the amplitude of oscillation is $N$ fringes, the intensity $I_0$ is reduced to $I_0/N^{\frac{1}{2}}$. In the Marton experiment $N \approx 100$, so that he could still observe 10% of the total effect. This explains why he had to expend such an effort to see any patterns at all and why his fringes were so ephemeral when seen, being subject to the gradual changing of his stray fields, all in sharp contrast to the beautiful fringe patterns obtained in the later experiments.

Finally it should be mentioned that a completely correct tracing of rays should yield the same results that we have described. In fact there is an extra effect that enters into the Werner–Brill analysis that was not taken into account in their original paper. This is that when the beams are bent apart (Fig. 1) and then recombined they do not recombine exactly. They do not land in the same place along the last grating but one beam slides along the grating relative to the other. This effect does not give a negligible contribution but modifies the Werner–Brill analysis so as to cancel out the effect of the bending of the beams altogether, which agrees with our analysis (Brill *et al.* 1978).

The details of this calculation, as well as a more general discussion of gravity experiments with interferometers, are contained in Greenberger and Overhauser (1979).

## References

AHARONOV, Y. and BOHM, D. (1959). *Phys. Rev.* **115**, 485.
BAYH, W. (1962). *Z. Phys.* **169**, 492.
BRILL, D.R., GREENBERGER, D.M., NEIER, B., OVERHAUSER, A., and WERNER, F.G. (1978). To be published.
CHAMBERS, R.G. (1960). *Phys. Rev. Lett.* **5**, 3.
COLELLA, R., OVERHAUSER, A.W., and WERNER, S.A. (1975). *Phys. Rev. Lett.* **34**, 1472.
GREENBERGER, D.M. and OVERHAUSER, A.W. (1979). *Rev. Mod. Phys.* **51**, 43.
LANDAU, L.D. and LIFSCHITZ, E.M. (1958). *Quantum mechanics*, Addison-Wesley, Reading, Mass.
MARTON, L. (1952). *Phys. Rev.* **85**, 1057.
MARTON, L., SIMPSON, J.A., and SUDDETH, J.A. (1953). *Phys. Rev.* **90**, 490.
—— (1954). *J. sci. Instrum.* **25**, 1099.
MENDLOWITZ, H. (1960). Am. Phys. Soc. Jan. Meeting, post-deadline paper.
MÖLLENSTEDT, G. and BAYH, W. (1962a). *Naturwissenschaften* **49**, 81.
—— (1962b). *Phys. Bl.* **18**, 299.
WERNER, F.G. and BRILL, D.R. (1960). *Phys. Rev. Lett.* **4**, 344.

# 14. AMPLITUDE INTERFERENCE AND INTERFEROMETRY

MING CHIANG LI

*Department of Physics, Virginia Polytechnic Institute and State University, Blacksburg, Va., U.S.A.*

Experimentally one often relies on the observation of an interference pattern to demonstrate the coherence of two coherent beams. The fringe spacing in an interference pattern

$$d = \frac{\lambda}{2\sin(\phi/2)} \tag{1}$$

depends on the wavelength $\lambda$ and the intersecting angle $\phi$ of these coherent beams. With the exception of light, all coherent X-ray, neutron, and electron beams have a short wavelength. Experimental requirements in obtaining an interference pattern from short wavelength coherent beams are stringent. First the intersecting angle between the coherent beams must be small so that the fringe spacing is greater than the resolving power of the observing instrument; hence a very precise alignment is required. Secondly the same angle must be sufficiently large in order to offset the inherent mechanical instability of the instrument for the purpose of obtaining a steady interference pattern.

All the above difficulties are directly associated with the observation of an interference pattern. To demonstrate the coherence of two coherent beams it is not always necessary to rely on the interference pattern. This can be illustrated by considering a scattering process initiated by two coherent beams. These two beams have wave vectors $\mathbf{k}_1$ and $\mathbf{k}_2$ with

$$\mathbf{k}_1{}^2 = \mathbf{k}_2{}^2 = k^2 = \left(\frac{2\pi}{\lambda}\right)^2 \tag{2}$$

and are described by the wave function

$$\exp(\mathrm{i}\mathbf{k}_1\cdot\mathbf{r}) + a\,\exp(\mathrm{i}\mathbf{k}_2\cdot\mathbf{r}) \tag{3}$$

where $a$ denotes the relative phase and amplitude of these two beams. The scattered wave function satisfies the Schrödinger equation

$$-\nabla^2\psi(\mathbf{r}) + \frac{2m}{\hbar^2}V(\mathbf{r})\psi(\mathbf{r}) = k^2\psi(\mathbf{r}) \qquad (4)$$

and the asymptotic condition

$$\psi(\mathbf{r}) \underset{r\to\infty}{\sim} \exp(\mathrm{i}\mathbf{k}_1\cdot\mathbf{r}) + a\exp(\mathrm{i}\mathbf{k}_2\cdot\mathbf{r}) + F(\mathbf{k}_1,\mathbf{k}_2;\mathbf{n})\frac{\exp(\mathrm{i}kr)}{r} \qquad (5)$$

where $m$ is the mass of beam particles and $V(\mathbf{r})$ is the potential of a fixed centre. $\mathbf{n}$ is the direction of the scattered wave, which comes from the disturbance of incident waves by the scattering centre. The Schrödinger equation (4) and the incident waves in equation (3) are both linear. The coherent scattering amplitude can be expressed as

$$F(\mathbf{k}_1,\mathbf{k}_2;\mathbf{n}) = f(\mathbf{k}_1;\mathbf{n}) + af(\mathbf{k}_2;\mathbf{n}) \qquad (6)$$

where $f(\mathbf{k}_i;\mathbf{n})$ for $i = 1, 2$ is the conventional scattering amplitude of a single incident beam. Amplitude $F(\mathbf{k}_1,\mathbf{k}_2,\mathbf{n})$ is a macroscopic quantity and is related to the experimentally observed differential cross-section

$$\mathrm{d}\Sigma(\mathbf{k}_1,\mathbf{k}_2;\mathbf{n}) = (1+|a|^2)^{-1}|F(\mathbf{k}_1,\mathbf{k}_2;\mathbf{n})|^2\mathrm{d}\Omega_{\mathbf{n}} \qquad (7)$$

where $\mathrm{d}\Omega_{\mathbf{n}}$ is a solid angle. This differential cross-section depends on the relative phase factor arg $a$, which inhibits the coherence of the two coherent beams in equation (2). In the illustration the coherence information is contained in the amplitude (6) of the scattered wave. Such an interference will be referred to as the amplitude interference. The interference pattern of two coherent beams arises from the difference on their wave fronts; this type of interference will be referred to as the wave-front interference.

The amplitude interference can be accomplished through the Fraunhofer diffraction of two coherent beams by a crystal. An electron interferometer (Ming Chiang Li 1978a,b) based on the amplitude interference is shown in Fig. 1.

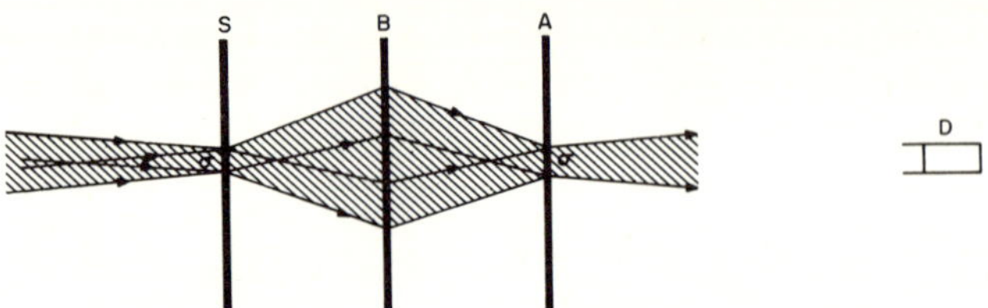


FIG. 1. The modified Marton electron interferometer. A convergent electron beam traverses three equally spaced thin crystalline laminae S, B, and A in succession. All irrelevant diffracted beams are omitted and the angles are exaggerated. In the actual scale the incident electron beam has an angular spread of $10^{-4}$ rad. The diffraction area $\sigma$ is about $5 \times 10^{-12}$ $m^2$. The distance between the interferometer and the solid state detector $D$ should be larger than 1 m.

A convergent electron beam traverses three equally spaced thin crystalline laminae S, B, and A in succession. Through Fresnel diffractions the electron beam is split by the lamina S and bent by the lamina B. The Fraunhofer diffraction of two coherent beams occurred at lamina A. The diffraction beam is measured by a solid state detector D which is mounted at a distance from the interferometer.

Owing to the symmetric arrangement the diffraction areas $\sigma$ on the beam splitter S and the analyser A are the same. To achieve a Fraunhofer diffraction at analyser A the linear dimension of the diffraction area $\sigma$ should be

$$\sigma \leqslant \lambda L \tag{8}$$

where $L$ is the distance between the interferometer and the detector $D$. By taking $L$ = 1 m and $\lambda$ = 0.05 Å, the upper limit of the diffraction area $\sigma$ is about $5 \times 10^{-12}$ $m^2$. By applying experimental techniques used in the electron microscope, an electron beam with an angular spread of $10^{-4}$ rad can be focused onto an area of these dimensions.

For an electron interferometer based on the observation of an interference pattern, the detector should be able to resolve the small fringe spacing. A photographic plate with the aid of a conventional microscope could be used as a detector. However, photographic plate has a poor temporal resolution. In order to be photographed the interference fringes should be kept steady for a relatively long period of time. The presence of a stray 60-Hz field and an inherent small mechanical vibration of the apparatus has made the task of stabilizing the interference fringes very difficult.

In the proposed interferometer the experimental objective is to measure the total intensity of a diffracted beam. A solid state device which has an excellent temporal resolution is employed as a detector. With the aid of fast electronics the solid state device can exclude the effects of the stray 60-Hz field and small mechanical vibration of the apparatus. The stringent stability requirement over a long period of time is no longer necessary.

For an electron interferometer based on the observation of an interference pattern, the original incident beam has to be well collimated and the allowed misalignment is small. A convergent electron beam can be used for the proposed interferometer. The allowed misalignment increases with the angular spread of the incident electron beam and thus the experimental difficulty of alignment can be reduced. An electron interferometer based on amplitude interference is more advantageous.

## References

MING CHIANG LI (1978a). *Z. Phys. B* **19**, 161.
—— (1978b). *Phys. Rev. A* **18**, 773.

# 15. FIZEAU EFFECTS FOR THERMAL NEUTRONS

M.A. HORNE† and A. ZEILINGER‡

*Massachusetts Institute of Technology, Cambridge, Mass. 02139, U.S.A.*

Fizeau's classic experiment (Fizeau 1851) on the propagation of light in moving water, although carried out over half a century prior to the advent of Special Relativity, is usually described today as a direct confirmation of Einsteinian velocity addition. An alternative, but equivalent, description is given here in terms of phase invariance. This approach, which is convenient for treating photon and neutron cases simultaneously, shows that with existing neutron interferometers an experiment analogous to Fizeau's can be done with thermal neutrons.

Consider, for simplicity, an arrangement as used for light by Zeeman (1914, 1915) where the moving medium is solid instead of fluid. A general arrangement of this type is shown schematically in Fig. 1. A parallel-faced plate of thickness $L$ and index of refraction $n(k)$ is inserted in one beam path of either a photon or a neutron interferometer. If initially the plate is at rest, the relative phase of the two interfering beams (1-2) is

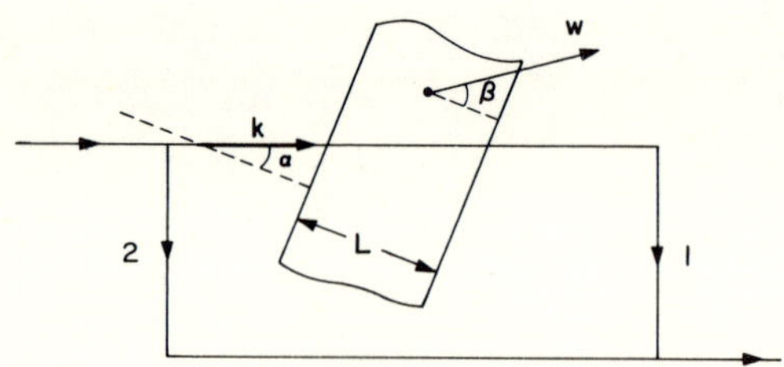


FIG. 1. Fizeau experiment employing a parallel-faced plate in one beam path.

---

† Permanent address: Stonehill College, North Easton, Massachusetts, U.S.A.

‡ Permanent address: Atominstitut der Oesterreichischen Universitaeten, Vienna, Austria.

$$\phi(\mathbf{k}) = L[\{n^2(k)k^2 - k_y^{\ 2}\}^{\frac{1}{2}} - k_x] \tag{1}$$

where $k$ is the magnitude of the incident wave vector $\mathbf{k}$ and $k_x(k_y)$ is the component perpendicular (parallel) to the plate surface. Suppose that the plate is now set in motion at a laboratory velocity $\mathbf{w}$. In the rest frame of the plate the relative phase will be $\phi(\mathbf{k}')$, where $\mathbf{k}'$ is the incident wave vector in that frame. However, since the relative phase is a relativistic invariant, $\phi(\mathbf{k}')$ must also be the relative phase in the laboratory frame. Thus the motion of the plate causes a phase shift, the Fizeau effect, given by

$$\Delta\phi = \phi(\mathbf{k}') - \phi(\mathbf{k}) \ . \tag{2}$$

For an alternative argument leading immediately to equation (2), one can imagine that instead of setting the plate in motion with velocity $\mathbf{w}$ the source of the radiation is set in motion with velocity $-\mathbf{w}$. In the case when the radiation is light and the plate is glass, equation (2) agrees with the result of familiar textbook derivations that appeal explicitly to velocity addition within the moving medium.

Now consider the case when the radiation is thermal neutrons and the material of the plate is non-absorbing. Equation (2) is relativistically exact and thus holds for both photons and neutrons; differences enter through $\mathbf{k}'$ and through the form of $n(k)$. For a slow neutron and low plate velocity $\mathbf{w}$ Galilean velocity addition holds and gives outside the moving medium

$$\mathbf{k}' = \mathbf{k} - \frac{m}{\hbar}\mathbf{w} \tag{3}$$

where $m$ is the neutron mass. For thermal neutrons in a non-absorbing plate the empirically determined index of refraction has the form

$$n(k) = (1 - 2A/k^2)^{\frac{1}{2}} \tag{4}$$

where $A$ is constant independent of $k$ and characteristic of the particular material. Thus from equation (2) the Fizeau effect for thermal neutrons is

$$\Delta\phi = -m\hbar^{-1}k^{-2}AL\omega(\cos\alpha)^{-2}\cos\beta \qquad (5)$$

to first order in $\omega$ and $A$. In terms of the density of nuclei $N$ and the scattering length $b$, the parameter $A$ is given by $A = 2\pi Nb$. The angles $\alpha$ and $\beta$ are defined in Fig. 1.

The $\beta$ dependence of $\Delta\phi$ deserves discussion. Independent of the angle of incidence $\alpha$, $\Delta\phi$ vanishes when $\beta = \pi/2$, i.e. when the plate moves parallel to its surface. This feature is easily understandable when one notes that a plate describable for neutron propagation by an index of refraction of form (4) is necessarily describable by a constant potential energy $U = \hbar^2 A/m$ in the Schrödinger equation and vice versa: a potential well, here the plate, moving parallel to its boundaries is indistinguishable from the well at rest. In addition, the fact that $\Delta\phi$ does not vanish for light in glass (Macek, Schneider, and Salamon 1964) can be directly traced to differences between equation (4) and the form of $n(k)$ for optical photons in glass. It is noteworthy that with the substitution $m \to E/c^2 = \hbar k/c$ equation (5) does apply to X-rays in, say, Lucite. This follows because with the same substitution equation (3) is the classical Doppler formula and because form (4) is correct for X-rays in Lucite.

Looking ahead to an actual experiment, we have considered several practical arrangements for moving metal plates at high speeds within a standard three-crystal neutron interferometer (Rauch, Treimer, and Bonse 1974). One design, shown schematically in Fig. 2, employs an aluminium rotor with diamond-shaped cogs. For this design equation (5) predicts a $2\pi$ phase shift at a rotor angular speed of 2940 rev min$^{-1}$. This suggests that an actual experiment for neutrons is feasible despite the limited available space in current interferometers. For X-rays the shift is smaller by a factor of $10^5$ and hence not easily observable.

Experiments, completed or in progress, have utilized four other methods of producing phase shifts in the radiation within a neutron interferometer: passage through stationary materials, terrestrial rotation of the interferometer, gravitational interaction with the radiation, and spin rotation of the radiation. All of these effects have a common feature which is also shared

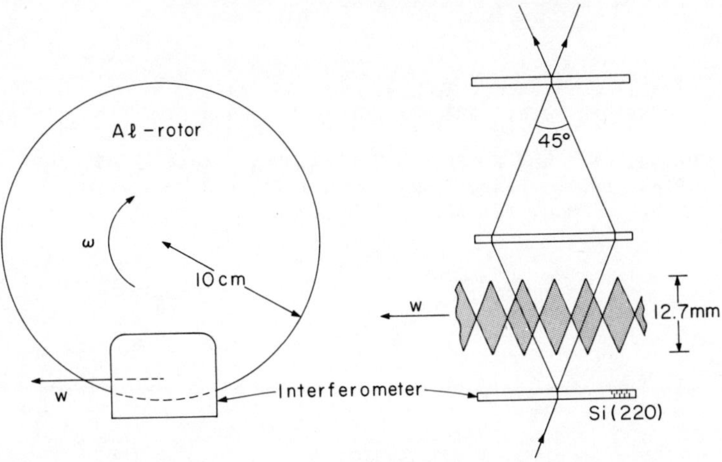


FIG. 2. Arrangement for a Fizeau experiment in a three-crystal interferometer. The design shown assumes a Bragg angle of 22.5° and employs an aluminium rotor with diamond-shaped cogs.

by the Fizeau effect: the shift is unambiguously predictable from fundamental principles. Each of the other effects has another common feature which is not shared by the Fizeau effect: the shift can be predicted by first calculating the difference $\Delta t$ in particle travel time via the two routes, converting this into an effective optical path difference $\Delta S = v\Delta t$, and obtaining the phase shift from $\Delta\phi = k\Delta S$. The reason that this simple argument fails for the Fizeau effect is that, in contrast to all previous experiments, the neutron will possess a different frequency during part of the path, i.e. while inside the moving medium.

## Acknowledgments

We wish to thank A.P. French, whose lecture at the Massachusetts Institute of Technology on the optical experiments led us to contemplate a neutron analogue, and C.G. Shull for discussions and encouragement. The research reported here was supported by grants from the National Science Foundation of the U.S.A. and the U.S. Department of Energy.

## References

FIZEAU, H.L. (1851). *C.R. Acad. Sci. (Paris)* **33**, 349.
MACEK, W.M., SCHNEIDER, J.R., and SALAMON, R.M. (1964). *J. appl. Phys.* **35**, 2556.
RAUCH, H., TREIMER, W., and BONSE, U. (1974). *Phys. Lett. A* **47**, 369.
ZEEMAN, P. (1914). *Proc. r. Acad. (Amsterdam)* **17**, 445.
—— (1915). *Proc. r. Acad. (Amsterdam)* **18**, 398.

# 16. LAUE-CASE DYNAMICAL NEUTRON DIFFRACTION WITH PERFECT NON-MAGNETIC CRYSTALS IN MAGNETIC FIELDS

A. ZEILINGER†

*Massachusetts Institute of Technology, Cambridge, Mass. 02139, U.S.A.*

## 1. Introduction

The predictions of the dynamical theory of magnetic neutron diffraction which has been developed in recent years (see Rauch and Petrascheck 1978 and references therein) cannot yet be tested owing to the unavailability of perfect magnetic crystals. However, the use of non-magnetic crystals bathed in magnetic fields can lead to some dynamical diffraction effects where both magnetism and the neutron spin play a crucial role. It is the purpose of the present paper to show some examples of the results to be expected in experiments of this kind.

## 2. Directions of neutron currents

The spin-dependent refraction of neutrons by a magnetic field influences the diffraction process at a non-magnetic crystal bathed in that field. In the present paper we restrict our attention to the symmetric Laue case and assume that in the absence of a magnetic field the incoming neutron beam satisfies the exact Bragg condition. This implies that within the crystal the neutrons propagate parallel to the lattice planes (for fundamentals of dynamical diffraction theory and notation details see Rauch and Petrascheck (1978)). Thus with a slit in front of the crystal much wider than the Pendellösung length the neutrons leave the crystal from a slit-like region along the lattice planes. If we b-the the crystal in a homogeneous magnetic field $B$ arranged in such a way that the neutrons enter it normally, i.e. there is no refractive bending, their wavelength

† Permanent address: Atominstitut der Österreichischen Universitäten, Vienna,

for field strengths which are not too large changes according to

$$\frac{1}{\lambda^{\pm}} = \frac{1}{\lambda}\left(1 \pm \frac{\mu B}{2E}\right) \tag{1}$$

where $E$ is the energy of the neutron, $\mu$ is its magnetic moment, $\lambda$ is its wavelength outside the magnetic field, and $\lambda^{\pm}$ are the wavelengths of the two spin states in the magnetic field. This wavelength change leads to a change of the propagation direction within the crystal which can be described using the parameter $y$ where

$$y_{\pm} = \mp \frac{\mu B}{2|V_G|} \tan\theta_B \sin(2\theta_B) \;. \tag{2}$$

$V_G$ is the Fourier component of the crystal lattice potential and $\theta_B$ is the Bragg angle ($\theta_B^{+} \approx \theta_B^{-}$). The linear parameter $\Gamma = \tan\Omega/\tan\theta_B$, where $\Omega$ is the angle between the direction of neutron propagation and the lattice planes, describing the position where the beams leave the interferometer if we again use an entrance slit is

$$\begin{aligned} \Gamma_{+} &= \frac{\mp y_{+}}{(1+y_{+}^{2})^{\frac{1}{2}}} \\ \Gamma_{-} &= \frac{\mp y_{-}}{(1+y_{-}^{2})^{\frac{1}{2}}} = \frac{\pm y_{+}}{(1+y_{+}^{2})^{\frac{1}{2}}} \end{aligned} \tag{3}$$

where the $\pm$ signs in the numerators refer to the two different wavefields. Thus it follows that neutrons of either polarization state will be split into two components which travel in directions symmetric to the lattice planes. The radiation following one path contains both polarization states as different wavefields. Thus, using again an entrance slit, the neutrons leave the crystal from two separated slit-like regions. This change of propagation direction can be obtained with unpolarized neutrons.

## 3. Polarization effects

As the neutron travels along each path as a superposition of the two spin states one can expect polarization effects. To demonstrate this, we assume that the incoming neutrons show a coherent superposition of spin-up and spin-down states with equal amplitudes, i.e. they are polarized in a direction perpendicular to the direction of the magnetic field experienced by the crystal. Thus, using again an entrance slit, both forward beams at the backface of the crystal can be written as

$$\begin{aligned} |I\rangle_0 &= \frac{1}{\sqrt{2}}(|\uparrow\rangle_1 + |\downarrow\rangle_2) \\ |II\rangle_0 &= \frac{1}{\sqrt{2}}(|\uparrow\rangle_2 + |\downarrow\rangle_1) \end{aligned} \tag{4}$$

where, for example, $|\uparrow\rangle_i$ denotes the neutron spin-up state travelling as wavefield $i$. It is obvious that these states experience different phase shifts within the crystal and thus the neutron spin is rotated. To demonstrate the kind of results to be expected we restrict our considerations to the forward components of beam I; their wavefunctions are

$$\begin{aligned} {}^1_{\mathrm{I}}\psi_0^{+} &= \frac{u_0^{+}}{2}\left\{1+\frac{y_+}{(1+y_+^2)^{\frac{1}{2}}}\right\}\exp(\mathrm{i}\mathbf{K}_1^{+}\cdot\mathbf{r}) \\ {}^2_{\mathrm{I}}\psi_0^{-} &= \frac{u_0^{-}}{2}\left\{1+\frac{y_+}{(1+y_+^2)^{\frac{1}{2}}}\right\}\exp(\mathrm{i}\mathbf{K}_2^{-}\cdot\mathbf{r}) \quad . \end{aligned} \tag{5}$$

where $u_0^+$ and $u_0^-$ ($|u_0^+|=|u_0^-|$) are the initial amplitudes and

$$\mathbf{K}_{1,2}^{+,-} = \mathbf{k}^{+,-} + \frac{\pi}{\Delta_0^{+,-}}\left\{\mp y_+ \pm (y^2+1)^{\frac{1}{2}}\right\}\hat{\mathbf{n}} - \frac{mV_0}{\hbar^2 k^{+,-}\cos\theta_B}\,\hat{\mathbf{n}} \ . \tag{6}$$

Here $\mathbf{k}^{+,-}$ are the neutron wave vectors outside the crystal but in the magnetic field, $\Delta_0^{+,-}$ are the spin-dependent Pendellösung lengths, $\hat{\mathbf{n}}$ is a unit vector normal to the crystal surface, $m$ is the neutron mass, and $V_0$ is the crystal potential for forward scattering. To calculate the polarization change by the crystal we first notice that the amplitudes of both waves in equation (5) are the same, which implies that the polarization vector is still confined to the plane orthogonal to the magnetic field, and secondly we recall that for a spinor wave function

$$\psi = \begin{pmatrix} \psi^+ \\ \\ \psi^- \end{pmatrix} = \frac{1}{N}\begin{pmatrix} e^{i\chi} \\ \\ e^{i\xi} \end{pmatrix} \tag{7}$$

the polarization vector with the usual choice of co-ordinate axis directions is

$$\mathbf{P} = [\cos(\xi-\chi), \sin(\xi-\chi), 0] \ . \tag{8}$$

Thus the spin-rotation angle upon traversing the crystal is obtained as the change in phase difference between spin-up and spin-down state which is

$$\phi = -\frac{\mu B}{E}\mathbf{k}\cdot\mathbf{r}_B + \frac{2mV_G}{\hbar k}\left\{y_+ - (y_+{}^2+1)^{\frac{1}{2}}\right\}D_{eff} - \frac{\mu B m V_0}{E\hbar k}D_{eff} \tag{9}$$

where $\mathbf{r}_B$ is the position vector of the exit point of the ray when the origin is chosen to coincide with its entrance point. To describe the thickness of the crystal we chose $D_{eff} = D/\cos\theta_B$ rather than its geometric thickness $D$. This effective thickness is the optical path length qualitatively interpretable as an 'infinitesimal zig-zag' path (Kato 1968) of the neutron in the crystal. The first term in equation (9) is a Larmor precession term. The last term is a refraction correction to Larmor precession and since it is smaller by a factor of about $10^5$ than the others, it can be omitted in our considerations. The second term $\phi_{dyn}$ is a spin rotation due to dynamical diffraction. To demonstrate the significance of this term we evaluate it for small $y(y^2 \ll 1)$ for which case we obtain

$$\phi_{dyn} \approx -\frac{\mu B}{E}kD_{eff}\sin^2\theta_B - \frac{V_G}{E}kD_{eff} \tag{10}$$

which combined with the first term of equation (9) leads us to

$$\phi \approx -\frac{\mu B}{E}kD_{eff} - \frac{V_G}{E}kD_{eff} \ . \tag{11}$$

Thus the spin rotation experienced by the neutron when traversing the crystal is composed of a term due to Larmor preccession in the magnetic field and a nuclear interaction term. Both of them are proportional to the effective thickness rather than the physical thickness of the crystal and, for example for silicon (220) reflection and a magnetic field of 1 T, of comparable magnitude. The nuclear interaction term is due to the property that the two spin states travel as different wave fields which again see a different crystal potential because one wave field has its nodes and the other its antinodes at the lattice planes.

## 4. Conclusion

Neutron diffraction in perfect crystals in magnetic fields provides the possibility of measuring some effects due to both dynamical diffraction and the spinor nature of the radiation. One of these effects is related to the direction of the neutron current in a perfect crystal which can be changed considerably by moderate magnetic fields. An experiment with polarized neutrons will show a spin rotation due to nuclear interaction besides the usual Larmor precession with both rotational effects demonstrating that the optical path length through the crystal is different from its geometrical thickness.

We finally note that the experiments proposed explicitly here can readily be done using a two-crystal neutron interferometer (Zeilinger *et al.* 1978) with the first crystal plate acting as a crystal collimator (Shull 1969, 1973) and the second being placed in a suitable magnetic field.

## Acknowledgments

I wish to acknowledge encouraging discussions with C.G. Shull and M.A. Horne. This work was supported by grants from the National Science Foundation of the U.S.A. and the U.S. Department of Energy.

## References

KATO, N. (1968). *J. appl. Phys.* **39**, 2225.

RAUCH, H. and PETRASCHECK, D. (1978). Dynamical diffraction on perfect crystals and its application in neutron physics. In *Neutron Diffraction* (ed. H. Dachs), *Topics in Current Physics*, Springer Verlag, Berlin.

SHULL, C.G. (1969). *Proc. Summer School on Neutron Physics, Alushta,* p. 395. JINR, Dubna, USSR.

—— (1973). *J. appl. Crystallogr.* **6**, 257.

ZEILINGER, A., SHULL, C.G., HORNE, M.A., and SQUIRES, G.L. (]978). This Conf., paper I.3.

Note added in proof:

In the meantime some of the effects predicted here have been verified experimentally (A. Zeilinger and C.G. Shull: *Phys. Rev.* **B19** (1979) 3957).

# PART III RELATED TECHNIQUES OF INTERFEROMETRY

# 1. ELECTRON INTERFEROMETRY

G. MÖLLENSTEDT and H. LICHTE

*Institut für Angewandte Physik, Universität Tübingen, Auf der Morgenstelle 12, 7400 Tübingen, Federal Republic of Germany*

## 1. Diffraction of electron waves by macroscopic objects

Diffraction phenomena of electron waves by macroscopic objects can be observed whenever they are illuminated with coherent electron beams. In the following some examples are given.

### 1.1. *Diffraction by an edge*

Figure 1 shows an example of diffraction by an edge as it is usually observed in a standard electron microscope.

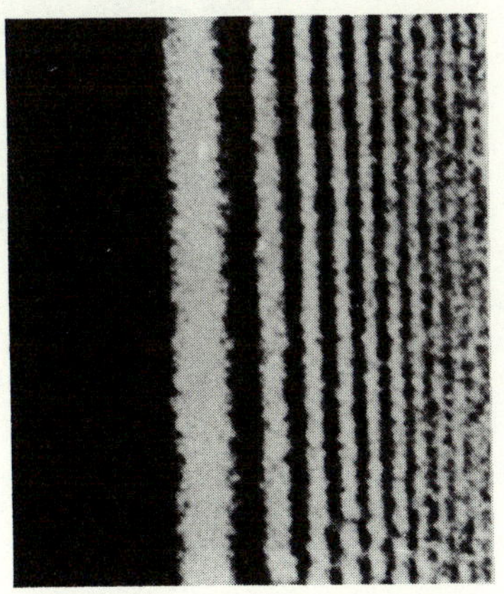

FIG. 1. Diffraction of electron waves by an edge taken in a 100 kV electron microscope (λ = 0.037 Å).

### 1.2. *Diffraction by one or several slits*

The left-hand side of Fig. 2 shows slits in a copper foil 5000 Å thick. If these slits are illuminated with coherent electrons the electron diffraction pattern shown on the right-hand side of Fig. 2 is observed (Möllenstedt and Jönsson 1959, Jönsson 1961). A grating of lines 100 Å wide spearated by 700 Å was also drawn on the foil using an electron microrecorder (Holl 1969). The upper part of Fig. 3 shows the grating and the lower part the electron diffraction pattern (note that there are 10 orders of diffraction in the pattern).

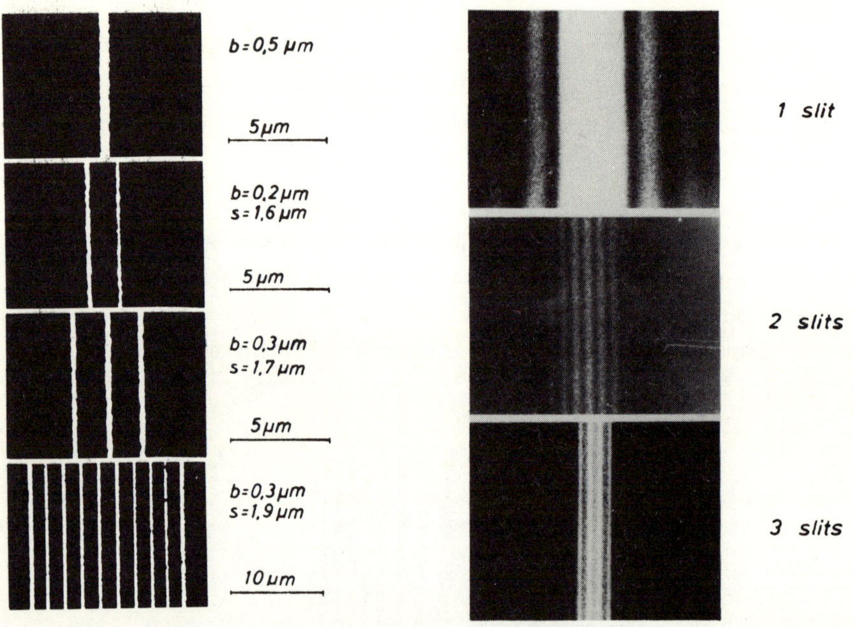


FIG. 2. Left-hand side: narrow slits in a copper foil 5000 Å in thickness; $b$, width of the slits; $s$, separation of the slits. Right-hand side: diffraction pattern produced with electron waves at the slits shown in the left-hand side; the pattern was magnified by using electron optics.

1.3. *Diffraction by two holes (Young's double-hole experiment)*
The diffraction of electron waves in an electron microscope by two holes 1 μm in diameter and separated by 2 μm yields a two-hole diffraction pattern (Fig. 4).

1.4. *Diffraction by a sinusoidal grating*
Diffraction of electrons by a sinusoidal grating produced using an electron microrecorder is feasible (Holl, unpublished). Only

the first positive and negative orders of diffraction appear in the pattern (Fig. 5).

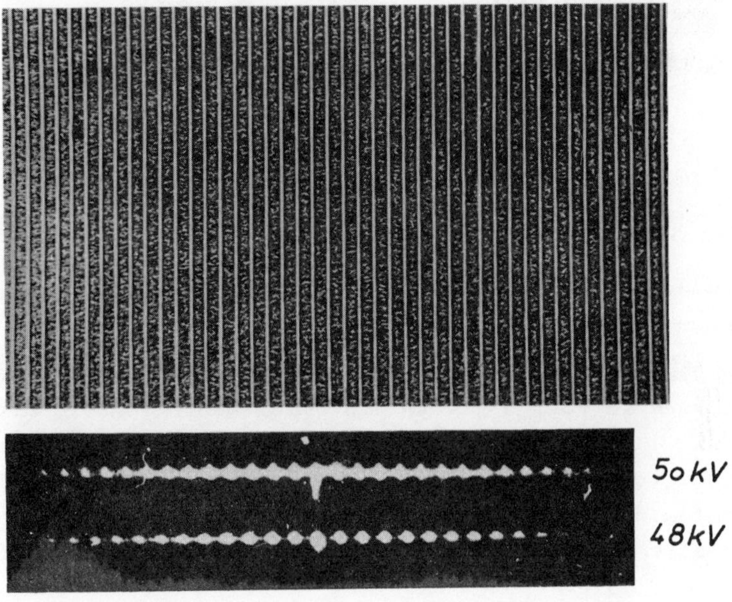


FIG. 3. Top: a grating with grating constant 700 Å and line width 100 Å drawn with an electron beam. Bottom: electron diffraction at the grating shown above for two different energies.

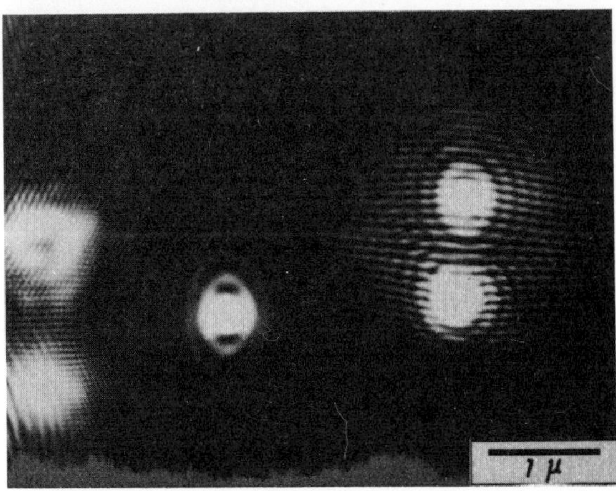


FIG. 4. Young's experiment with electrons: electron diffraction by two holes obtained using an electron microscope. (From Ohtsuki and Zeitler 1977.)

FIG. 5. Diffraction pattern obtained by diffraction of electron waves ($U_b$=50 kV) by a sinusoidal grating (grating constant 400 Å).

## 2. Amplitude splitting of electrons using a single crystal and interference of the coherent partial waves

The principle of amplitude splitting by a thin single crystal is illustrated in Fig. 6(a): the incoming monochromatic electron wave $\mathbf{K}_0$ is split into the undiffracted wave $\mathbf{K}_0$ and the diffracted wave $\mathbf{K}_1$ (corresponding to Bragg's law $2d \sin\theta/2 = n\lambda$)

The following experiment can then be realized in a high-quality electron microscope. The direct beam $\mathbf{K}_0$ and the diffracted beam $\mathbf{K}_1$ are superposed by an electron optical lens. The interference of $\mathbf{K}_0$ and $\mathbf{K}_1$ results in an image of the lattice plane. The ability to produce an image of a lattice plane with a lattice constant of about 0.6 Å is evidence for the high stability of the electron microscope (Fig. 6(b)).

## 3. Electron interferometer

In an electron interferometer an incoming wave is split into two coherent partial waves by amplitude division or by wave-front division and the two partial waves are superposed to give an interference pattern.

### 3.1. *Amplitude division by means of single crystals*

Marton (1952) suggested that an incoming electron wave could be diffracted by a set of three single crystals which would divide its amplitude into the different orders of diffraction (Fig. 7(a)). Limitation of the number of contributing rays by apertures leads to the ray path shown in Fig. 7(b).

The realization of this type of interferometer is very

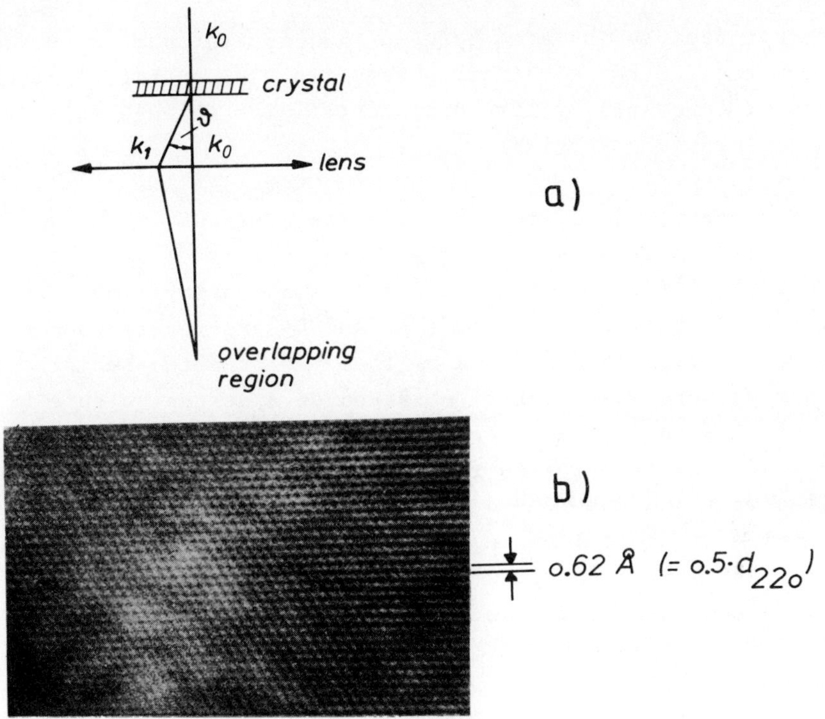


FIG. 6. (a) The partial wave $\mathbf{K}_1$ of the monoenergetic electron wave $K_0$ is deflected by the angle θ in a single crystal according to Bragg's law. $\mathbf{K}_0$ and $K_1$ are superimposed by the lens. (b) Lattice image of Ni (220) planes taken with a field emission electron microscope. (Courtesy of Hitachi Ltd, Japan.)

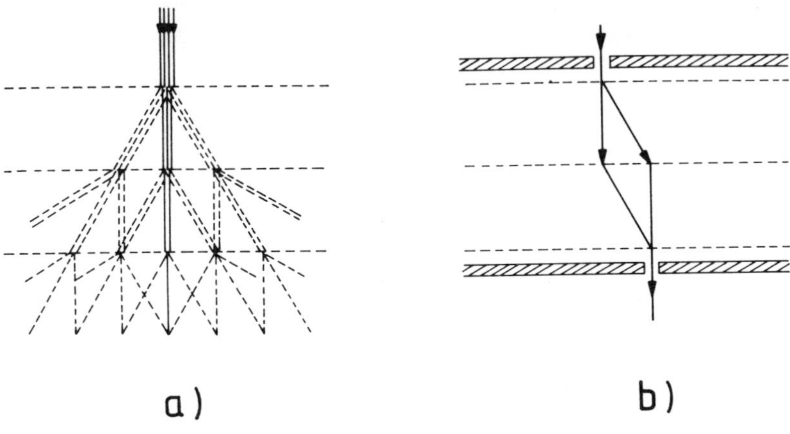


FIG. 7. (a) Schematic illustration of the generation of partial waves and their superposition by diffraction at three single crystals. (From Marton 1952.) (b) By insertion of two apertures only two partial waves are superposed.

difficult because the area between the coherent partial waves is too great. Stray alternating magnetic fields that result in a loss of coherence by time-dependent fluxes through this area exist in all laboratories.

### 3.2. *Wave-front division by an electron biprism*

By way of analogy we show in Fig. 8 the effect of a light optical biprism on the path of light waves coming from a source L: two coherent virtual sources L′ and L″ are generated and superposing the waves emitted by them yields a two-beam interference pattern which can be observed on a screen. Figure 9(a) demonstrates this analogy between the biprism for light (left-hand side) and the biprism for electron waves (right-hand side) (Möllenstedt and Düker 1954, 1956). The electrons are deflected by a filament about 1 μm in diameter which carries a charge of several volts with respect to the earthed electrodes at the sides. Since the field strength decreases according to $1/r$, where $r$ is the distance from the filament, in the region surrounding the filament, the wavefront of each partial wave is deflected by the same angle regardless of the distance from the filament. This means that the electron biprism also produces two coherent virtual sources.

While each biprism in light optics is characterized by its fixed angle of deflection, this angle of the electron biprism can be changed over a wide range by variation of the filament voltage $U_F$. Changing the voltage continuously from $U_F$ = 0 V to

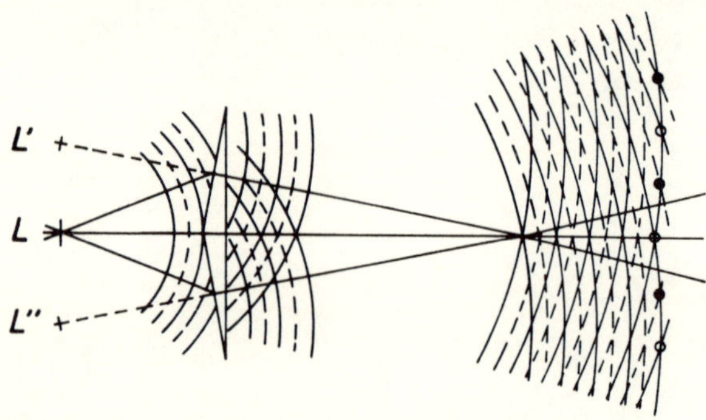


FIG. 8. Schematic representation of light optical biprism interference.

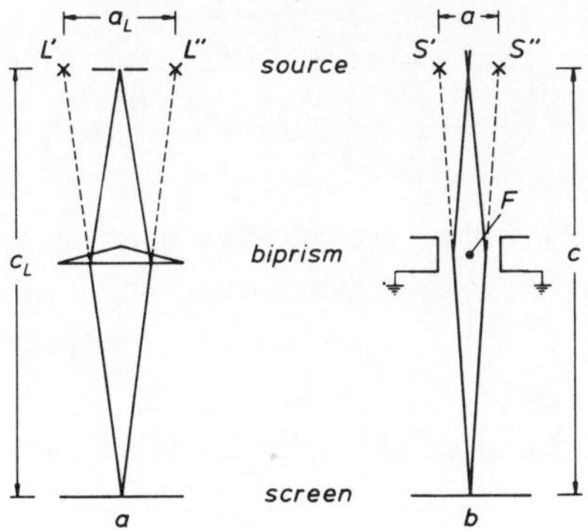


FIG. 9. Ray paths: (a) light optical biprism; (b) biprism for electrons.

$U_F$ = 7 V makes the Fresnel diffraction fringes come closer and overlap to produce the well-known biprism interference pattern in the observation plane (Fig. 10).

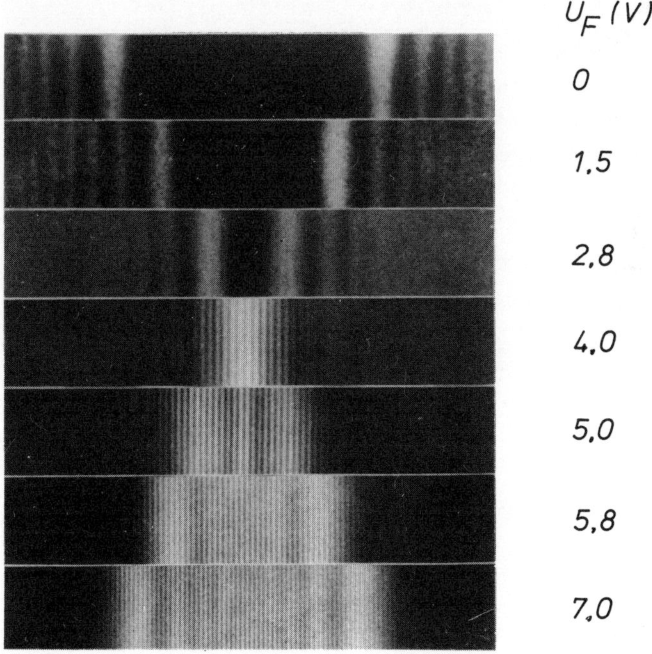


FIG. 10. Superposition of the partial waves and fringe spacing of the biprism interference as a function of the filament voltage $U_F$.

## 4. Experiments using the electron biprism interferometer

### 4.1. *Measurement of the mean inner potential of thin films*

Keller (1961) constructed the interferometer with the optical path drawn in Fig. 11: the two partial waves transmitted through an object with two thicknesses $D_1$ and $D_2$ are superimposed on each other. The interference fringes are shifted by an amount $\Delta s$ owing to the phase shift of one partial wave with respect to the other, and the mean inner potential $\phi_i$ can be calculated using

$$\phi_i = \frac{h}{e} \frac{v \Delta s}{\mathrm{d}s}$$

where $d = D_1 - D_2$, $v$ is the velocity of the electrons, $s$ is the spacing of the interference fringes and $\Delta s$ is the fringe shift.

Figure 12 illustrates the way to adjust the two partial waves and the steps of the objects and what to expect in the observation plane.

As an experimental result, Fig. 13 shows an interferogram allowing the calculation of the mean inner potential. This method has been shown to work for different materials and for accelerating voltages up to 300 kV (Kerschbaumer 1967, Schaal 1971).

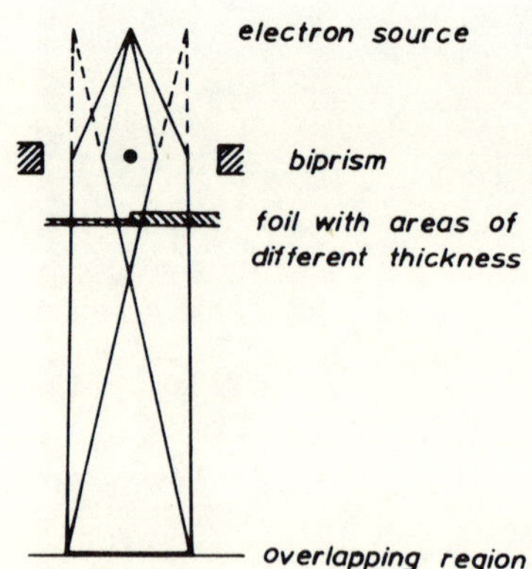


FIG. 11. Ray paths in the electron interferometer for measuring the mean inner potential.

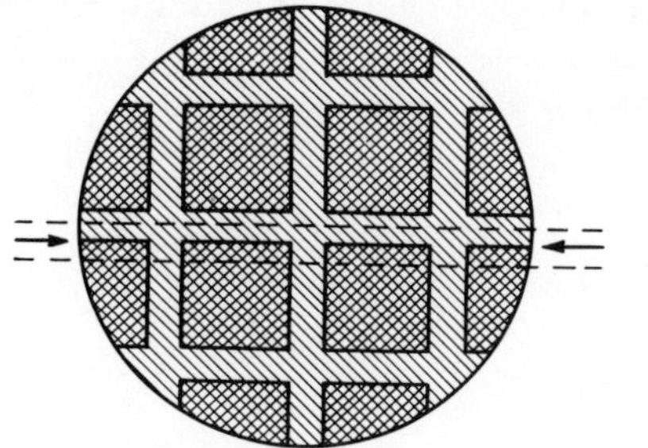

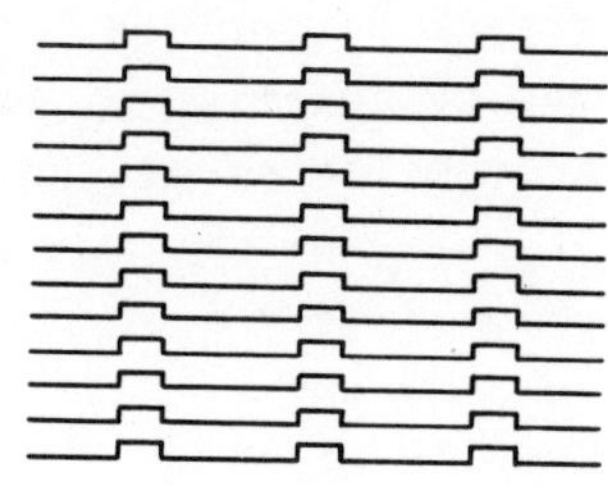

FIG. 12. Measurement of the mean inner potential: left-hand side, adjustment of the coherent partial waves with respect to the areas of different thickness; right-hand side, expected interference fringes.

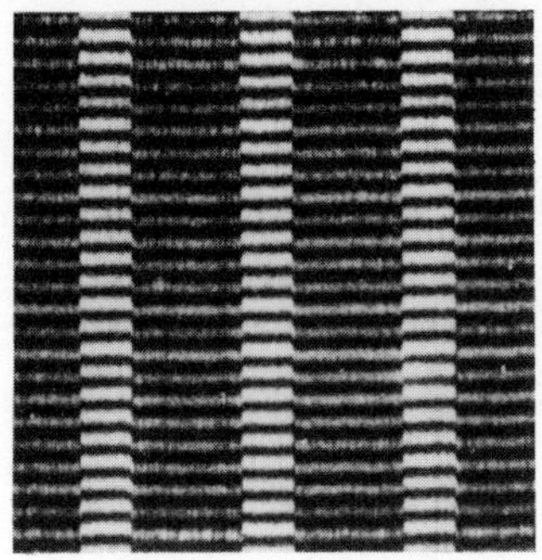

FIG. 13. Example of an interference pattern for measuring the mean inner potentials: $U_b$ = 60 kV; material, aluminium; difference in thickness, 100 Å.

## 4.2. *Measurement of the contact potential*

The deflection of the electrons by the biprism filament is strongly affected by the potential difference between the filament and the earthed electrodes beside it. The contact potential between the filament and the earthed electrodes influences the spacing of the interference fringes if they are made of different materials. Variation of the filament metal produces a change in fringe spacing, the amount of which allows one to determine the difference of the contact potentials.

Figure 14(a) shows two distinct metals evaporated side by side onto the filament. Metal A is in the area a-a and metal B is in the area b-b. In the interference pattern (Fig. 14(b)) the variation of the fringe spacing due to the difference of the contact potentials between A and B can be seen. This permits the calculation of the contact potential difference (Krimmel, Möllenstedt, and Rothemund 1964).

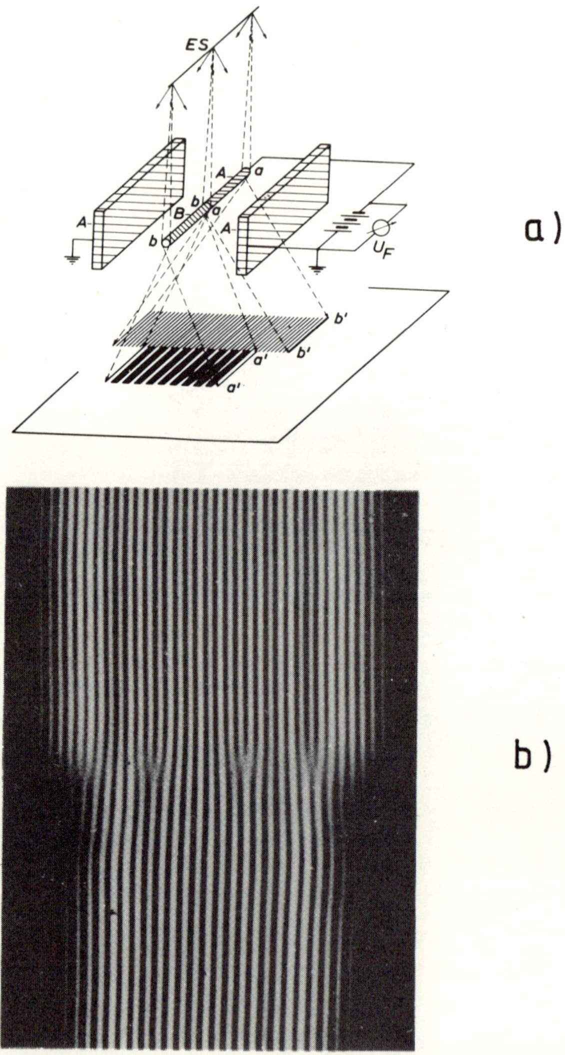


FIG. 14. (a) Experimental arrangement: the filament a-a and b-b has been metalized with different metals by vacuum evaporation. Owing to the potential difference with respect to the plane electrodes the interference fringes show different spacing. (b) Variation of fringe spacing due to the change of contact potential difference along the biprism filament as shown in Fig. 14(a).

Another way to perform the experiment is to evaporate material B onto the filament when the fringe spacing with material A on the filament is already determined, and to measure the change of spacing that occurs for use in calculating the difference of contact potential between A and B (Brünger 1972). By using the Moiré method illustrated in Fig. 15 this measurement can be made very precisely.

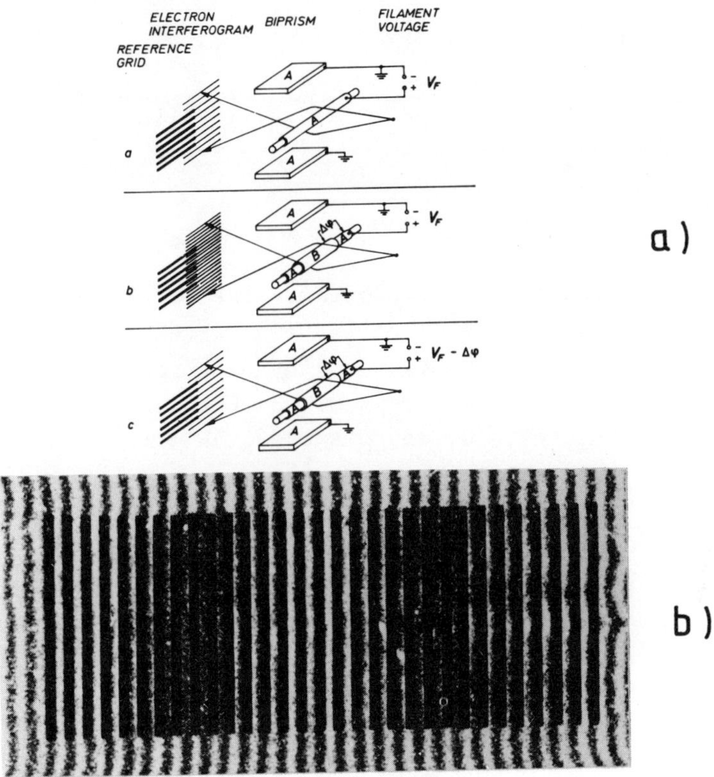


FIG. 15. (a) The filament originally covered with material A is metalized with another material B under ultra-high vacuum. The interference fringes form a moiré pattern with a fixed lattice cut into the surface of the fluorescent screen. The potential difference to be measured is identical with the correction voltage needed to remove the moiré pattern. (b) Moiré pattern phenomenon created by superposing the biprism interferences and the reference grid on the screen.

### 4.3. *Measurement of the phase shift by the magnetic vector potential*

By placing both partial waves in a uniform magnetic field (Fig. 16(a)) the whole interference pattern is shifted without changing its structure because of the phase shift produced by the magnetic vector potential **A** (Fig. 16(b)). In the presence of a magnetic field **B** = rot**A** the electron optical index of refraction is given by

$$n(\mathbf{r},\mathbf{s}) = \{2em_0U(1+eU/2m_0c^2)\}^{\frac{1}{2}} - e(\mathbf{A},\mathbf{s}) \ .$$

The phase of both partial beams is mutually shifted by the amount

$$\Delta\phi = -\frac{e}{\hbar} \oint \mathbf{A}\,\mathrm{d}\mathbf{s} = -\frac{e}{\hbar}\Phi$$

where $\Phi$ is the magnetic flux enclosed by the partial waves. Strangely enough, the phase of the partial waves is shifted if they enclose a magnetic flux even if the electrons move in regions without any magnetic field **B** and thus do not experience a Lorentz force $\mathbf{K} = -e\mathbf{v} \times \mathbf{B}$. In this case the position of the

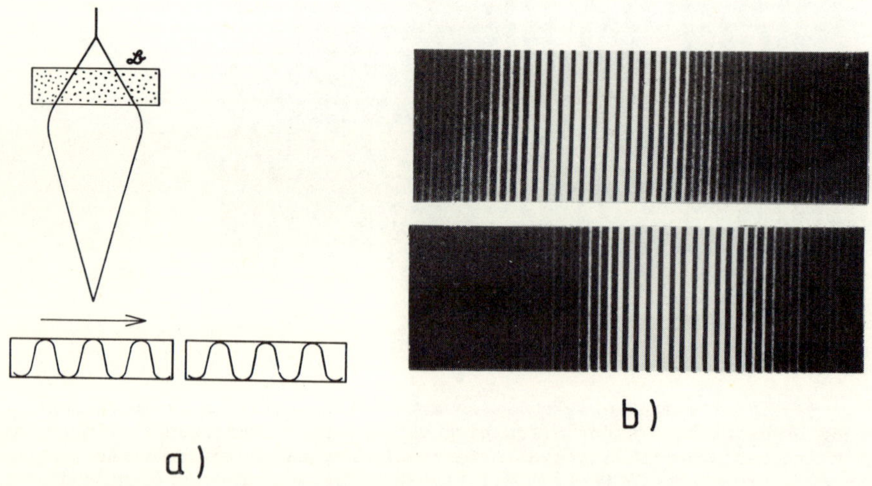


FIG. 16. (a) Top: a magnetic field is superposed upon the coherent partial waves. Bottom: the interference pattern is entirely shifted aside. (b) Experimental result corresponding to Fig. 16(a).

interference field itself stays fixed, while the interference fringes move sideways owing to the phase shift (Fig. 17(b)).

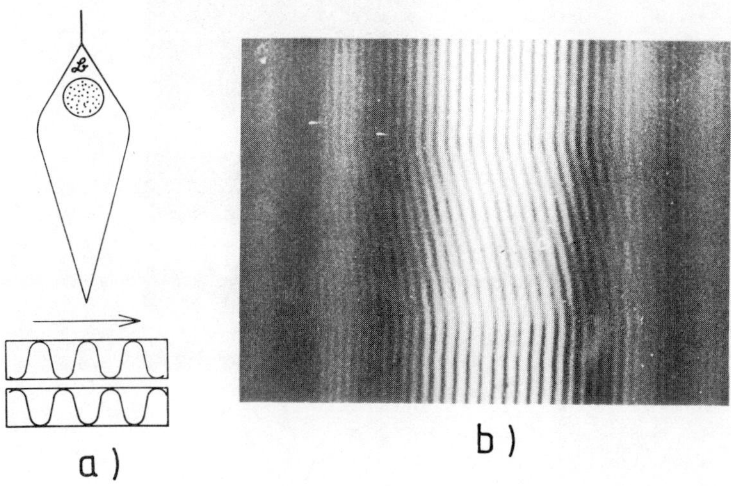


FIG. 17. (a) Top: phase shift of electron wave in a field-free space by the magnetic vector potential. Bottom: the interference region stays fixed while the interference fringes are shifted aside. (b) Reproduction of the motion of interference fringes with respect to the interference region by a continuously changed magnetic flux.

This was proved experimentally by Bayh (1962) who used the arrangement sketched in Fig. 18(a). The fabrication of small-diameter magnetic coils turned out to be a difficult problem but by means of special equipment, coils of diameter 4.7 μm were finally produced (Fig. 18(b)). The electric current through the coil was changed continuously and the fringe shift thus produced was measured as a function of the magnetic flux. The best result for one fluxon $\Phi_0 = h/e$ (i.e. the magnetic flux needed to shift the phase by $2\pi$) was (Schaal, Jonsson, and Krimmel 1966)

$$\Phi_0 = 4.15 \times 10^{-7} \text{ G cm}^2 \pm 1.5\%$$

in good agreement with theoretical predictions (Glaser 1952, Ehrenberg and Siday 1949, Aharanov and Bohm 1959).

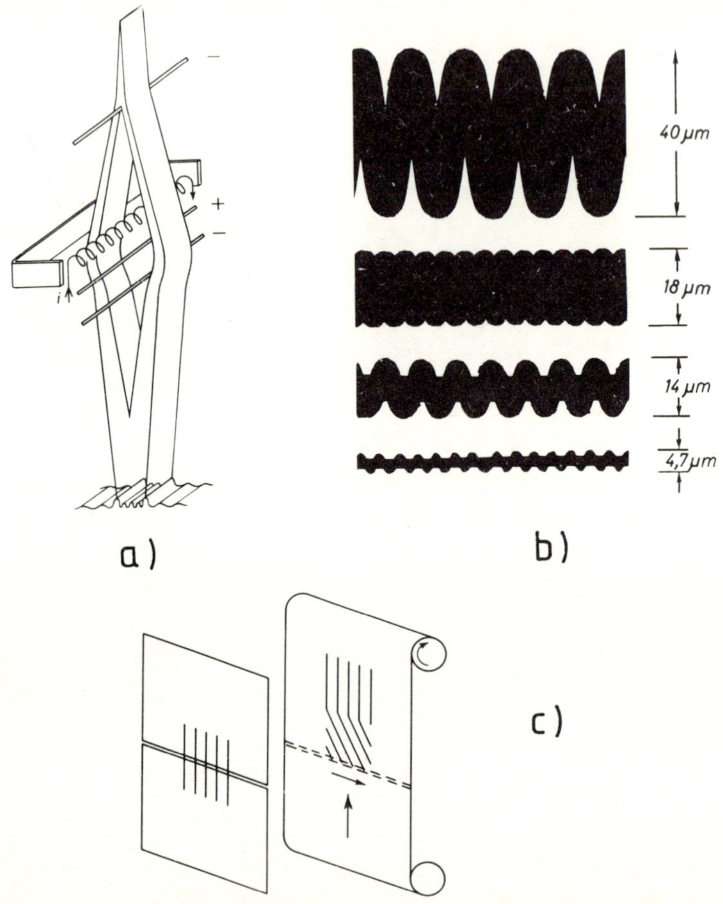


FIG. 18. (a) Principal set-up of the electron interferometer constructed by Bayh (1962): the upper biprism with a negative voltage makes the partial waves diverge and then converge again by the action of the second biprism with a positive voltage. The third biprism is used to create a small angle of superposition in the observation plane. (b) Microcoils for the generation of the magnetic flux to be inserted between the partial waves. (c) Illustration of the method used for recording the interference fringes: the fringes are recorded on a moving photographic film through a fine slit.

### 4.4. *Measurement of the fluxoid in a superconducting hollow cylinder*

Small magnetic fluxes play an important role in the field of superconductivity where the theory predicts that the 'frozen-in' flux in superconducting hollow cylinders is quantized in multiples of $\Phi_0/2$. Doll and Näbauer (1961) and Deaver and Fairbank (1961) confirmed this experimentally.

Wahl (1970a,b) and Boersch and Lischke (1970) measured the quantization of the flux by means of an electron interferometer. Wahl used an apparatus similar to that of Bayh (1962) shown in Fig. 18(a) where instead of a coil he inserted a superconducting hollow cylinder between the partial waves. The subsequent shift of the interference fringes permitted the calculation of the 'frozen-in' flux. These measurements also verified the theoretical predictions.

### 4.5. *Measurement of the source width using an electron interferometer*

Applying the method Michelson (1890) used in his stellar interferometer, the intensity distribution in an electron source can be determined interferometrically (Braun 1972). According to Born and Wolf (1959) the intensity distribution in the interference pattern is given by

$$J = J_1 + J_2 + 2(J_1J_2)^{\frac{1}{2}}|\gamma_{12}|\cos(Ky+\beta_{12})$$

where $J_{1,2}$ is the intensity of the partial waves, $\gamma_{12} = |\gamma_{12}|\exp(i\beta_{12})$ is their mutual coherence function, $K = 2\pi/s_F$ is the spatial frequency of the interference fringes and $y$ is the coordinate in the observation plane perpendicular to the fringes. If the intensities equal each other (i.e. $J_1 = J_2$) $|\gamma_{12}|$ is the visibility (i.e. contrast) of the interference fringes. If the electrons are sufficiently monoenergetic $\gamma_{12}$ is only dependent on $K$, and $\gamma_{12}(K)$ and the intensity distribution $I(x)$ in the electron source are Fourier transforms of each other. For this reason the intensity distribution can be determined by measuring the dependence of the visibility on the spatial frequency $K$ and calculating its Fourier transform (Braun 1972).

In Fig. 19(a) interference patterns are reproduced with different spacings $s_F$ generated using a field emission gun. The visibility measured by means of a microdensitometer trace (Fig. 19(b)) can be used to obtain the intensity distribution $I(x)$ of the field emitter (Fig. 19(c)) by Fourier transformation (Speidel and Kurz 1977).

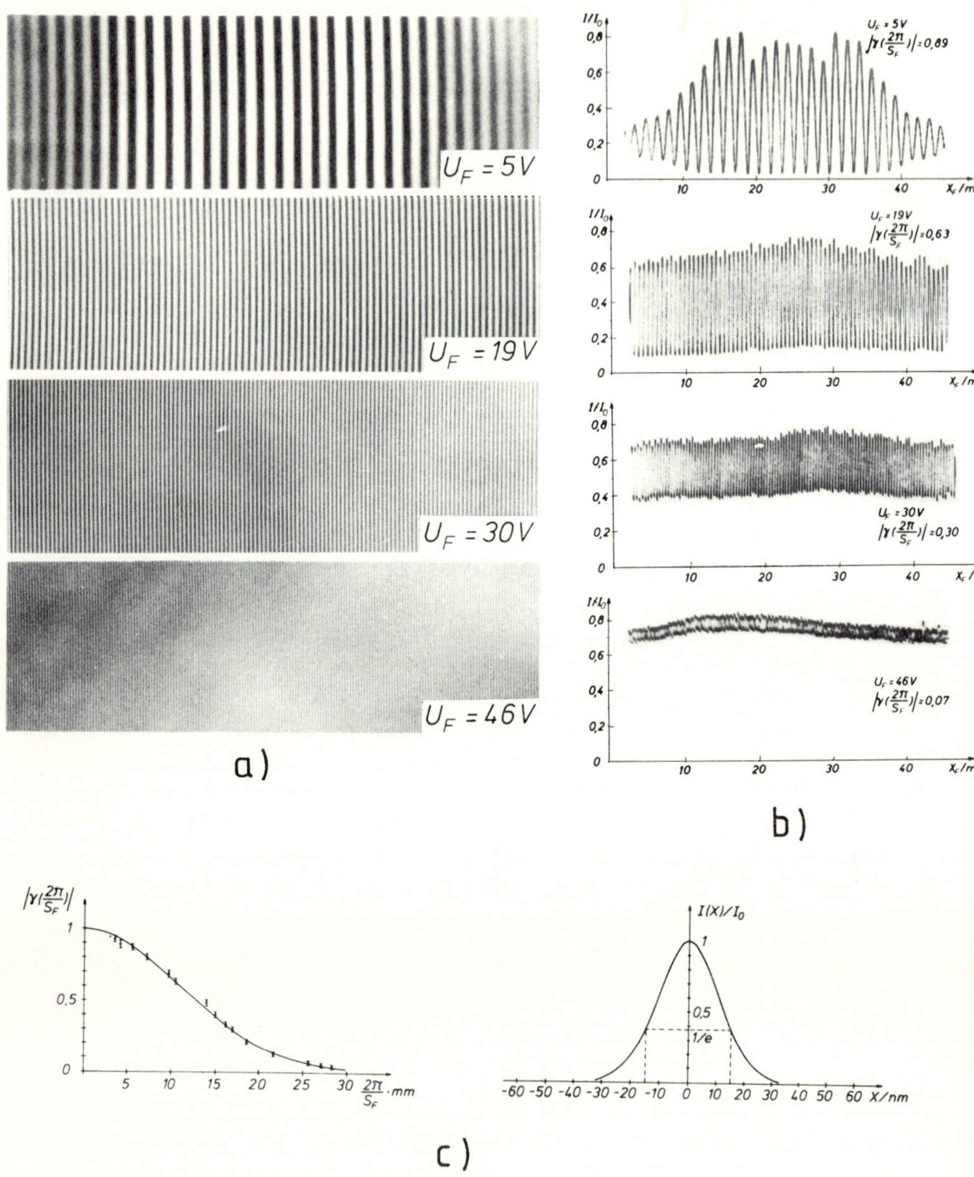


FIG. 19. (a) Electron biprism interferences with decreasing fringe spacing $s_F$ obtained using a field emission gun. (b) Microdensitometer traces for evaluation of the visibility of the fringes in Fig. 19(a). (c) Visibility $|\gamma_{12}(K)|$ as a function of $K = 2\pi/s_F$ (left-hand side) and intensity distribution $I(x)$ in the field emission electron source as a Fourier transform of $\gamma_{12}(K)$ (right-hand side).

## 5. Interference microscopy with electrons

Combining the principles of an electron microscope and of an electron interferometer produces an electron interference microscope. Such a microscope forms an image of the object and makes the phase modulation by the object visible in an interference pattern in the image plane, affording a high lateral resolution.

### 5.1. *Electron interference microscope in the transmission mode*

Figure 20 shows the path of rays in the interference microscope in the transmission mode devised by Buhl (Möllenstedt and Buhl 1957, Buhl 1959). The microscope is used for measuring the mean inner potential of very small extended areas in the object (Fig. 21). Another arrangement was used by Faget and co-workers (Faget and Fert 1957, Fert *et al.* 1962). Merli, Missiroli, and Pozzi (1974), Pozzi and Missiroli (1973), and Tonomura (1972) demonstrated the ability to measure electric and magnetic structure by means of electron interference microscopy.

Wahl (1974, 1975) used an interference microscope for recording image plane holograms with electrons. Reconstructing the holograms using a He–Ne laser he achieved a lateral resolution of about 20 Å which indicates the great potential for applica-

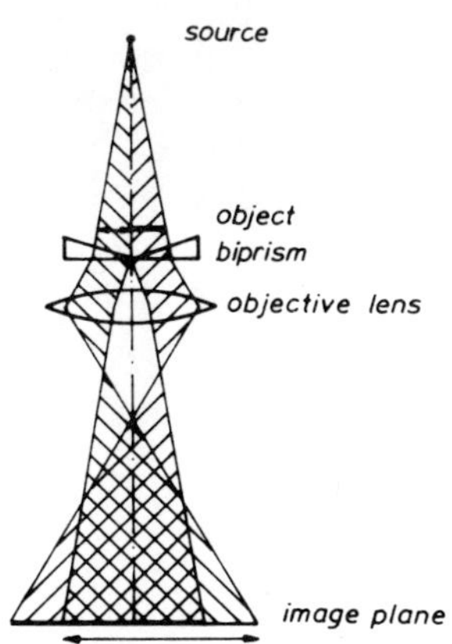


FIG. 20. Ray paths in the electron interference microscope constructed by Buhl (Möllenstedt and Buhl 1957, Buhl 1959).

tion of holographic interferometry in the field of electron holography (Fig. 22).

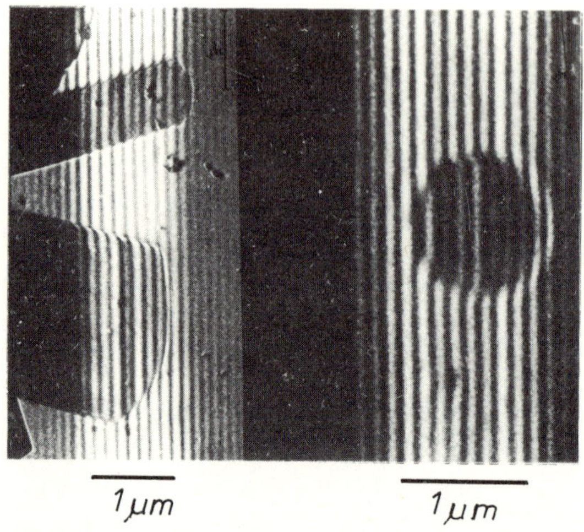


FIG. 21. Electron interference micrograph of $MoO_3$ crystals (left-hand side) and of a small aluminium disc about 100 Å thick (right-hand side) taken with the microscope shown in Fig. 20.

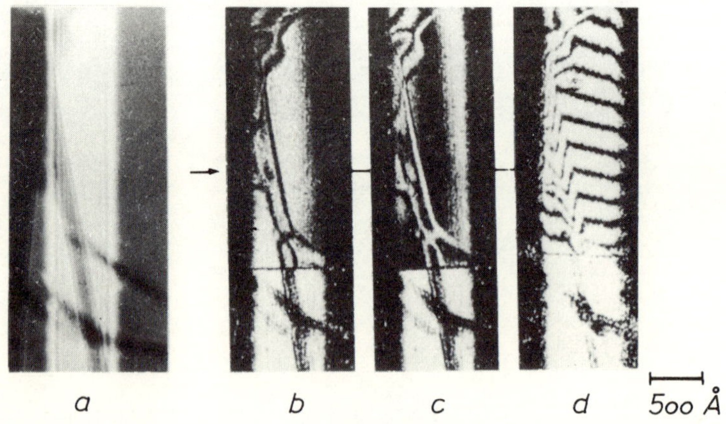


FIG. 22. (a) Image plane hologram taken with electrons b–d. The reconstructions were made using laser light. The upper parts were reconstructed using different reference waves to represent the phase distribution in the object (holographic interferometry).

### 5.2. *Electron interference microscope in the reflection mode*

In an electron mirror interference microscope a highly coherent 25 keV electron beam from a field emission gun with a coherence width of 10 μm is directed against a decelerating field. The beam is reflected close to a conducting 'mirror' electrode at a potential slightly negative with respect to the cathode.

While it is being reflected in the electron mirror the phase of the electron wave is modulated according to the electrical or topographical structure of the mirror electrode. An electron mirror interference microscope with the schematic arrangement illustrated in Fig. 23 has been operated successfully (Lichte, Möllenstedt and Wahl 1972, Lichte and Möllenstedt 1977); it allows one to measure topographical surface structures with a height of only a few tenths of an Ångström. In Fig. 24 an interferogram that was taken after the reflection of the electron wave at the object drawn in Fig. 24(a) is reproduced. It shows clearly that the phase is shifted by $2\pi$ by this object for every 5 Å in height difference.

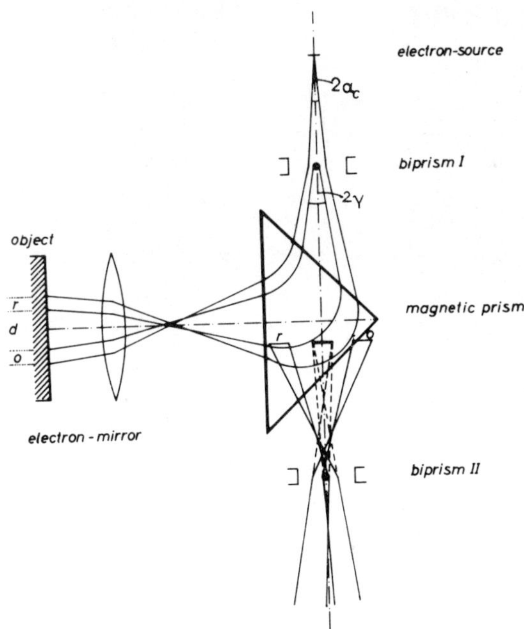


FIG. 23. Schematic set-up of the electron mirror interference microscope.

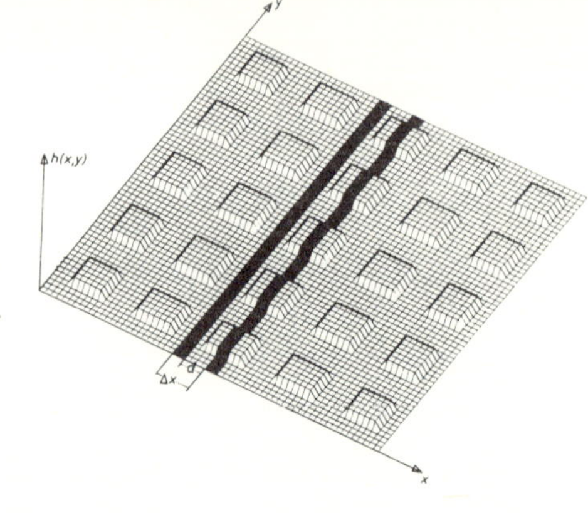


a)

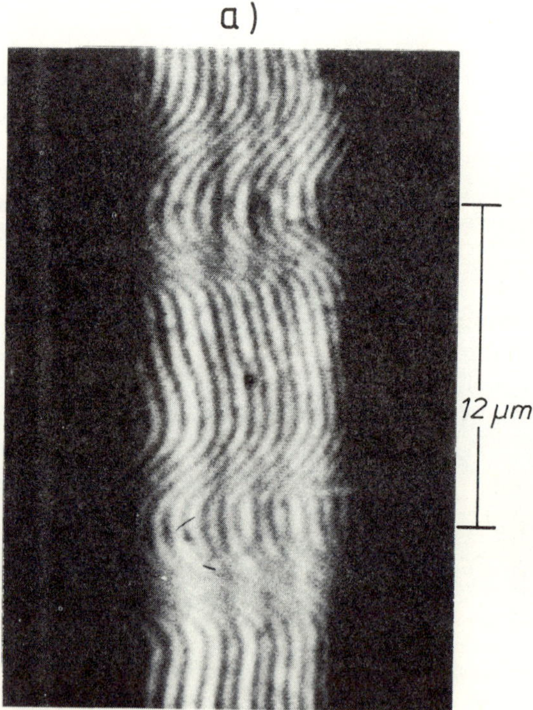


b)

FIG. 24. (a) Specimen for emphasizing the height resolution of the electron mirror interference microscope. Gold squares 25 Å in height and with a lateral spacing of 12 μm are evaporated onto a polished glass plate covered with gold. (b) Interference pattern produced by the partial beams r and o reflected at the black areas in Fig. 24(a).

This extremely high sensitivity is obviously useful for measuring surface structures of minimal height as for example the roughness of highly polished glasses. In Fig. 25 interference patterns of some different areas of a surface are reproduced where the mean roughness was measured to be 2.7 ± 0.9 Å. Even this slight roughness produces large phase shifts.

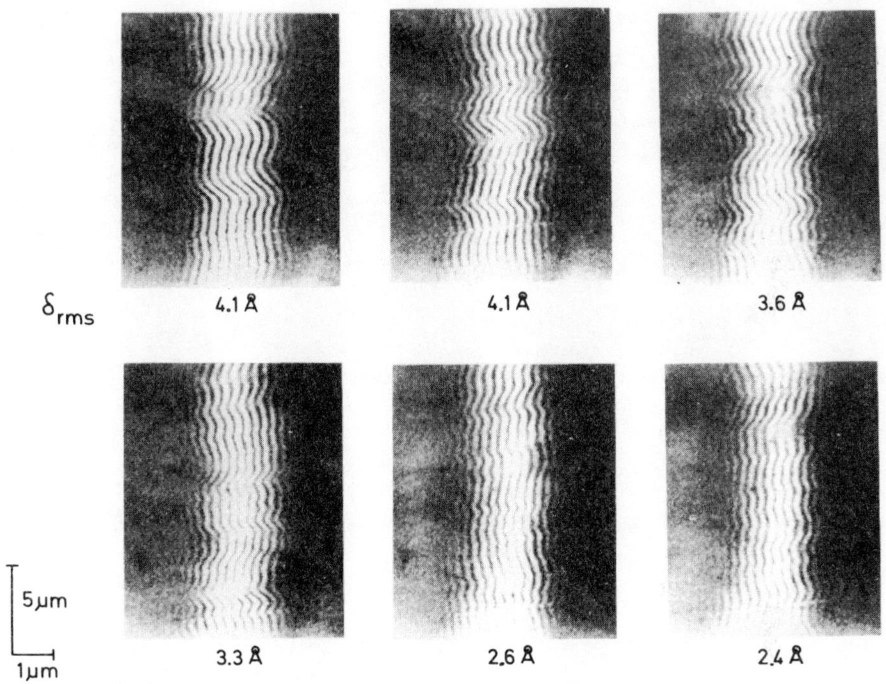


FIG. 25. Interferograms of several areas of a highly polished glass plate covered with gold recorded in the electron mirror interference microscope. By evaluation of 15 such interferograms the mean roughness of the surface was determined to be $\delta_{rms}/\sqrt{2} = 2.7 \pm 0.9$ Å.

## 6. Physical applications of the electron interferometer

### 6.1. *Beating electron waves produced by the microbetatron*

It was experimentally shown by Bayh (1962) that the phase is shifted by the magnetic vector potential or, identically, by the enclosed flux. A continuous change of this flux produces an electrical vortex **E** around the flux tube (Fig. 26). Measuring this changing flux and the surrounding vortex by electron interferometry, the electron passing on the right-hand side (path 1) is accelerated by **E** while it is decelerated by **E** passing on the left-hand side (path 2). This gives rise to two distinct wavelengths of the electron wave differing only by a very small amount which produces a beat of the electron wave observable in the interference pattern. The beat frequency can be calculated as follows.

After acceleration or deceleration the energy of the electron is given by

$$E_1 = m_0c^2 + e\left(U_b + e\int_{a \text{ path } 1}^{b} \mathbf{E}\cdot d\mathbf{s}\right)$$

and

$$E_2 = m_0c^2 + e\left(U_b + e\int_{a \text{ path } 2}^{b} \mathbf{E}\cdot d\mathbf{s}\right)$$

respectively. Therefore the difference in energies is

$$\Delta E = E_1 - E_2 = e \oint \mathbf{E}\cdot d\mathbf{s} = -e\dot{\Phi}$$

and remembering that $E = h\nu$ one obtains the beat frequency $\Delta\nu$ as

$$\Delta\nu = \frac{\Delta E}{h} = -\frac{e}{h}\dot{\Phi} \ .$$

Changing the flux continuously by one fluxon $\Phi_0 = h/e$ per second induces a voltage $U_{\text{ind}} = \oint \mathbf{E}\cdot d\mathbf{s} = 4.135 \times 10^{-15}$ V and produces a beat frequency of 1 Hz.

### 6.2. *Doppler shift of electron waves*

Lichte has obtained similar beating of electron waves when he produced a Doppler effect for electrons by reflecting an electron wave at a slowly moving mirror (Möllenstedt and Lichte 1978a). In the electron mirror interference microscope (Fig. 23) the mirror surface is tilted slowly using a cylindrical piezoceramic spacer ($\varepsilon = 6$ Å $\text{V}^{-1}$) (Fig. 27(a)) which produces a tilt angle of $\varepsilon U_{\text{piezo}}/s$. The piezovoltage is increased linearly with time so that two points separated by a lateral

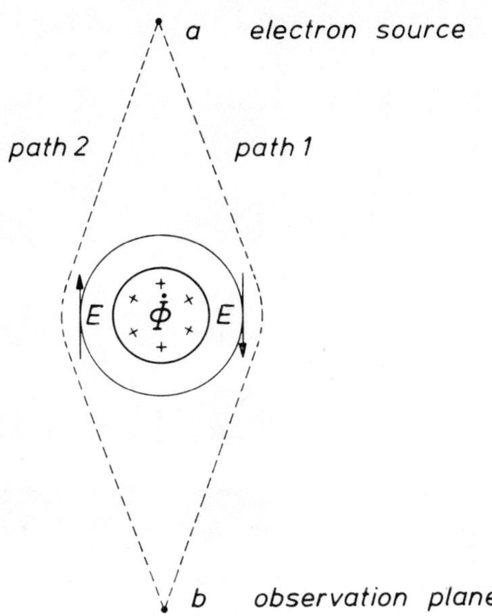


FIG. 26. Illustration of the generation of beats in a microbetatron.

distance $\Delta x$ on the mirror surface move with velocities perpendicular to the mirror which differ from each other by $\Delta v = \varepsilon(\mathrm{d}U/\mathrm{d}t)\Delta x/s$. The two partial waves reflected at the mirror at a lateral distance $\Delta x$ from each other suffer a Doppler frequency shift $\Delta\nu_D = \Delta E/h$ where $\Delta E = (4/3)(h/\lambda)\Delta v$ is the difference in the energies taken up by the moving mirror. Superposition of the partial waves yields a time-dependent intensity distribution (Fig. 27(b)). In the recording plane a sufficiently narrow slit is arranged parallel to the interference fringes. The current passing through this slit varies with the beat frequency $\Delta\nu_b$. Figure 27(c) shows the result of this experiment: the upper curve shows the variation of the piezovoltage and the lower curve the resulting intensity beats. The calculated Doppler frequency and the measured beat frequency agree closely.

### 6.3. *Bimirror interferences*

The optical bimirror interference experiment by Young and Fresnel which has been known for much more than a century has been realized with electrons (Möllenstedt and Lichte 1978b).

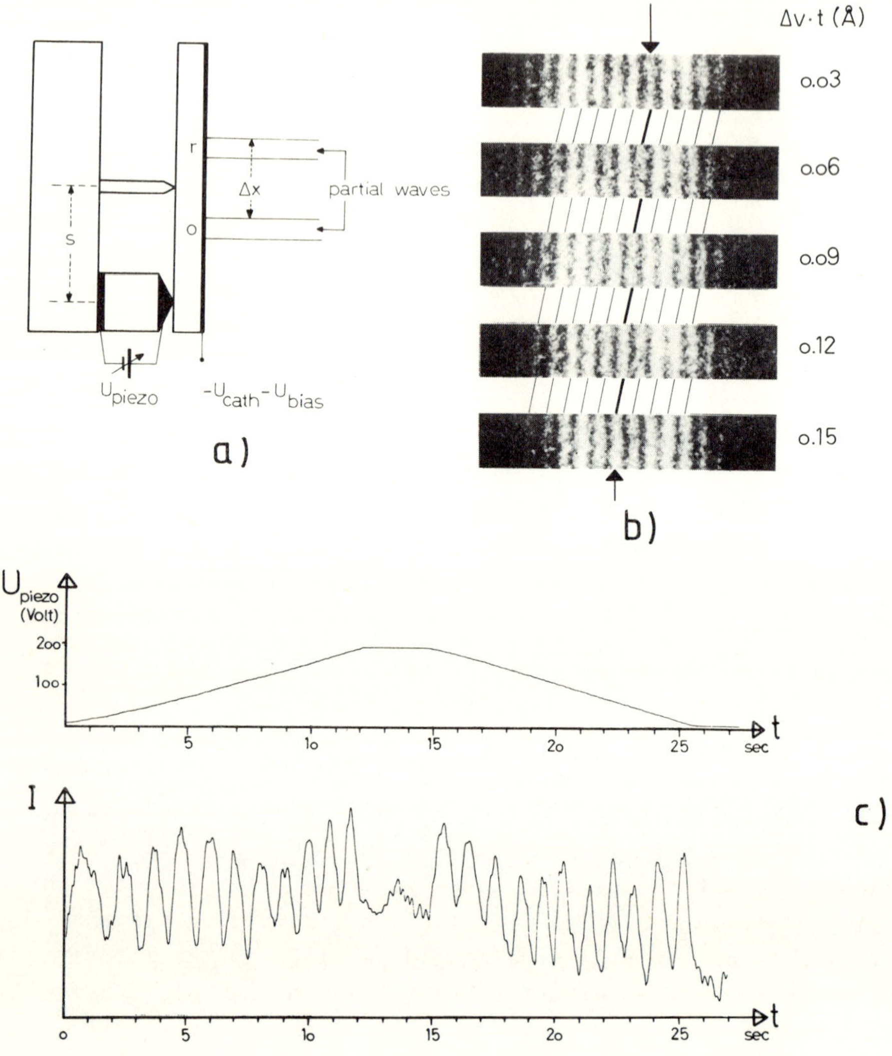


FIG. 27. (a) Mirror surface tiltable by a piezoceramic spacer. (b) Motion of the interference fringes due to Doppler shift. (c) Experimental result: simultaneously recorded piezovoltage (upper curve) and intensity beats (lower curve).

In the electron mirror interference microscope an echelette grating with a lattice constant of 1.6 μm is used as a mirror electrode. Owing to the periodic structure of the grating the inclination of the equipotentials as well as the phase modulation of the reflected wave vary with the same spacing (Fig. 28). Consequently, without the use of a biprism interference phenomena in the reflected wave are observed in the region of each grating element which can be approximately understood as interference of the partial waves originating from two coherent virtual sources (Fig. 29).

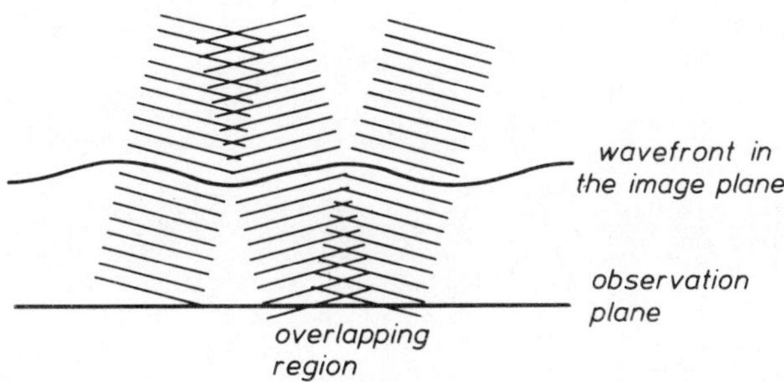


FIG. 28. Illustration of the formation of interference fringes by reflection of the electron wave at an echelette grating. If the observation plane is out of focus with respect to the image plane of the mirror surface each grating element produces one overlapping region in the observation plane.

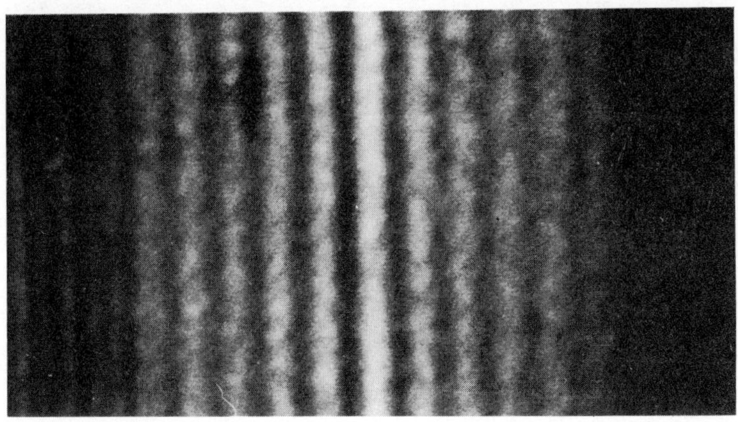

FIG. 29. Interference pattern produced by wavefront division using an echelette grating element as a wavefront divider.

**Acknowledgment**

The authors wish to thank Prof. W.D. Gregory for his assistance in translating the manuscript of this paper into English.

**References**

AHARANOV, Y. and BOHM, D. (1959). *Phys. Rev.* **115**, 485.
BAYH, W. (1962). *Z. Phys.* **169**, 492.
BOERSCH, H. and LISCHKE, B. (1970). *Z. Phys.* **237**, 449.
BORN, M. and WOLF, E. (1959). *Principles of optics.* Pergamon Press, Oxford.
BRAUN, K.J. (1972). Diplomarbeit, Tübingen.
BRÜNGER, W. (1972). *Z. Phys.* **250**, 263.
BUHL, R. (1959). *Z. Phys.* **155**, 395.
DEAVER, B.S. and FAIRBANK, W.M. (1961). *Phys. Rev. Lett.* **7**, 43.
DOLL, R. and NÄBAUER, M. (1961). *Phys. Rev. Lett.* **7**, 51.
EHRENBERG, W. and SIDAY, R.E. (1949). *Proc. phys. Soc. B* **62**, 8.
FAGET, J. and FERT, C. (1957). *Cah. Phys.* **83**, 285.
FERT, C., FAGET, J., FAYOT, M. and FERRÉ, J. (1962). *J. Microsc.* **1**, 1.
GLASER, W. (1952). *Grundlagen der Elektronenoptik*, Springer Verlag, Wien.
HÖLL, P. (1969). *Optik* **30**, 116.
JÖNSSON, C. (1961). *Z. Phys.* **161**, 454.
KELLER, M. (1961). *Z. Phys.* **164**, 274.
KERSCHBAUMER, E. (1967). *Z. Phys.* **201**, 200.
KRIMMEL, E., MÖLLENSTEDT, G., and ROTHEMUND, W. (1964). *Appl. Phys. Lett.* **5**, 209.
LICHTE, H. and MÖLLENSTEDT, G. (1977). *Proc. 8th Int. Congr. X-Ray Optics and Microanalysis, Boston*, p. 20, Science Press, Princeton, New Jersey.
LICHTE, H., MÖLLENSTEDT, G., and WAHL, H. (1972). *Z. Phys.* **249**, 456.
MARTON, L. (1952). *Phys. Rev.* **85**, 1057.
MERLI, P.G., MISSIROLI, G.F., and POZZI, G. (1974). *J. Microsc. (Paris)* **21**, 11.
MICHELSON, A.A. (1890). *Phil. Mag.* **5**, 1.
MÖLLENSTEDT, G. and BUHL, R. (1957). *Phys. Bl.* **13**, 357.
MÖLLENSTEDT, G. and DÜKER, H. (1954). *Naturwissenschaften* **42**, 41.
—— (1956). *Z. Phys.* **145**, 377.
MÖLLENSTEDT, G. and JÖNSSON, C. (1959). *Z. Phys.* **155**, 472.
MÖLLENSTEDT, G. and LICHTE, H. (1978a). *9th Int. Congr. Electron Microscopy, Toronto*, Vol. 1, p. 178, Microscopical Society of Canada, Toronto.
—— (1978b). *Optik.* **51**, 423.
OHTSUKI, M. and ZEITLER, E. (1977). *Ultramicroscopy* **2**, 147.
POZZI, G. and MISSIROLI, G.F. (1973). *J. Microsc. (Paris)* **18**, 103.
SCHAAL, G. (1971). *Z. Phys.* **241**, 65.
SCHAAL, G., JÖNSSON, C., and KRIMMEL, E.F. (1966). *Optik* **24**, 529.
SPEIDEL, R. and KURZ, D. (1977). *Optik* **49**, 173.
TONOMURA, A. (1972). *Jap. J. appl. Phys.* **11**, 493.
WAHL, H. (1970a). *Optik* **30**, 508.
—— (1970b), *Optik* **30**, 577.
—— (1974). *Optik* **39**, 585.
—— (1975). *Habilitationsschrift*, Tübingen.

# 2. MULTIPLE-WAVELENGTH INTERFEROMETRY WITH X-RAYS AND NEUTRONS

M. HART and D.P. SIDDONS
*Wheatstone Laboratory, King's College, London, United Kingdom*

## 1. Introduction

Apart from our previous work (Cusatis and Hart 1974, 1977, Hart and Siddons 1978) all other neutron and X-ray interferometers have attempted to use only single wavelengths. In this paper we describe two different systems which are currently in use which demonstrate the feasibility and some advantages of working simultaneously at more than one wavelength.

In two-wavelength X-ray interferometry the two photon wavelengths are separated in the solid state detector system so that the ratio of the phase shifts can be measured. An exact analogue would be a two-wavelength neutron interferometer used with a spallation neutron source and time-of-flight order sorting. Crystal interferometers can also be used as high speed phase grating choppers. Photon chopping with nanosecond time resolution has been demonstrated and the application to neutron-chopper spectroscopy is obvious.

Both of these crystal interferometer systems rely for their operation on freedom from long-term drift and on the excellent mechanical frequency response which has been achieved with monolithic elastic systems.

In experiments which measure phase shifts one usually generates fringe patterns and determines phase from intensity measurements. When designing spectrometers for use with either very low intensity sources or for very expensive sources such as the neutron and synchrotron radiation facilities, it is most important that efficient systems are produced and, preferably, that the only source of experimental error is the inevitable random error caused by counting statistics. As far as one can tell from the literature no other X-ray or neutron interferometer system has yet reached that fundamental limit of performance.

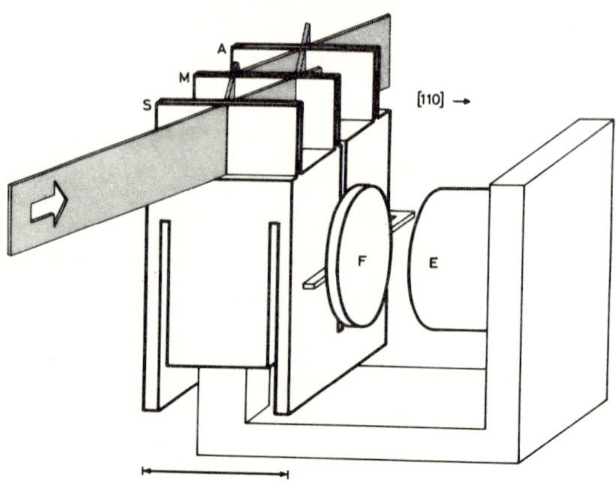


FIG. 1. Scanning interferometer constructed from a single crystal of silicon. S, M, and A are the beam splitter, mirrors and analyser respectively. Forces generated between the electromagnet E and the ferrite disc F cause rotation-free translations of the analyser parallel to the [110] direction.

Figure 1 shows the crystal interferometer. The crystal design is the same as that first described by Hart (1968) but the necessary force generator has since evolved from the original spring system through piezoelectric and capacitative elements to the present electromagnetic drive. The force between the electromagnet E and the ferromagnetic material F causes a rotation-free translation of the analyser A parallel to the [110] direction which is normal to the Bragg planes. Since the fringes so generated are a two-wave image of the Bragg planes, their spacing and profile is precisely determined. X-ray and neutron fringes may also be generated with suitable phase plates. Then the precision with which phase-shift measurements can be made may be limited by the quality of the profile of the phase plate. Because the electromagnetically scanned crystal interferometer is held at only one point it is drift free to the limit set by counting statistics. During the last six months the zero point (for a particular X-ray wavelength) has proved stable to a few millifringes, i.e. the drift rate is less than 1 pm per year. During that time the interferometer had been driven over more than $10^{10}$ fringes at frequencies between 0 and 100 kHz. Over short time scales the drift rate is difficult to quantify but we have shown that fringes which are present for less than 1 μs can be detected by integration over $10^8$ μs.

Earlier experiments (Cusatis and Hart 1977) showed that the stability of the interferometer was sufficiently high that phase-shift measurements could be made using only 10 photons $s^{-1}$. Integration times of a few hours suffice to determine phase shifts to within a few millifringes.

## 2. Two-wavelength interferometry

In principle any number of wavelengths *n* determined by Bragg's Law

$$2d \sin\theta = n\lambda$$

can be simultaneously used in a crystal interferometer. In practice the available laboratory X-ray generators have restricted us to two simultaneous wavelengths. The experimental

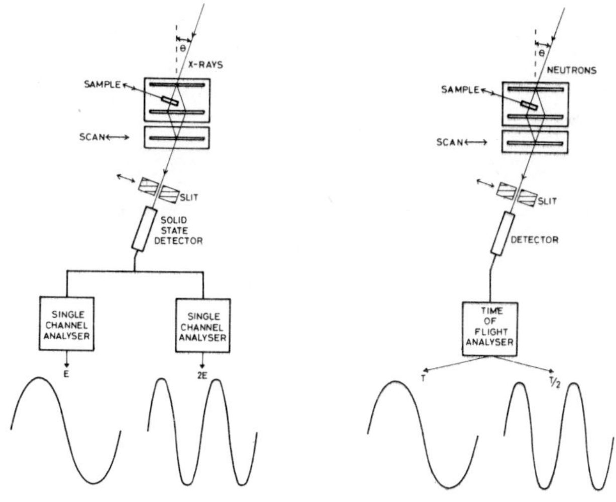


FIG. 2. Schematic arrangement of the scanning crystal interferometer for multiple-wavelength spectroscopy of X-rays (a, left) and neutrons (b, right). The X-ray system is already running with a laboratory bremsstrahlung source and will in due course be transferred to the U.K. Storage Ring Source. The neutron system appears to be well suited to the proposed spallation neutron sources.

arrangement is shown in Fig. 2(a) in which the whole instrument is controlled by a Hewlett–Packard model 9825 calculator.

On selecting the desired wavelength from the continuum the calculator positions the interferometer angle θ and the diffracted beam collimator slit and sets the channel positions in the pair of single-channel pulse height analysers. Fringes are then measured automatically, with and without the sample, for both wavelengths simultaneously and after profile analysis the two sample phase shifts are determined. Typical fringe patterns are shown in Fig. 3 which was obtained in a total time of about 5 h.

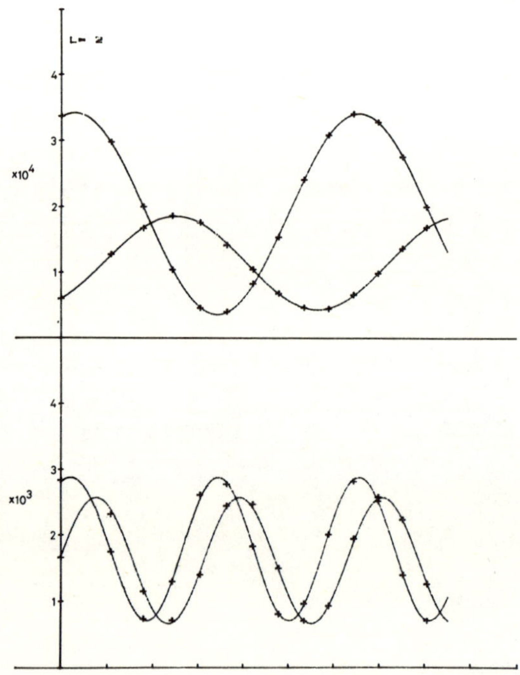


FIG. 3. Typical fringe patterns obtained with the X-ray interferometer system sketched in Fig. 2(a).

The upper fringes were obtained at 0.5442 Å wavelength with the 220 Bragg reflection and the lower fringes were the 440 harmonic at 0.2721 Å wavelength. In each case the more intense fringes relate to the empty interferometer and one sees that the images of the 220 and 440 Bragg planes coincide as expected. Since the mean intensity in the fundamental is approximately 20 000 photons, the standard deviation is approximately 140 and it can be seen in Fig. 3 that the scatter of data points about the least-squares sinusoid is consistent with the statistical fluctuations. The standard error in the phase determination is 0.0054 fringes. With characteristic radiation, which is available with higher intensity, the phase can be determined to approximately 0.001 fringes in a few hours. With synchrotron radiation we expect enough intensity for this precision to be routinely available at any wavelength in the 0.1–2 Å band. Figure 4 shows the results so far obtained for niobium and indicates that the precision and energy resolution are quite competitive (though much slower) with those obtained using film techniques and synchrotron radiation sources (Bonse and Materlik 1974).

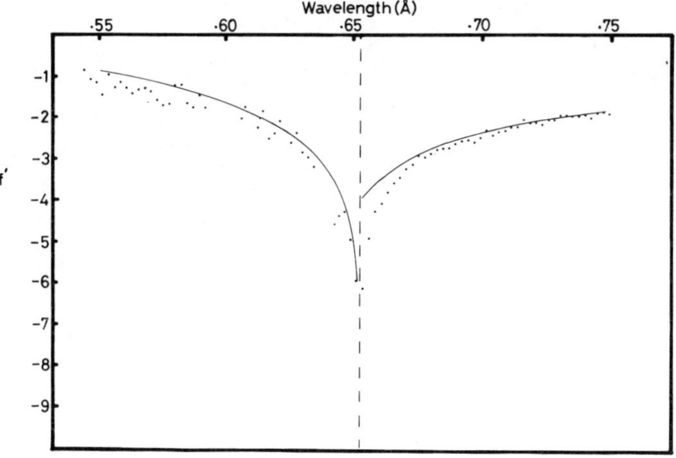


FIG. 4. Dispersion corrections determined for niobium near the K absorption edge ($E_k$ = 18.99 keV). The solid curve is calculated from Cromer and Liberman's theory and there are no adjustable parameters. Note that the two-wavelength method gives absolute measurements of $f'$ near the absorption edge.

Figure 2(b) shows how the present X-ray system might be adapted for neutron interferometry. With time-of-flight energy analysis, measurements exactly analogous to these X-ray measurements could be made. With $n$ simultaneous channels the system would of course be $n$ times faster than a single-wavelength interferometer. Such a system would be most attractive in the epithermal energy range with a spallation neutron source.

### 3. Signal-averaging techniques

The fringes shown in Fig. 3 were not obtained by simply counting for a sufficiently long time at each position in sequence. Earlier measurements (Cusatis and Hart 1977) were made in that way and show fluctuations which are larger than the expected statistical fluctuations. All of the neutron interferometer fringe patterns so far published show fluctuations which are up to four or five times greater than expected on statistical grounds (see, for example, Bauspiess *et al.* 1974, Bauspiess 1977, and these Proceedings). After normalization some data sets obtained with neutrons do practically reach the limit set by counting statistics (Bauspiess 1977, p. 84).

Provided that the interferometer itself does not drift conventional signal-averaging techniques can be applied. For example, in the present experiments we measured the whole fringe pattern spending only a few seconds at each point and then repeated the whole measurement cycle about 100 times. In the average of those 100 data sets the influence of external disturbances such as source intensity fluctuations is reduced ten-fold. To avoid losing experimental time in the signal-averaging process it must be possible to scan rapidly across fringes and that is rather conveniently done in the elastic scanning interferometer. An incidental advantage of the signal-averaging method is that no incident beam or source monitor is necessary.

In measurements of phase shifts with characteristic radiation we have sometimes adopted another method of operation which may prove useful with synchrotron radiation sources. The output intensity is modulated by oscillating the analyser over a range of a few hundredths of an ångström unit at an arbitrary frequency, usually a few tens of Hertz. The output is then analysed with a digital phase-sensitive detector and the derivative signal is used to servocontrol the interferometer position. Phase-shift measurements are made by recording the servo loop control current at adjacent fringe positions rather than by intensity measurement. In principle this same X-ray system could be used to servocontrol the position of a neutron interferometer, thereby eliminating the elastic deformation which has been troublesome in gravitational phase-shift experiments (Colella, Overhauser, and Werner 1975), or to eliminate mechanical drifts.

## 4. High-speed neutron choppers

Commonly used neutron choppers are either massive rotating objects which rely on neutron absorption (Feshback and Shelton 1977) or they are crystals whose diffracting power is modulated by magnetic fields (Badurek and Westphal 1975) or dynamic strains (Mikula, Michalec, and Vavra 1977). All of these devices suffer from practical limitations of speed and have a poor duty cycle which entails loss of intensity. In principle a phase-sensitive chopper could be made by moving a phase plate in one of the beam paths in a neutron interferometer.

The elastic scanning interferometer which we have already described also acts as a phase-sensitive chopper, directing the flux into either the forward or deviated direction as it is scanned over a distance of 1.92 Å for the 220 reflection. Since the necessary amplitude is so small this scanning interferometric chopper will operate at far higher frequencies than other systems. Since the amplitude of vibration for a harmonic oscillator is proportional to $\omega^{-2}$ and since the stored energy is proportional to the square of the amplitude we can work either at a low frequency with a large amplitude or at a high frequency with a small amplitude to achieve a given chopping rate. For a chopping period of 100 ns the mean velocity of the crystal analyser is only 2 mm $s^{-1}$.

Figure 5 shows the experimental arrangement used to obtain the chopper fringe patterns in Fig. 6. The scanning interferometer is driven by a sinusoidal current at frequency $\omega$. Each photon detected produces a spot on the oscilloscope ($z$ modulation) which lasts for 500 ns, which is the time necessary for

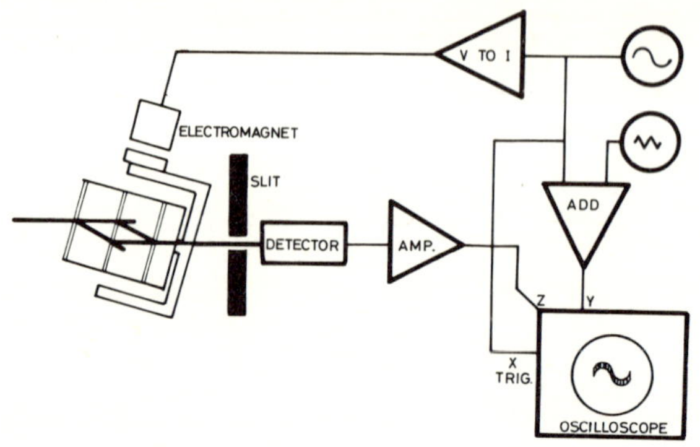


FIG. 5. Experimental arrangement used to display chopper fringe patterns with X-rays (see text).

detection on Polaroid 3000 speed film. The time base is triggered by the sinusoidal driving voltage which is displayed and broadened by an incommensurate triangular voltage in the $y$ direction. Fringes with spacings down to the 500 ns limit set by the oscilloscope and camera are clearly seen in Fig. 6. Only 15 mW of power is necessary for this performance and several watts could be used if necessary. We estimate that 1 ns chopping times could be obtained if necessary with the existing system.

In practice neutron-chopping interferometers may preferably be operated under constant-velocity or zero-velocity conditions rather than with the present sinusoidal drive. We have driven the chopper with many different waveforms and have found that its amplitude, phase, and frequency response follow closely that required for any simple harmonic oscillator.

## 5. Conclusions

We have developed a monolithic scanning interferometer system for neutrons and X-rays which is drift-free to the limit set by counting statistics. Phase-shift errors are determined only by counting statistics. The interferometer can work simultaneously at several different wavelengths, given suitable sources and detectors of X-rays and neutrons, and can also work as a

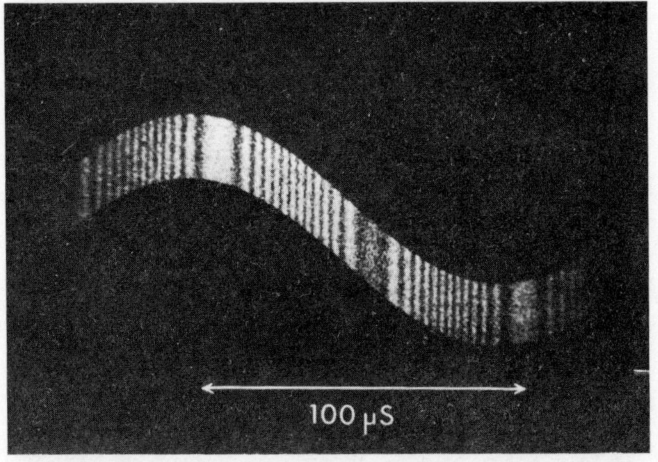


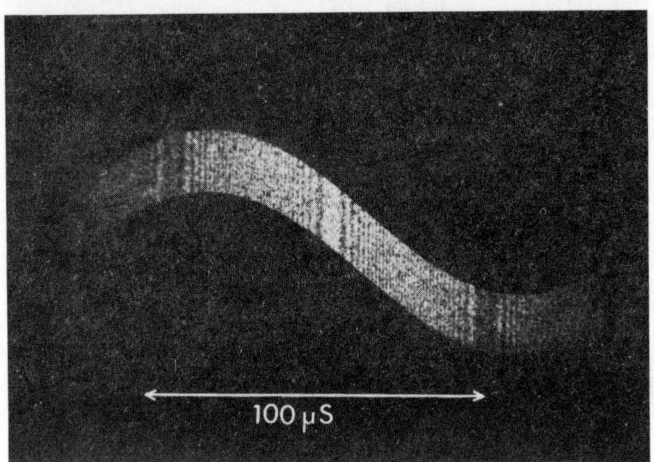


FIG. 6. Examples of fringe patterns obtained with the arrangement shown in Fig. 5.

multiplexed chopper.

Preliminary experiments which use rudimentary signal averaging techniques and servocontrol at the picometre length scale have demonstrated worthwhile gains in performance. We look forward to the more general application of these techniques so that the use of costly facilities such as neutron sources and synchrotron radiation sources may become more efficient and so that the precision of measurement in interferometry will approach more closely that determined by counting statistics.

## References


BADUREK, G. and WESTPHAL, G.P. (1975). *Nucl. Instrum. Methods*, **128**, 315–321.

BAUSPIESS, W. (1977). Ph.D. thesis, University of Dortmund.

BAUSPIESS, W., BONSE, U., RAUCH, H., and TREIMER, W. (1974). *Z. Phys.* **271**, 177.

BONSE, U. and MATERLIK, G. (1974). *Z. Phys. B* **24**, 189.

COLELLA, R., OVERHAUSER, A.W., and WERNER, S.A. (1975). *Phys. Rev. Lett.* **34**, 1472.

CUSATIS, C. and HART, M. (1974). *Anomalous scattering*. Copenhagen, Munksgaard.

—— (1977). *Proc. R. Soc. (Lond.) A* **354**, 291–302.

FESHBACK, H. and SHELDON, E. (1977). *Phys. Today* **30**, 40–51.

HART, M. (1968). *J. Phys. D* **1**, 1405–1408.

HART, M. and SIDDONS, D.P. (1978). To be published.

MIKULA, P., MICHALEC, R.T. and VAVRA, J. (1977). *Nucl. Instrum. Methods* **128**, 315–321.

# 3. X-RAY INTERFEROMETRY: THE OPTICAL TO GAMMA RAY CONNECTION

RICHARD D. DESLATTES

*U.S. National Bureau of Standards, Washington, D.C. 20234, U.S.A.*

## 1. Introduction

In the X-ray and γ-ray region there are two classes of spectra for which accurate calculations are possible and where testing of these theoretical estimates is particularly interesting. The first class to have received intense recent study is that of the 'optical' transitions of the bound negative muon. These muonic X-ray spectra are most often measured relative to γ-ray reference lines whose accuracy clearly influences the quality of possible comparisons with theory. The second class of spectra is that of the K emission spectra of mid- to high-*Z* atoms. Although 'direct' measurements are possible and, indeed, are in preparation, a specific historical circumstance makes indirect values obtained via γ rays more readily available at the time of writing. Both muonic spectra and high-*Z* electronic spectra have been considered at various times as testing grounds for renormalized QED calculations. In the most recent past, however, muonic spectra have dominated, especially as regards tests of the vacuum-polarization estimates. At present it would seem that departures from theory in high-*Z* electronic spectra are more likely to be occasioned by the inadequacy of the average field approximation itself than by inadequacy of current QED calculations.

In either case, and regardless of what one is looking for, theoretical results contain the lepton masses $m_e$ and $m_\mu$, a rather small reduced-mass type correction for the finite nuclear mass, the fine-structure constant α, and the Rydberg constant $R_\infty$. None of these constants is sufficiently uncertain to broaden theoretical estimates perceptibly. The problem is, instead, dominated by measurement imprecision and, until our recent work, a still larger inaccuracy in the reference lines (Kessler *et al.* 1978). What is at issue here is not some vaguely metrological connection to conventional 'standards' but the evidently needed connection to the Rydberg constant itself. Clearly, without this no comparisons other than those involving scaling behaviour within each class of spectra are possible.

Past connecting chains (which will not be reviewed here) have in common that they were complex, i.e. involved many steps, and were uncertain to the extent of nearly 20 ppm which is comparable to or greater than projected comparison levels in many cases. To describe the new connection scheme it is first noted that the hydrogen Rydberg was in fact most recently determined in an interferometric comparison of the Balmer $\alpha$ with a molecularly ($^{129}I_2$) stabilized laser (Hänsch *et al.* 1974). We used an effectively identical stabilized laser to measure the repeat distance of a particular silicon crystal in a scanning Laue case X-ray interferometer. Next, the calibrated silicon crystal was used in non-dispersive transfer measurements to calibrate other specimens having sizes and shapes suitable for $\gamma$-ray diffraction. These latter crystals were then used in a transmission two-crystal instrument to measure Bragg–Laue angles. The two-crystal instrument had a high resolving power ($\sim 10^5$) and was equipped with angle-measuring interferometers which were calibrated from first principles by summing to closure the external interfacial angles of a 72-sided optical polygon.

In our most recently published work (Kessler *et al.* 1978) the results have an estimated uncertainty (1 $\sigma$) of 0.37 ppm while nowadays somewhat better values nearer 0.25 ppm are being obtained. Such results have now been accumulated for approximately 40 $\gamma$-ray transitions in the range $0.06 < E < 1.1$ MeV.

The plan of this report is to describe briefly the four steps from the hydrogen Balmer $\alpha$ to a typical $\gamma$-ray line. These descriptions will be quite schematic and brief since a general treatment has already been given (Deslattes 1979) and a more thorough technical discussion is in a late stage of preparation. Next, an inventory of our measurements to date is given and the results from several recent closure tests in $^{169}$Yb are indicated. As examples of applications, data are presented both from recent muonic X-ray spectroscopy and from sone high- and mid-*Z* electronic X-rays. The concluding part indicates our plans for work at still higher energies.

## 2. Experimental procedures

### 2.1. *The Rydberg measurement*

Doppler-free spectroscopy of the hydrogen Balmer α has been reported (Hänsch *et al.* 1974). The various fine-structure components were compared interferometrically with a $^{129}I_2$ stabilized He–Ne laser locked to a peak conventionally labelled n (Knox and Pao 1971). In previous work we had already determined the interval between the feature n and another peak labelled B which we found more convenient to use. This interval was measured by direct counting of the beat frequency between two lasers locked to these absorption features (Schweitzer *et al.* 1973). Thus, by application of the Dirac equation, the Balmer α frequency (vacuum wavenumber) was interpreted as a value for the hydrogen Rydberg $R_H$ and thereafter via a reduced mass correction as the Rydberg constant $R_\infty$. Through the beat measurement our $^{129}I_2$ (B) laser was therefore calibrated in units of $R_\infty$. Detailed analysis and tests of $I_2$-stabilized lasers indicate an upper limit of error for this connection below 0.01 ppm which does not limit our accuracy at the present time (Layer, Deslattes, and Schweitzer 1976).

### 2.2. *Optical determination of a silicon (110)-lattice period*

The $^{129}I_2$ (B) stabilized laser illuminated a pair of high-reflectivity mirrors arranged as a short (0.3 mm) hemispherical cavity. As is shown in Fig. 1 one of these was attached to the stationary platform of a mechanical contrivance (Fig. 1(a)) whilst the other was fixed upon a moving part. The (00*l*) laser mode was matched into that of the optical cavity (Fig. 1(c)) by a single lens and steering mirrors. Optical feedback was reduced by a standard quarter-wave plate plus a linear polarizer decoupler followed by a neutral attenuator having an approximately 20-fold value.

Also mounted on the stationary and moving parts of the mechanical contrivance were the separated parts of a symmetri-

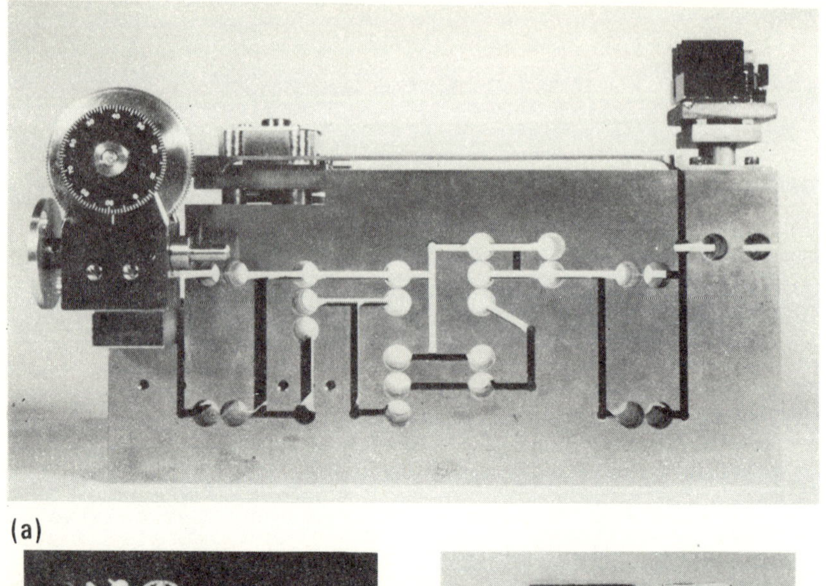

(a)

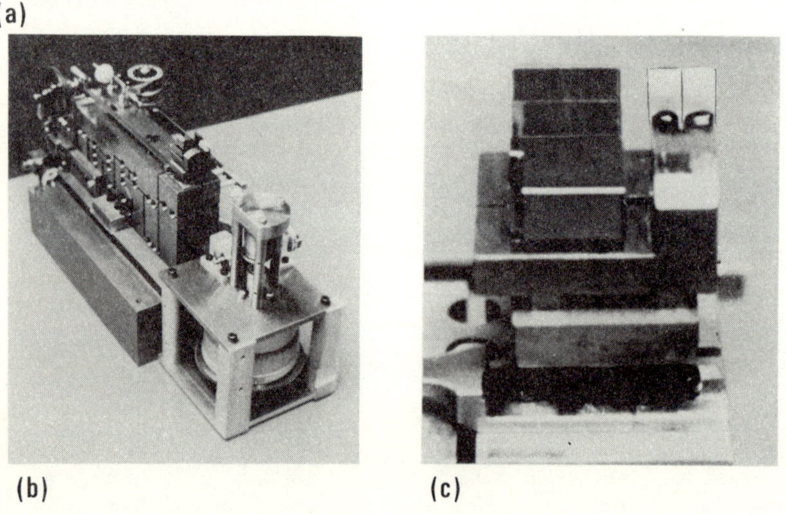

(b) (c)

FIG. 1. Photographs of the mechanical stage for the X-ray/optical interferometry: (a) the principal features of the parallelogram structure with its internal lever reducer; (b) overall view of the apparatus showing the input hydraulic motor; (c) details of the mirrors and crystals.

cally cut Laue case interferometer of the type introduced into X-rays by Bonse and Hart (1965) (Fig. 1 (c)). As suggested by the presence of slightly weakened necks in the structure near the platforms, differential bending devices permitted the required crystal alignments to be obtained by small rotations about two orthogonal axes. Drive to the scanning table was furnished by a motorized micrometer followed by a fluid-filled bellows reducer acting on a pin which pressed the input lever (Fig. 1(b)). The pin contained a piezoelectric ceramic which, through the lever reducer shown, was quite effective until well below the picometre (0.01 Å) level.

In operation the system was brought to a number of optical transparency maxima and the phase of the X-ray fringe signal established at each. By choice of optical order intervals we determined the fractional X-ray order excess by successive refinements as in the method of exact fractions. Apart from one early report using a Lamb dip laser and an early X-ray interferometer (Deslattes and Henins, 1973), all subsequent measurements (Deslattes *et al.* 1974) were of the type described above, using $I_2$-stabilized lasers.

The results obtained in these measurements exhibited an observational standard deviation $\sigma_m$ of the mean of about 0.03 ppm for approximately 150 observations. There are, however, several uncompensated sources of variance which must be accounted for as well as two corrections, each of which is associated with a significant uncertainty. Contributions at the level of 0.01–0.02 ppm arise from temperature measurement, together with uncertainties on thermal expansion and bulk modulus. The two corrections, one for path curvature (Deslattes 1976, Deslattes *et al.* 1976) and one for wavefront curvature, each contribute at the 0.02 ppm level. Taken together, these lead to an overall estimated standard deviation $\sigma_T$ near 0.1 ppm. (In earlier publications a laser problem was taken into account twice in estimating the error budget with a resultant uncertainty quoted as 0.15 ppm.) Our current estimate of the repeat distance for the interferometer silicon sample is $d(220)$ = 192.01707 pm (0.1 ppm) referred to 25 °C. This result was obtained from measurements in the range 22–23 °C using $\alpha = 2.56 \times 10^{-6}\ K^{-1}$.

### 2.3. *Crystal to crystal transfers*

As suggested in §1, large and rather perfect crystals are needed for the γ-ray diffraction measurements (see below). These need to be compared with the crystal sample used in the optical/X-ray interferometer in a way which does not add a significant error burden. To this end we used quasi-non-dispersive methods as will be described. All these methods have the common feature that they yield an estimate of the difference in lattice period between two nearly equal specimens, so that the result depends only weakly on the value of X-ray wavelength used in the transfer measurement. The small-angle measurements were carried out on a single-axis double-crystal device described previously (Deslattes 1967). It was equippped with a two-beam Michelson-type angle interferometer (Marzolf 1964) which could be read to about 0.01″ and had been calibrated to about 10 ppm. As will be seen, differences in $d$ spacings studied were mostly of the order of $10^{-4}$. Thus the 10 ppm calibration uncertainty had no perceptible effect except in one early measurement where a difference of $2 \times 10^{-2}$ was encountered.

We carried out transfer measurements on one sample of germanium and three of silicon. For the germanium comparison we made use of the approximate equality of $d$(Si 355) and $d$(Ge 800) (Baker and Hart 1975) in the geometry of Fig. 2(a). This arrangement is a single-source variant of the double-source quasi-non-dispersive procedure described by Hart (1969). Normalization of the Si 355 crystal was obtained by extraction of 220 samples from the parent boule. These were then compared with the interferometer crystal in the same geometry but with a long Si 220 in the first crystal position.

For other silicon to silicon comparisons, most notably for the case of a Si 111 pair used for a W $K\alpha_1$ measurement, we attempted to make use of theprocedure suggested by Ando, Bailey, and Hart (1978) for comparison of similar crystals in surface reflection. Unfortunately we found an instability in this procedure which may be associated with humidity effects on surface layers or other disturbing processes. Subsequently we realized a version of the Ando *et al.* procedure which used a symmetric transmission geometry as shown in Fig. 2(b). This was successful and obviated the need for an exceptionally long specimen as was required for the transfer measurement shown in Fig. 2(a).

Results of these several lattice parameter transfer measurements have been obtained with uncertainty estimates in the

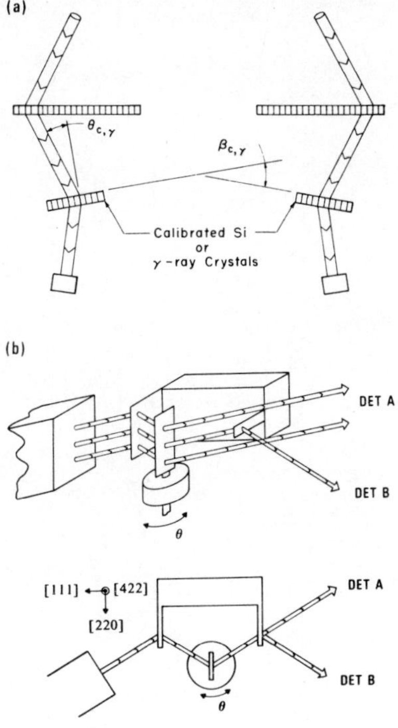


FIG. 2. Lattice parameter comparisons were carried out by the two forms of quasi-non-dispersive transfer shown. (a) The single source method used for most work including especially the silicon to germanium comparisons. (b) The transmission geometry used for certain silicon to silicon comparisons.

range 0.2–0.02 ppm. It is clear that any of these specimens are capable, given an adequate signal-to-noise ratio, of producing γ-ray reference values below the 1 ppm level. It is not, however, so clear that such comparisons can be pressed significantly beyond the 0.01 ppm level which *a priori* seemed easy. In any case we look forward to more diligent efforts to make sure that the quality of γ-ray data is not degraded in this, the easiest step in the measurement chain.

2.4. γ-*ray determinations*

Pairs of crystals calibrated as described in §2.3 were used in the double-crystal diffraction geometry shown in Fig. 3. For radiation parallel to the plane of dispersion, the Bragg relation $\sin\theta_0 = \lambda/2$ is exact. When a finite vertical divergence

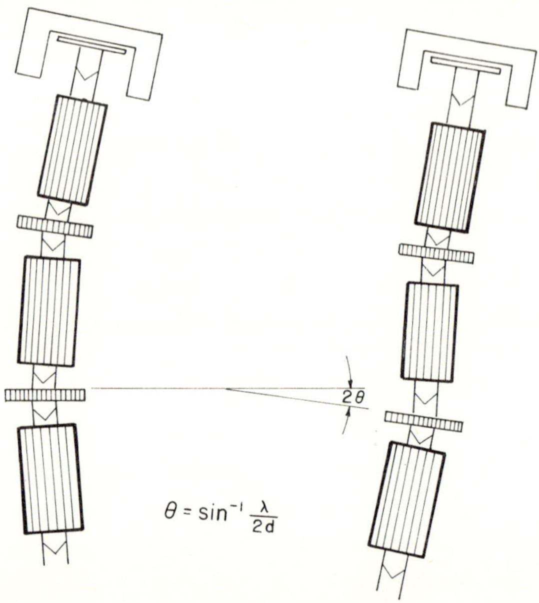


FIG. 3. Geometry of the transmission two-crystal γ-ray instrument. For rays parallel to the plane of dispersion the Bragg—Laue equation is burdened by no known error.

is permitted (as is necessary) the crystal diffraction pettern is, in the (n, +n) configuration, convolved with a singular vertical divergence function to form the Doniach—Sunjić integral (Doniach and Sunjić 1970). The resulting model functions for the two diffraction profiles are fitted to the data and the unshifted $\theta_0$ is extracted.

The general features of the γ-ray instrument are shown in Fig. 4. The instrument chassis is of grey cast iron and is about 1 m long (Fig. 4(a)). The two axes shown are 50 cm apart and are defined by bearings of high precision. The axes are approximately parallel with a bias of 1″. Also visible in Fig. 4(a) is a plate with optical elements kinematically supported from the chassis. These optical elements, together with four corner cubes mounted by pairs on each axis, constitute a pair of angle interferometers having considerable precision and accuracy (see below). Figure 4(b) is a closer view of one of the corner-cube arms while Fig. 4(c) shows a crystal, one of its reference mirrors, and the entrance of an intercrystal

(a)

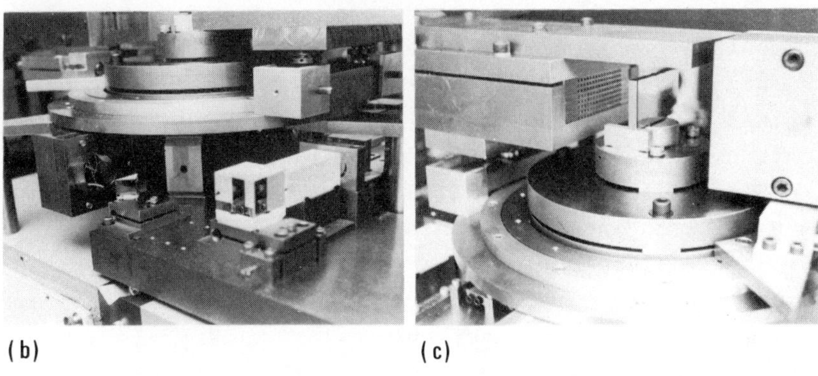

(b) (c)

FIG. 4. Several views of the National Bureau of Standards transmission two-crystal instrument: (a) an overview of the main chassis and angle interferometer plate; (b) a closer view of one of the cube-corner reflectors and nearby steering and beam-folding optics; (c) a crystal with one of its reference mirrors and the entrance to an intercrystal collimator.

collimator. These angle interferometers have considerable sensitivity (about $50 \times 10^{-6}$ arcsec) and, after calibration, a high accuracy as well. Internal to the cast iron chassis are micrometer-driven sine arms 60 cm long. The prime drive is from external motors and each microcmeter has a piezoelectric vernier. In this version the driverange is limited to 5°, while a successor instrument will have a range of 30°.

The diffractometer is normally surrounded by an acoustic and thermal shield. This, the input collimator, and the detector rest on a large iron plate, of area 1.2 m × 2.5 m, whose mass is about 5000 kg. The whole is supported on air springs whose decoupling from a rather hostile environment is essential. Sources, typically 0.1–10 kCi, are prepared in the National Bureau of Standards reactor and brought out of the containment vessel to an assembly area for measurement. Here we have built an environment chamber within which the instrumentation is housed. Sources are held in 800 kg shields and are brought to the entrance port by a floor-mounted jack. All operating controls are external to the housing whose temperature is held to ± 0.2 K. Within the thermal and acoustic inner shield we noted slow drifts of $10^{-2}$ K over several hours.

Calibration of the (sine-measuring) interferometers required two steps. First, a zero of angle was obtained by fitting the arcsine measurement function to data obtained from successive measurements of the interval given by a small-angle gauge in the neighbourhood of mechanical zero. Next, a 72-sided quartz polygon was mounted on top of an axis and the interval between each pair of adjacent faces was determined in interferometer units and fractions thereof. Each unit corresponds to approximately 0.03″ with interpolation by a null-seeking polarizer which is angle encoded. The closure condition, namely $\Sigma\alpha_i = 2\pi$, then fixed the interferometer constant. Such calibrations carried out every 3–6 months over the past several years show a slow growth (about 0.7 ppm per year) of the invar arms separating the corner cubes. The value of the interferometer constant for a particular γ-ray measurement has been obtained by interpolation between earlier and later calibrations with an accuracy of approximately 0.1 ppm.

Figure 5 shows typical rocking curves as obtained from silicon and germanium crystals. Observed widths are not in conspicuous disagreement with elementary theory down to approximately 0.1″. The particular widths shown in Fig. 5 correspond to resolving powers of 40 000 and 100 000 for germanium and silicon respectively. The (n, +n) curves are slightly broadened and made asymmetric by the vertical divergence effect noted above. This aside, the equality indicated between the widths of (n, +n) and (n, -n) follows from the spectral narrowness of the

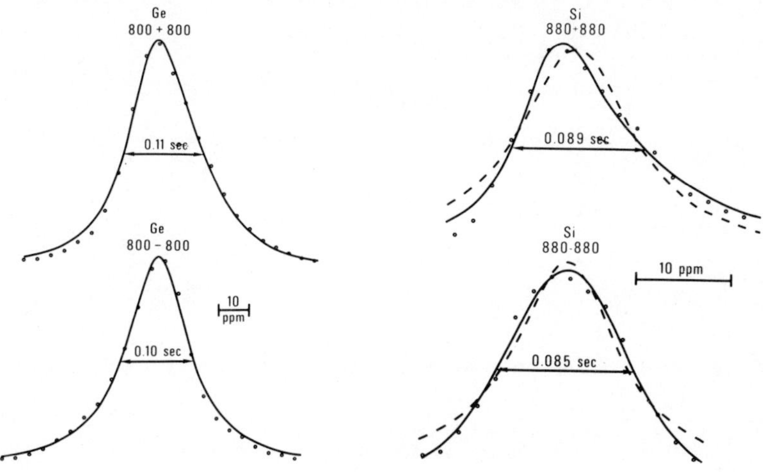


FIG. 5. Sample rocking curves for silicon and germanium crystals. Note the approximate equality of the (+) and (-) position widths as is appropriate for the narrow γ-lines. The asymmetry in the (+) position curves is due to finite vertical divergence and is a well-known analytic function represented by the solid curves. The symmetric curves shown by the broken curves are included for visual reference only and are not used in data reduction.

γ-line.

Results have been reported thus far on a number of γ-ray lines including $^{198}$Au 411 and 675 keV together with $^{192}$Ir 205, 295, 308, 316, 468, 484, 588, 604 and 612 keV (Kessler *et al.* 1978). In addition, tungsten K and thulium K X-ray lines have been recorded together with $^{170}$Tm 84 keV and $^{169}$Yb 63, 93, 110, 119, 130, 177, 198, 261 and 307 keV γ rays which have been reported elsewhere (Kessler *et al.*, 1979). Wavelengths have been determined in picometres under the condition that the $^{129}I_2$ (B) = 632990.074 (1) pm which is equivalent to $^{129}I_2$ (n) = 632991.341 (1) pm to which the Rydberg measurement was referred. Energy values were obtained in a conventional sense, i.e. by assuming that the voltage wavelength conversion factor is exactly $1.2398520 \times 10^{-6}$ eV m (Taylor and Cohen, 1973). If any of these results need to be compared with an electrically determined energy, then the *V*–λ uncertainty of 2.6 ppm must be considered. Several groups of three lines each are connected by cascade-crossover relations. Closure can be tested in such a case after correcting for nuclear recoil. The results of two such tests are shown in Table 1.

TABLE 1
*Tests of closure in $^{169}Yb$*

| Level differences (eV) | | Sum values (eV) | $\Delta E/E$ (ppm) |
|---|---|---|---|
| $^{169}$Yb | 197958.00 | | |
| $^{169}$Yb | 109779.91 | 307737.92 | |
| $^{169}$Yb | 307737.88 | | 0.13 ± 0.53 |
| $^{169}$Yb | 130523.74 | | |
| $^{169}$Yb | 177214.12 | 307737.85 | |
| $^{169}$Yb | 307737.88 | | 0.10 ± 0.37 |

An impression of what these new results do to the γ-ray energy scale can be formed by considering the particular case of the 411 keV line from the mercury daughter of $^{198}$Au. Many γ-ray and mesic X-ray measurements have been referred to this line. Until our work the best estimate of its energy was 411.794 keV (17 ppm) (Murray, Graham, and Geiger 1965). Our result is 411.80441 (0.37 ppm). This represents a shift of 25 ppm which may be expected to operate on any line whose ratio to the 411 keV line was well established.

## 3. Applications

### 3.1. *Outer-orbit muonic spectra*

Beginning with an initial set of high-precision measurements (Dixit *et al.* 1971), comparisons between muonic spectra and QED calculations were unsatisfactory and somewhat confused by certain experimental conflicts. On the other hand, the problem appeared persistent with discrepancies increasing with $Z$, perhaps quadratically. A number of papers too large to list here addressed this problem from both theoretical and experimental points of view. The results of this turbulent period are summarized in two substantial review articles (Watson and Sundareson 1974 and Rafelski *et al.* 1974). In these it was concluded that further improvement in QED calculations could not reconcile the experimental situation. These review articles then considered the possibility that the discrepancies might signal the existence of a (~ 10 MeV mass) meson of the type permitted (required?) by renormalizable gauge theories.

Subsequent to these review papers, comparison of mesic X-ray lines with γ-ray references has improved in observational precision and analytical modelling. In addition, our work has improved (and changed) the values of the γ-ray reference standards. The effect of combining these two efforts is indicated in Fig. 6.

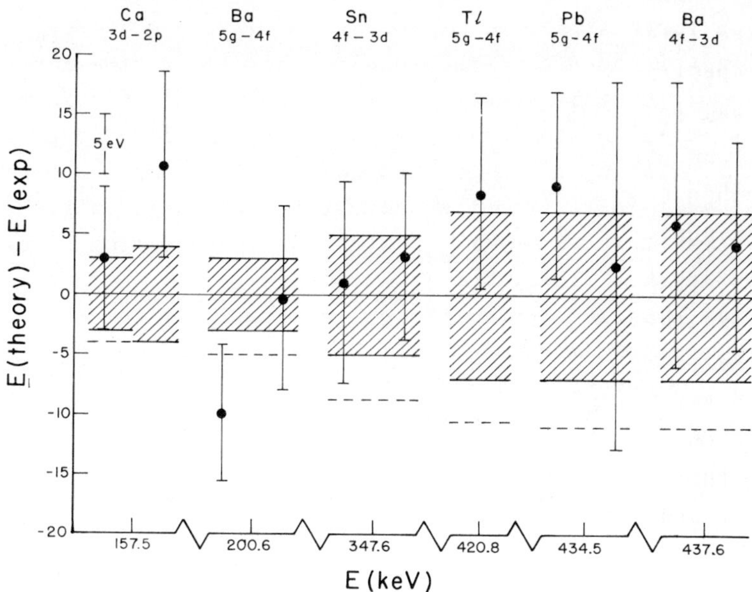


FIG. 6. A comparison of recent muonic X-ray data theory. The hatched bands represent theoretical uncertainties. The data appear to be consistent with theory as shown. In the absence of the revised γ scale, the theory bands would be centred on the broken lines and some case might be made that significant disagreement persisted.

Here we have made use of a recent re-analysis of the experimental data (Hargrove *et al.* 1977) together with our re-adjustment of reference γ-ray energies. The hatched bands are plotted so as to represent estimates of theoretical uncertainties (Vuilleumier *et al.* 1976). Figure 6 clearly suggests that agreement between theory and experiment is at present satisfactory. This result is consistent with additional new data and analysis reported by Dubler *et al.* (1978).

Were it not for the energy scale adjustment, Fig. 6 would appear with the hatched regions centred on the broken lines indicated. This would have the effect of distinctly separating theory from experiment, thereby lending a certain credence to the 10 MeV meson. Thus, conversely, we can assert that revision of the $\gamma$-ray reference scale as required by the measurements reported here has the effect of eliminating any requirement for the introduction of this extra particle.

### 3.2. *Electronic spectra*

A further consequence of the results that led to the improved situation for muonic spectra is to be found in the case of electronic spectra. Historically, high-$Z$ X-ray spectra have been an occasional testing place for QED calculations. On the other hand, if we take it from the above that QED is in satisfactory shape, then what, if anything, might be learned from high-$Z$ K-series spectra? The issue is, it seems to us, whether or not the 'average-field' approximation is itself sufficiently robust to survive these improved tests. We have therefore begun an effort to re-determine a representative sample of mid- and high-$Z$ K-series spectra. Clearly, a main focus of this work is to apply the same crystal spectroscopy described above for $\gamma$ rays to the X-ray problem. Such an effort is planned but not yet fully underway; however, some results are already at hand.

A few K-series spectra (tungsten and thulium) have already been directly determined using the present $\gamma$-ray instrumentation while there are equivalent estimates for molybdenum and copper using other instrumentation (Deslattes and Henins 1973). Another group of results is available from combining the high-accuracy X-ray to $\gamma$-ray ratio measurements (Borchert 1976) for thulium, thorium, uranium, plutonium and tungsten with our re-determination of the $\gamma$-ray reference values. On the theoretical side new relativistic Hartree–Fock–Slater calculations that have been carried out to considerable accuracy are available (Huang *et al.* 1976). These are compared with the adjusted experimental data for $K\alpha_1$ transitions described above in Fig. 7.

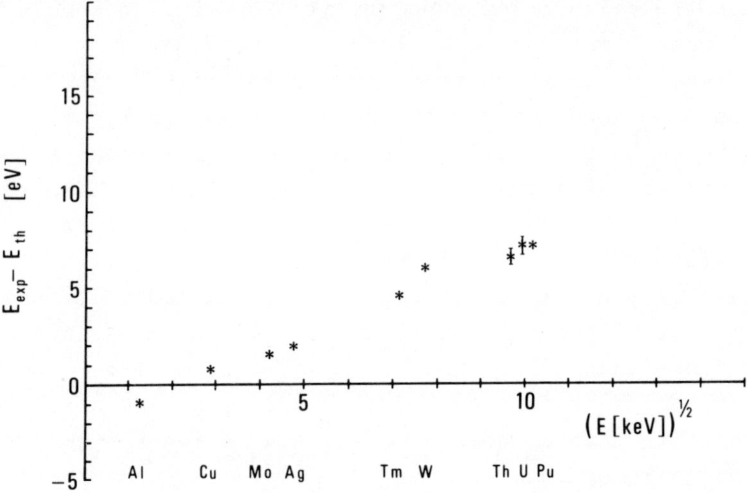


FIG. 7. Comparison of relativistic Hartree–Fock–Slater calculations with experiment for $K\alpha_1$ transition in selected elements. The X-ray data come from a few direct measurements but mostly via the γ-ray connection. This approximately linear Moseley plot is suggestive of a screening-type problem.

The approximately linear behaviour of the discrepancy versus $\sqrt{E}$ is suggestive of a screening-type error. Such a term may in fact arise from the approximation procedure used by Huang *et al.* (1976) for the effect of screening on self-energy terms. They make some reasonable estimates of screening effects on (Coulomb) self-energy contributions for the case of *s* states. Thereafter it was assumed that the result for $2s_{1/2}$ could be transferred to $2p_{1/2}$ while this effect on $2p_{3/2}$ was taken to be negligible. It must be emphasized that this is a very tentative and at least incomplete analysis. Further experimental data obtained in this indirect fashion are available. Also, a systematic programme of direct measurements is planned.

## 4. Future plans

In addition to a programme of measurement on X-ray spectra, we plan to extend γ-ray measurements toward higher energies. This will require use of relatively prompt neutron-capture sources in a tangent port of the National Bureau of Standards reactor.

Aside from the calibration type utility, there is one possibly timely outcome which can be anticipated. Consider a measurement, by means of mass spectroscopy, of the mass difference $\Delta M$ between $^{15}NH_3$ and $^{14}NH_2D$. The corresponding energy difference can be reckoned as the difference $\Delta E = hc/\lambda$ between the neutron-capture cascade in $^{14}N$ and the neutron–proton capture $\gamma$. If $\Delta Mc^2$ is then identified with $\Delta E$ and the result manipulated slightly (Taylor 1976) it follows that $F_{NBS} = (2e/h)_{NBS}c(\Delta M)\lambda/2$ where the subscripts indicate local as-maintained units and $\lambda$ is the reciprocal of the term value difference between a neutron bound in deuterium and one bound in $^{15}N$. Since the electrochemical Faraday constant $F$ is one of the more troubling of the constants, this independent route to its evaluation seems to be of significant current interest.

## Acknowledgments

The local group working in this area includes principally E.G. Kessler, Jr. with the partial involvement of A. Henins and W.C. Sauder. The present report and planned efforts for the future contain significant contributions from two guest workers, Ludo Jacobs (Mol Laboratories and the Catholic University of Leuven, Belgium) and Wolfgang Schwitz (Physics Department, University of Fribourg, Switzerland).

## References

ANDO, M., BAILEY, D., and HART, M. (1978). *Acta Cryst. A* **34**, 484.
BAKER, J.F.C. and HART, M. (1975). *Acta Crystallogr. A* **31**, 364.
BONSE, U. and HART, M. (1965). *Appl. Phys. Lett.* **7**, 99.
BORCHERT, G.L. (1976). *Z. Naturforsch. A* **31**, 102.
DESLATTES, R.D. (1967). *Rev. sci. Instrum.* **38**, 815.
—— (1976). *Atomic masses and fundamental constants*, Vol. 5 (eds. J.H. Sanders and A.H. Wapstra), p. 552. Plenum Press, New York.
—— (1979). *Proc. Int. School of Physics, 'Enrico Fermi', Course 68, Metrology and Fundamental Constants, July 1976*. To be published.
DESLATTES, R.D. and HENINS, A. (1973). *Phys. Rev. Lett.* **31**, 972.
DESLATTES, R.D., HENINS, A., BOWMAN, H.A., SCHOONOVER, R.M., CARROLL, C.L., BARNES, I.L., MACHLAN, L.A., MOORE, L.J., and SHIELDS, W.R. (1974). *Phys. Rev. Lett.* **33**, 463.
DESLATTES, R.D., HENINS, A., SCHOONOVER, R.M., CARROLL, C.L., and BOWMAN, H.A. (1976). *Phys. Rev. Lett.* **36**, 898(C).

DIXIT, M.S., ANDERSON, H.L., HARGROVE, C.K., McKEE, R.J., KESSLER, D., MES, H., and THOMPSON, A.C. (1971). *Phys. Rev. Lett.* **27**, 878.
DONIACH, S. and SUNJIĆ, M. (1970). *J. Phys. C* **3**, 285.
DUBLER, I., KAESER, K., ROBERT-TISSOT, B., SCHALLER, L.A., SCHELLENBERG, L., and SCHNEUWLY, H. (1978). *Nucl. Phys. A* **294**, 397.
HÄNSCH, T.W., NAYFEH, M.H., LEE, S.A., CURRY, S.M., and SHAHEN, I.S. (1974). *Phys. Rev. Lett.* **32**, 1336.
HARGROVE, C.K., HINCKS, E.P., McKEE, R.J., MES, H., CARTER, A.L., DIXIT, M.S., KESSLER, D., WADDEN, J.S., ANDERSON, H.L., and ZEHNDER, A. (1977). *Phys. Rev. Lett.* **39**, 307.
HART, M. (1969). *Proc. R. Soc. A* **309**, 289.
HUANG, K.-N., AOYAGI, M., CHEN, M.H., CRASEMANN, B., and MARK, H. (1976). *At. Data nucl. Data Tables* **18**, 243.
KESSLER, E.G., Jr., DESLATTES, R.D., HENINS, A., and SAUDER, W.C. (1978). *Phys. Rev. Lett.* **40**, 171.
KESSLER, E.G., Jr., JACOBS, L., SCHWITZ, W. and DESLATTES, R.D. (1979). *Nucl. Inst. Meth.* **160**, 435.
KNOX, J.D. and PAO, Y.H. (1971). *Appl. Phys. Lett.* **18**, 360.
LAYER, H.P., DESLATTES, R.D., and SCHWEITZER, W.G., Jr. (1976). *Appl. Opt.* **15**, 734.
MARZOLF, J.G. (1964). *Rev. sci. Instrum.* **35**, 1212.
MURRAY, G., GRAHAM, R.L., and GEIGER, J.S. (1965). *Nucl. Phys.* **63**, 353.
RAFELSKI, J., MÜLLER, B., SOFFARD, G., and GREINER, W. (1974). *Ann. Phys. (N.Y.)* **88**, 412.
SCHWEITZER, W.G., Jr., KESSLER, E.G., Jr., DESLATTES, R.D., LAYER, H.P., and WHETSTONE, J.R. (1973). *Appl. Opt.* **15**, 734.
TAYLOR, B.N. (1976). Private communication.
TAYLOR, B.N. and COHEN, E.R. (1973). *J. phys. chem. ref. Data* **2**, 663.
VUILLEUMIER, J.L., DEY, W., ENGFER, R., SCHNEUWLY, H., WALTER, H.K., and ZEHNDER, A. (1976). *Z. Phys. A* **278**, 109.
WATSON, P.J.S. and SUNDARESON, M.K. (1974). *Can. J. Phys.* **52**, 2037.

# 4. A TWO-CRYSTAL X-RAY INTERFEROMETER FOR PRECISION MEASUREMENTS OF LATTICE PARAMETERS: PRESENT STATE AND PROSPECTS

P. BECKER

*Physikalisch-Technische Bundesanstalt, D 3300 Braunschweig, Federal Republic of Germany*

## 1. Introduction

As a result of the proposals of Bonse and Hart (1965) the Physikalisch-Technische Bundesanstalt (PTB) and other research institutes, initiated work on a project for a very precise measurement of the (220) lattice plane spacing in a silicon crystal. These investigations remain important even after the publication of Deslattes' successful precision measurements (see for example Deslattes *et al.* 1976) carried out at the National Bureau of Standards (NBS) because there is an urgent need for an independent determination of this important parameter which is required for a new definition of the mole and the kilogram and for fixing a number of atomic constants. Since 1972 the PTB's Annual Reports (PTB-Jahresberichte) have given information on this interlaboratory project[†] to which Professor Bonse is giving his advice and support. In order to convey an idea of the concept and the experimental difficulties connected with the desired relative measuring uncertainty of $10^{-6}$, the essential details of our set-up are presented. Subsequently the present state of the research and the further development of the project will be discussed (see also Kind 1977).[‡]

## 2. The X-ray interferometer

### 2.1. *The interferometer crystal*

The most important part of the instrument, the X-ray interferometer crystal, is shown in Fig. 1. This crystal has three thin

---

[†] The collaborators in this project are acknowledged in the headings of the individual sections.

[‡] A detailed discussion of the present state of the project will be given in the PTB-Report APh-i3 which will be published shortly.

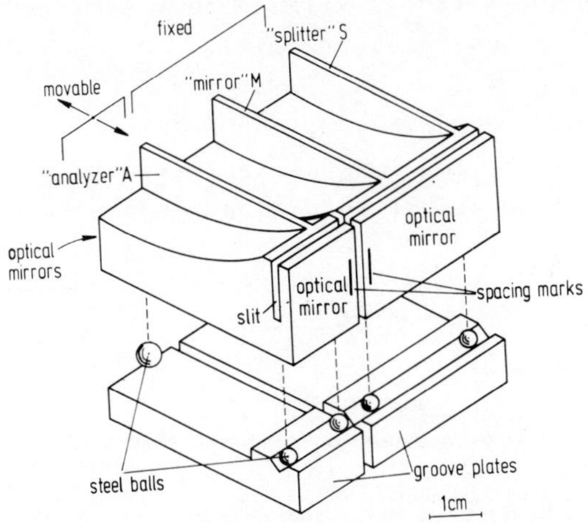


FIG. 1. The divided silicon interferometer crystal. The slit is to keep the strain originating from the mirror polish off the wafers.

wafers S, M, and A and is designed according to the proposals of Bonse and te Kaat (1968). Its shape is characterized by the following special features. Each crystal has been polished on the side facing the observer to form an optical mirror with a flatness of 1/20 of the wavelength of the visible light. With the aid of these mirrors the displacement of the analyser crystal A is measured by means of optical interferometry. The surfaces of the mirrors are parallel to the (220) lattice planes within an uncertainty of $10^{-4}$ rad. Thus the positions of the analyser crystal and the optical mirror are fixed with respect to each other. The attachment of separate optical mirrors is superfluous (Siegert 1974).

2.2. *Arrangement of the interferometer (Lehrke 1974, Becker)*
The crystal faces which face away from the observer have also been polished to form optical mirrors. With the aid of these and using an autocollimation telescope the completely separated crystal parts are adjusted with respect to one another. Three small steel balls are fixed under each crystal as feet. Two steel balls are placed in a groove. Spacing marks are etched on the surfaces of the mirrors by means of which the spacings

of the wafers can be exactly aligned to ± 1 μm with a reading telescope (Seyfried and Siegert 1974).

### 2.3. *The adjustment aids (Lauer, Becker)*

The adjustment of the two crystals relative to one another is done by means of the adjusting elements shown in Fig. 2.

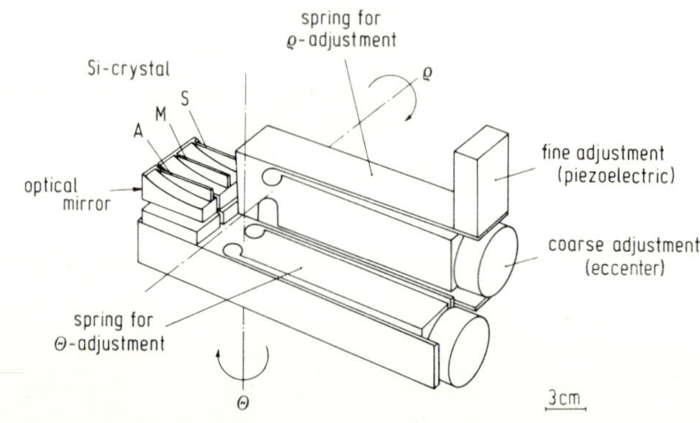


FIG. 2. The adjusting elements of the silicon crystal. The axes of rotation ρ (for the crystal part with wafers S and M) and θ (for the analyser crystal A) are perpendicular to one another. A photograph of this device is shown by Kind (1977).

With their aid the lattice planes of the two crystals are aligned parallel to one another with a relative uncertainty of less than $10^{-8}$ rad. The elements are similar in form to those used by Deslattes (1971). They consist of steel spring joints with one leaf spring each for coarse and fine adjustment of the crystals. The leaf springs have a resonance frequency above 200 Hz. A deflection of the leaf springs by ± 1 mm effects a turn of the crystal by $5 \times 10^{-5}$ rad. Coarse adjustment to $5 \times 10^{-6}$ rad was done with the eccenter shown in the figure and with the aid of an autocollimation telescope. The interference range of $1 \times 10^{-8}$ rad can be adjusted reproducibly and without vibration by means of a piezoelectric fine adjustment.

## 3. The displacement device

### 3.1. *The translation stage (Rademacher 1974, Hoffrogge and Rademacher 1973)*

During the interferometric measurements the analyser crystal A is displaced with a velocity of 0.01–500 nm $s^{-1}$ parallel to the fixed part of the interferometer. For this purpose translation stages with guiding errors of less than $10^{-8}$ rad are necesary. These requirements greatly exceed those usual in measuring techniques. Figure 3 shows the bi-parallel spring used in our set-up. Theoretically springs of this kind provide sufficiently good translation stages without backlash or friction. When assembling bi-parallel springs from individual leaf springs (Bonse, te Kaat, and Spieter 1971) guiding errors may occur caused by inexact mounting and inhomogeneities of the material. To avoid this the spring used was manufactured from a high-quality steel plate. The springs are similar in shape to tuning forks. Each reduced section of the springs, shown hatched in Fig. 3, acts as an elastic pivot. One part of each fork is connected to the external frame and the other to the movable central bar supporting the analyser crystal A. In order to keep undesirable relative motions between the crystal parts sufficiently small the stiffness of the bi-parallel spring must be very high. In this case it is 1500 N $mm^{-2}$.

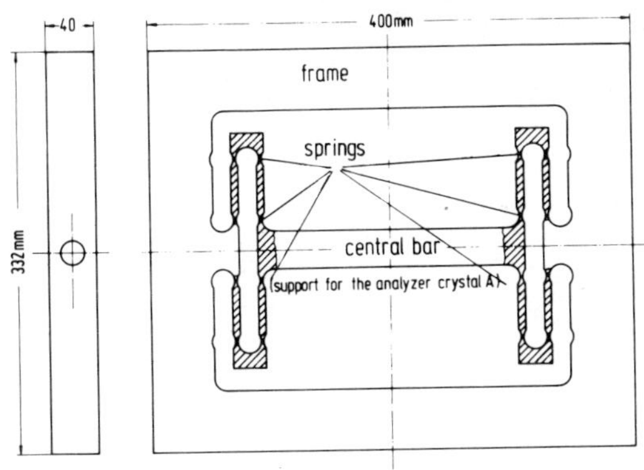


FIG. 3. The bi-parallel spring for lattice parameter measurements. The movable part has been hatched.

### 3.2. *The drive (Rademacher)*

Figure 4 shows the schematic arrangement of the apparatus. In the centre, parts of the bi-parallel spring and the interferometer crystal are shown. The drive for the central bar is shown schematically on the right of the figure. This bar can be displaced over larger distances by means of a step motor and a high-reduction gear. Displacements of up to 300 μm are possible so far a displacement of 40 μm has been made use of by means of a piezoelectric drive. The direction of the displacement can be reversed without noticeable backlash so that individual X-ray interference signals can be reproducibly scanned (see §6).

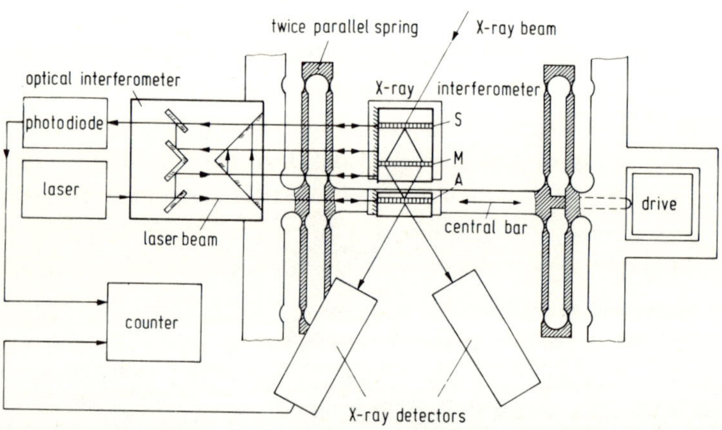


FIG. 4. Scheme of the X-ray scanning interferometer. The optical signal is recorded by means of a semiconductor photodetector; the output signal of the X-ray interferometer is recorded with the aid of two X-ray detectors. The X-ray interference fringes occurring during the displacement are electronically counted.

## 4. The optical interferometer

The scheme of the optical interferometer is shown on the left-hand side of Fig. 4. The optical interferometer measures the displacement of the analyser in meters. The main difficulties lie in achieving, over a relatively long measuring period of several hours, a measuring uncertainty of 0.01 nm which is a factor of $10^{-2}$ of that obtained with conventional interferometry.

4.1. *The design of the interferometer and ray tracing (Reim 1974, 1977, Dorenwendt)*

To solve these problems an interferometer of the Michelson type was used which had been developed by Curtis *et al.* (1971) and which had been used by Hart and his collaborators at the National Physical Laboratory (NPL) for the same purpose. Figure 5 shows the ray tracing and the technical arrangement of the interferometer built at the PTB. We decided in favour of the two-beam interferometer for the following reasons:

the silicon crystal can be directly polished to form optical mirrors;

the influence of misalignments and diffraction effects can be calculated more easily;

the interferometer signal is invariant with respect to a common rotation and translation of the crystals.

An incident polarized beam of known wavelength is divided as shown. By means of a half-wave plate the two partial beams are polarized perpendicularly to one another. The phase difference of the two beams caused by the displacement is determined by electro-optical phase modulation in a Pöckels cell and synchronous detection. Thus a measuring uncertainty of 0.01 nm is obtained.

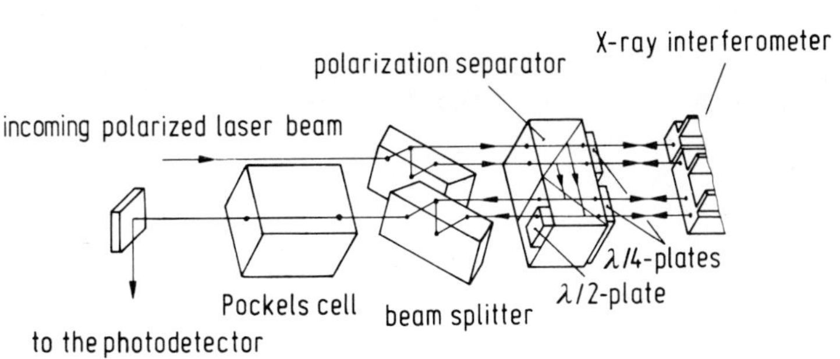


FIG. 5. The optical interferometer. A photograph of this device is shown by Kind (1977).

4.2. *Measures for the improvement of the long-time stability and the adjustment (Dorenwendt, Reim, Siegert)*

The long-time stability of the interferometer signal is better than about $10^{-2}$ nm $h^{-1}$. According to our experience it is reached after a running-in period of two or three weeks. The following measures were necessary in order to obtain these values.

The surrounding temperature was kept constant to $10^{-3}$ K (for thermostating methods see §5).

The optical interferometer was installed in an airtight aluminium housing and the beam was guided along a tube between the laser and the beam splitter in order to eliminate the influence of air streaks. Experience has shown that air streaks occurring in the unsplit beam have a surprisingly large influence on the phase difference of the two partial beams.

Equal optical paths are produced in both partial beams in order to eliminate the influence of air pressure on the phase difference.

The adjustment of theoptical interferometer set-up with regard to the laser beam is also very critical. In order to keep phase changes in the exit signal to within tolerable limits angular positions must be set with an accuracy to better than $10^{-4}$ rad and lateral positions must be set with an accuracy to better than 100 μm.

## 5. Installation of the X-ray and optical interferometer

The instrument so far described is arranged on a table with a granite plate about 600 kg in weight which is partly shown in Fig. 6. Unfortunately this plate and its legs constitute a vibration system (resonance frequency 4 Hz, five-fold step-up resonance) which is excited by the mechanical vibrations of the environment near the resonance frequency. Most importantly the low-frequency X-ray signal is affected by these vibrations. Almost all vibrations with amplitudes higher than 0.01 nm prevent the goal of determining the position of the analyser crystal exactly to a fraction of the lattice plane spacing being reached. Therefore it was necessary to devise methods of attenuating the mechanical vibrations.

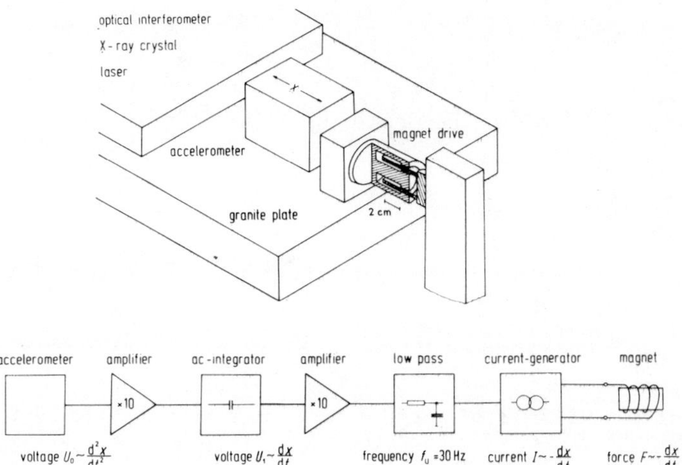


FIG. 6. Electronic device for attenuating the mechanical vibrations. The accelerometer and the iron core of the magnet have been fastened to the table plate. The coil extends the magnetic field of the air gap at both sides to such a degree that small movements in the displacement direction ($x$ direction) allow the Lorentz force to remain constant. Therefore the support of the coil on the ground need not be rigid. The lower part of the figure shows a block diagram of the control circuit. A maximum force of 1 N is sufficient to attenuate all disturbing vibrations.

5.1. *Measures for the attenuation of mechanical vibrations (mechanical meausres, Rademacher; electronic measures, Seyfried, Lucas, Siegert)*

Attenuation in the direction of displacement is produced by an electronic control circuit. An additional force counteracting the vibration component in the direction of displacement and proportional to the vibration velocity acts on the table. The force is converted by amplification, integration, and transformation into a current of the output signal of an accelerometer; this current feeds a magnetic drive mounted on the granite table in the direction of the displacement. The result of this measure can be seen in Fig. 7 which shows the vibration velocity of the table plate as a function of time for an exciting force of 1 N. In the upper part the case without feedback is shown, and in the lower part the case with feedback.

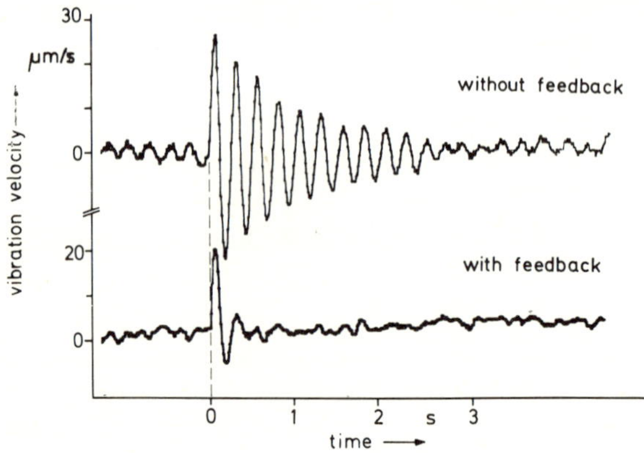


FIG. 7. Vibration velocity of the granite table. The vibration velocity was obtained from the signal of an accelerometer by single integration. At a time $t_0$ the table was suddenly stimulated to vibrate in the $x$ direction by a force of 1 N: (a) without feedback the movements of the table plate (resonance frequency approximately 4 Hz) fade only because of the low internal damping of the vibration system; (b) with feedback the vibrations of the table plate are damped almost aperiodically.

5.2. *Thermostating measures (Seyfried, Ebeling, Siegert)*

Figure 8 shows the thermostating measures which were necessary to guarantee a temperature stability of $10^{-3}$ K in the vicinity of the interferometer for several hours. The table was surrounded by a system of three heat-insulating chambers. Disturbing heat sources such as step motors were situated as far as possible from the instrument. The heat generated by the laser was removed in a controlled form via a water-circulating system. To reduce the thermal coupling of the granite plate to the environment the temperature of the steel legs was kept constant at half height by means of regulated additional heating. The heat conduction resistance between the granite plate and the temperature-regulating points yields, in connection with the thermal capacity of theplate, a time constant of about three days which attenuates short-time variations in a sufficiently strong manner.

## 6. Measurements

The following equation is used to describe the relationship between the lattice plane spacing $d$ and the optical wavelength $\lambda$:

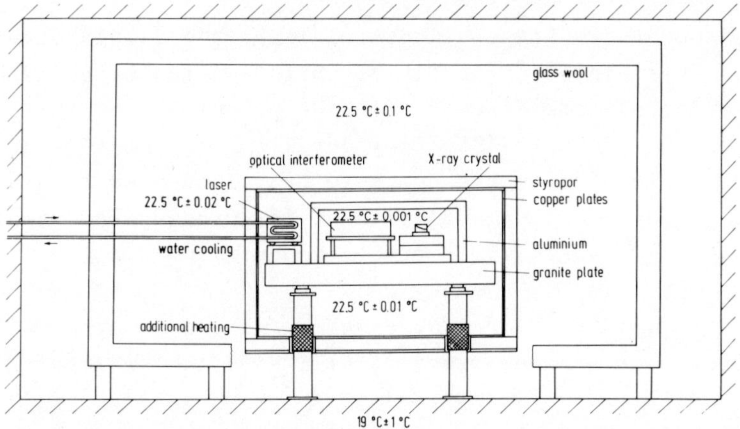


FIG. 8. Thermochambers of the interferometer. The granite plate which carries the whole instrument is surrounded by three chambers made from glass wool, polystyrene (trade name 'Stypropor') and aluminium. Going from the outer to the inner chamber, the temperature stability is improved by a factor of 10 in each chamber. The temperature gradients along the granite plate are measured by means of differential thermoelements and the temperatures for control purposes are measured by means of resistance thermometers. During all measurements the temperature was kept constant to nearly 22.5 °C. for future measurements in vacuum the aluminium chamber can be evacuated.

$$d = \frac{m}{n}\,\frac{\lambda}{2} \tag{1}$$

where $m$ is the number of the half optical wavelengths $\lambda$ contained in the total displacement and $n$ the number of the X-ray fringes. In order to achieve the desired measuring accuracy of $10^{-6}$ both the whole number $n_0$ of the X-ray fringes and the fraction $f$ must be determined to the highest accuracy:

$$\frac{n}{m} = n_0 - f_a \qquad \text{in air} \tag{2}$$

$$\frac{n}{m} = n_0 + f_v \qquad \text{in vacuum.} \tag{3}$$

All the numerical values obtained by measurements in air can be converted to vacuum conditions by using a wavelength-correction factor.

6.1. *Determination of* $n_0$ *(Becker, Lucas)*

The whole number of X-ray periods relating to a displacement of $\lambda/2$ was determined reproducibly, during 48 tests, by means of a phase-locked-loop circuit (for the principles of this circuit see e.g. Gardner 1966; for details see Best 1975 and Morgan 1975). It permits the noise to be balanced up to a displacement velocity of 1 nm $s^{-1}$ without influencing the result of the counting. The required number amounts to 1648.

6.2. *Definition of* f *(Seyfried, Becker, Lucas)*

Figure 9 demonstrates the reproducibility of the analyser displacement. Following the proposals made by Deslattes (1971) and Hart (1975) the X-ray signal is plotted against the optical signal near its zero level. The analyser crystal was made to

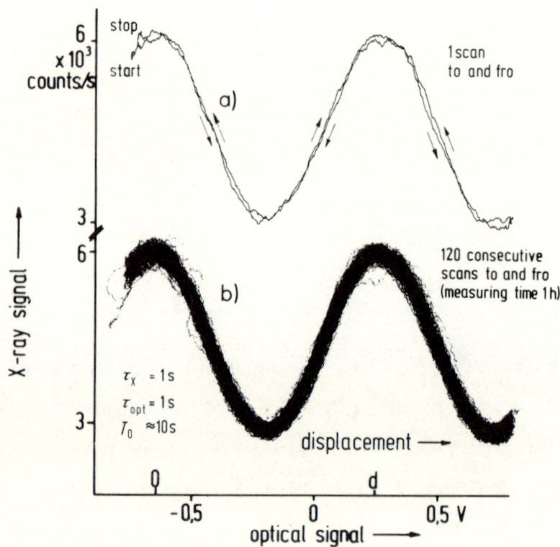


FIG. 9. X-ray signal plotted *versus* the optical signal in the region of its zero level. The optical signal is in this region proportional to the displacement. The analyser crystal was caused to wobble triangularly by 1.5 lattice plane spacings. The time constants $\tau_{opt}$ of the optical signal and $\tau_X$ of the X-ray signal were 1 s; the recording time $T_0$ of one signal period was about 10 s. The direction of the displacement (indicated by arrows) was reversed only once. The deviation of both branches is small compared with one-tenth of an X-ray fringe. The direction of the displacement was reversed 240 times. All curves are enclosed in an interval of one-tenth of an X-ray fringe. The curves are taken from a magnetic tape which recorded all scans during 8 h at night. The fine structure of the curves is caused by tape background noise.

wobble by not much more than one lattice plane spacing. The upper part shows the result of one scan. The two branches of the curve deviate from one another less than 0.03 *d*. In the lower part of the figure the result of 120 consecutive scans over a measuring time of 1 h is shown. During this time the displacement of the analyser crystal can be reproducibly measured to better than one-tenth of the lattice plane spacing.

Referring to the determination of the fraction *f* Fig. 10 shows two curves recorded at two consecurive optical orders *m*. The unknown fraction *f* can be determined from the relative displacement of these curves. As can be gathered from Fig. 10 *f* has been repeatedly determined by evaluating several X-ray fringes. When the conditions are good the individual values deviate from one another by less than 0.06.

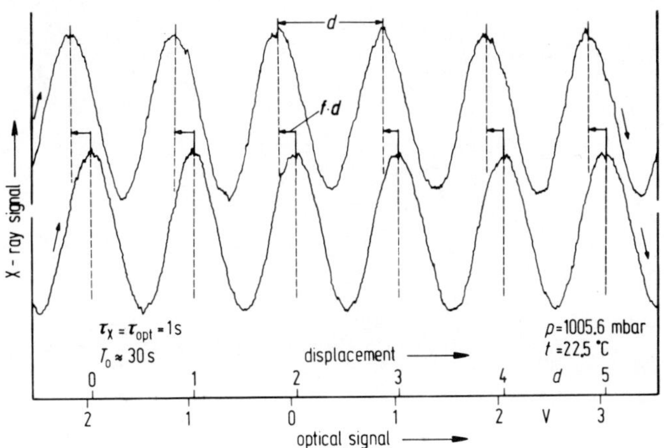


FIG. 10. X-ray signal as a function of the optical signal recorded at two consecutive optical orders. The distance between two maxima of the X-ray signal corresponds to a crystal displacement of one lattice plane spacing *d* (the displacement direction is indicated by arrows). All the relative variations are small compared with one-tenth of an X-ray period. From the relative displacement of the second curve with respect to the first one, the unknown fraction *f* is obtained as $f = 0.18 \pm 0.02$.

### 6.3. *Evaluation (Hanszen, Becker)*

When the number $n_0$ has been fixed, according to the findings of Deslattes and Henins (1973) it is only necessary to measure the fringe fractions $\Delta n$ between increasing optical intervals $m^*$.

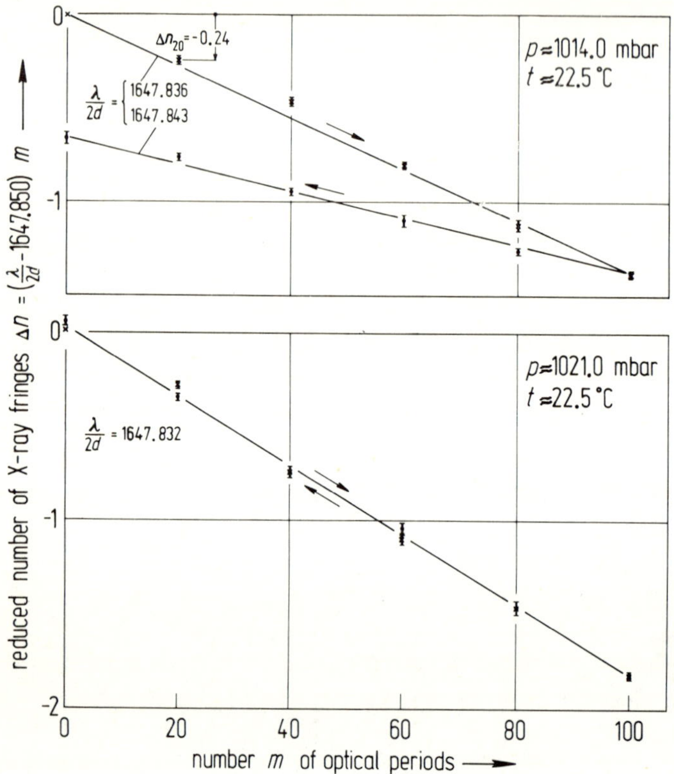


FIG. 11. Determination of the ratio 'optical wavelength to lattice plane spacing' (last step). Total displacement, 100 optical orders. The general evaluation formula reads $\Delta n = (\lambda/2d - N^*/m^*)m$ where $m^*$ is the sampling distance and $N^*$ is the smallest integer above $n(m^*)$. 1st step: $m^* = 1$; i.e. scanning at $m = 0, 1, 2, \ldots, 10$; $N^*_1 = 1648$. For $m^* = 3$ $N^*_3 = 4944$. 2nd step: $m^* = 3$; i.e. scanning at $m = 0, 3, 6, \ldots, 15$; $N^*_3$ was taken from the above step. For $m^* = 10$ $N^*_{10} = 16479$. 3rd step: $m^* = 10$, i.e. scanning at $m = 10, 20, 30, \ldots, 100$; see above. For $m^* = 20$ $N^*_{20} = 32957$. The next step is shown described in detail in the diagram.

Figure 11 shows the last step of a variant of this procedure applied in the PTB which we call the 'stroboscopical' procedure. At the twentieth optical order, for example the fraction $\Delta n_{20}$ was measured in comparison with the zeroth order. In the example given in the upper part of Fig. 11 the result is $\Delta n_{20} = -0.24$. The maximum deviations of the measured values have also been marked. At the fortieth and sixtieth order we proceeded in the same way. At the eightieth order the overflow over one total X-ray period must be taken into account.

Proceeding in this way an almost straight line results. From its slope the ratio $\lambda/2d$ was determined according to the equation given at the ordinate. The deviations of the measuring points from the straight line plotted would indicate a measuring uncertainty of less than $1 \times 10^{-6}$ when evaluating a total displacement of 100 optical orders.

Unfortunately there is an instrumental error which at present renders this accuracy still an illusion. Only in exceptional cases, see the lower part of Fig. 11, do the forward and backward displacements yield identical branches. At present the deviation of both branches from one another and the different slope of all measured curves actually limit the accuracy. The results obtained for $\lambda/2d$ have been marked on the curves. Even after a pressure correction these values do not coincide.

A total of 14 series of measurements, each with forward and backward displacements, have been carried out. At the bottom of Fig. 12 the results of these measurements corrected to vacuum conditions are given and compared with the results obtained by other authors. The old value for the lattice parameter determined by X-ray diffraction (Parrish 1960) is indicated at the top; underneath appears the value found by Deslattes *et al.* (1976) of the NBS which deviates from Parrish's value in the fourth digit behind the decimal point. The last line gives the result of our measurements. It is in agreement with the value published by Deslattes. From the result of the two recent measurements it can be concluded that the value for the X-ray wavelength used in former calculations was incorrect in the fourth digit. The fifth digit given by Deslattes, i.e. zero, was also confirmed by our measurements.

| | | |
|---|---|---|
| Internat. comparison | 1960: $0.543\ 0^{37}_{71}$ | nm |
| NBS | 1976: $0.543\ 106\ ^{30}_{62}$ | nm |
| PTB | 1978: 0.543 10█ ██ | nm |

FIG. 12. Lattice parameter $a_0$ (25 °C) of silicon. The small numbers in the last two digits indicate the confidential limits, given by the ± 1σ (standard deviation).

Unfortunately the sixth digit is still uncertain up to six units. The printed blockings give a hint as to the problem which must still be solved. We hope to be able to overcome the existing difficulties by improving the apparatus.

## Acknowledgments

I wish to express my gratitude to Professor K.-J. Hanszen, head of the project, for helpful discussions and suggestions during the preparation of this manuscript.

## References

BEST, R. (1975). *Elektroniker* **14** (6), 9–15; see also the subsequent articles.
BONSE, U. and HART, M. (1965). *Z. Phys.* **188**, 154–164.
BONSE, U. and TE KAAT, E. (1968). *Z. Phys.* **214**, 16–21.
BONSE, U., TE KAAT, E., and SPIEKER, P. (1971). In *Precision measurement and fundamental constants* (ed. D.N. Langenberg and B.N. Taylor). *Nat. Bur. Stand. U.S.A. Spec. Publ.* 343, 291–293.
CURTIS, I., MORGAN, I., HART, M., and MILNE, A.D. (1971). In *Precision measurement and fundamental constants* (ed. D.N. Langenberg and B.N. Taylor). *Nat. Bur. Stand. (U.S.A.) Spec. Publ.* 343, 285–289.
DESLATTES, R.D. (1971). In *Precision measurement and fundamental constants* (ed. D.N. Langenberg and B.N. Taylor). *Nat. Bur. Stand. U.S.A. Spec. Publ.* 343, 279–283.
DESLATTES, R.D. and HENINS, A. (1973). *Phys. Rev. Lett.* **31**, 972–975.
DESLATTES, R.D., HENINS, A., SCHOONOVER, R.M., CARROLL, C.L., and BOWMAN, H.A. (1976). *Phys. Rev. Lett.* **36**, 898–900.
GARDNER, F.H. (1966). *Phaselock techniques*. Wiley, New York.
HART, M. (1975). *Proc. R. Soc. A* **346**, 1–22.
HOFFROGGE, CHR. and RADEMACHER, H.-J. (1973). *PTB-Mitt.* **83**, 79–82.
KIND, D. (1977). *P.-V. Seances Com. Poids Mes. Ser. 2* **45**, 23–33.
LEHRKE, K. (1974). See Mewes *et al.* 1974, pp. 38–52.
MEWES, E.-R., SIEGERT, H., LEHRKE, K., RADEMACHER, H.-J., SEYFRIED, P., and REIM, G. (1974). *PTB-Bericht APh-8*, Physikalisch-Technische Bundesanstalt, Braunschweig.
MORGAN, D.K. (1975). In *RCA SSD 203C Databook Series*, pp. 471–8. RCA Solid State, Somerville, U.S.A.
PARRISH, W. (1960). *Acta Crystallogr.* **13**, 838–850.
*PTB-Jahresbericht* (1972–1977). Physikalisch-Technische Bundesanstalt, Braunschweig.
RADEMACHER, H.-J. (1974). See Mewes *et al.* 1974, pp. 66–74.
REIM, G. (1974). See Mewes *et al.* 1974, pp. 75–95.
—— (1977). *PTB-Bericht Me-13*, pp. 139–155. Physikalisch-Technische Bundesanstalt, Braunschweig.
SEYFRIED, P. and SIEGERT, H. (1974). See Mewes *et al.* 1974, pp. 60–65.
SIEGERT, H. (1974). See Mewes *et al.* 1974, pp. 24–37.

# 5. USE OF SYNCHROTON RADIATION IN X-RAY INTERFEROMETRY

G. MATERLIK

*Deutsches Elektronen-Synchrotron DESY, Notkestr. 85, D-2000 Hamburg 52, Federal Republic of Germany*

## 1. Introduction

During the past five years the outstanding characteristics of synchrotron radiation have found rapidly increasing application to problems in X-ray spectroscopy and diffraction. Among these applications were the first tests with X-ray interferometers (Bonse and Materlik 1975a,b) who used DESY in Hamburg as a radiation source and showed that the very high spectral and angular luminosity of this source, combined with the low divergence of the emitted beam, its wavelength tunability and the well-defined polarization state harmonizes ideally with the small intrinsic angular acceptance of single crystals. This experience led to the first continuous measurement of the anomalous dispersion correction on each side of an X-ray absorption edge (Materlik 1975, Bonse and Materlik 1976) and to the first successful operation of a dispersive three-beam X-ray interferometer (Graeff 1976, Graeff and Bonse 1977).

The profitable utilization of synchrotron X-rays was in each case realized by specially adapting the interferometric set-up to the properties of the synchrotron source which are quite different to those of conventional X-ray tubes. This difference made substantial changes of the goniometer arrangement necessary, e.g. a vertical reflection plane was used for the dispersion correction measurement. New concepts also had to be developed to overcome problems caused by the nature of this radiation. For example, the elimination of higher Bragg reflected orders became essential. These appear because of the continuous wavelength distribution from the visible down to 0.1 Å $\gamma$ rays. Of course the high intensity is also responsible for a drastic increase in the scattered radiation background and for a strong heat load on the first diffraction element which can exceed several W $cm^{-2}$.

The solution of these problems not only made it possible to carry out the interferometer experiments mentioned above, but also opens the way for several other studies of interferometer devices and interferometer applications. The easy accessibility of the short-wavelength region below 0.5 Å allows, as in the case of neutrons, the study of diffraction moirés from 'thin' interferometers for which absorption is small and Pendellösung effects became dominant. Diffraction focusing (Indenbom, Slobodetskii, and Truni 1975) can be investigated as in the neutron case, and furthermore a well-defined photon polarization state can be applied to reduce the number of excited wavefields and to study the polarization dependence of Pendellösung effects for 'thin' interferometers. The problem of X-ray resonators (Deslattes 1968) and X-ray Fourier spectroscopy (Hart 1975) can then also be experimentally attacked by making use of the high intensity and a suitable wavelength fit which is best for highly dispersive interferometer designs. This last property might also be useful for certain applications of interferometric phase contrast microscopy. The number of dispersive electrons can then be changed by tuning into an absorption edge.

It is therefore useful to describe the interferometer set-up which was adapted to synchrotron X-rays and to discuss in detail in which way special source properties affected the arrangement and which methods were used to make full use of the wavelength tunability and the angular collimation.

## 2. Interferometric set-up for use with synchrotron X-radiation

The diffractometer is shown schematically in Fig. 1 for two different angles of incidence $\theta_1$ and $\theta_2$. The respective beam paths are also inserted. The stationary entrance slit is located about 30 m away from the tangent point of the electron orbit. The radiation is monochromatized by a grooved monochromator (or 'channel cut') which fulfils several tasks. In the first place it cuts a very small wavelength region out of the white incident spectrum which is given by $\Delta\lambda/\lambda = \Delta\theta \cot\theta$. The value of $\Delta\theta$ is determined by the divergence coming from

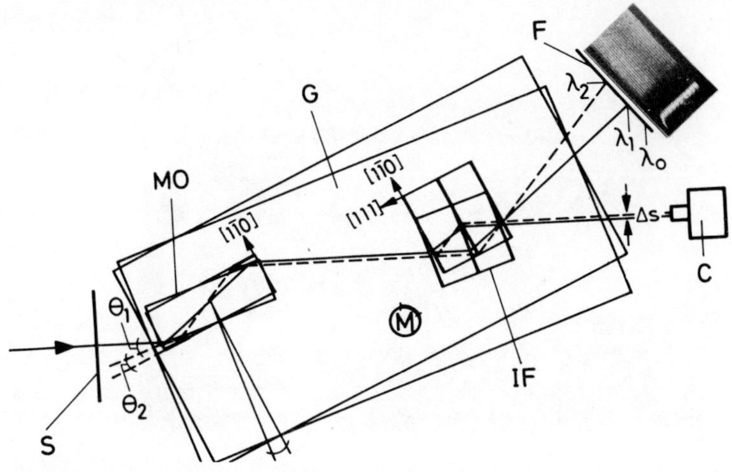


FIG. 1. Interferometric set-up: S, stationary slit; MO, monochromator; G, goniometer; M, monitor counter; IF, interferometer; C, counter; F, film.

the source and is usually about $10^{-4}$–$10^{-3}$ for each reflected order. The singly diffracted beam therefore already has the monochromacy of a characteristic X-ray line or even better. Secondly by closely shielding the crystal as shown in Fig. 2 the radiation background can be suppressed because the spot where the primary beam is stopped by the first mirror is no longer in line with the doubly reflected beam. Thirdly if the first order Bragg mirror is heated by the intensive radiation, one can design the groove in such a way that the relative angle between the Bragg mirrors can be changed by elastic deformation of a weak link cut in the base of the groove (Materlik and Kostroun 1979). Since different $d$ spacings for the first and second reflection shift the $\theta$ position of the respective single-crystal reflection ranges, this is then corrected by adjusting the Bragg angle for the second reflection so that the overlap is again maximized.

The two interference beams generated by the LLL interferometer respond quite differently to a change of the Bragg angle $\theta$. As illustrated in Fig. 1 rotation of the spectrometer through an angle $\Delta\theta$ results in a sweep of the H beam across the stationary film while the O beam, being redirected parallel to the primary beam, stays at virtually the same spatial position. In

FIG. 2. Channel-cut monchromator with lead shielding (the lead cover is removed).

the case shown $\Delta s$ = 0.5 mm for $\Delta\theta$ = 3° and $\lambda$ = 1.54 Å. This beam is therefore very well suited for use with detectors while the H direction produces a wavelength scale on a film or on another position sensitive device such as a TV camera or position-sensitive detector.

The fringe moiré of a wedge placed in one of the interference beams is displayed in the inset of Fig. 1. A section pattern has been recorded at the position $\lambda_0$. The wavelength resolution is then determined by the source divergence and the distance of the groove to the tangent point. In the region from $\lambda_1$ to $\lambda_2$ the diffractometer was driven continuously and the resolution thus also depends on the distance of the film to the interferometer and can be varied by this means.

The LLL interferometer is remotely adjustable parallel and perpendicular to the Laue mirrors as are the three independent rotations $\Delta\theta$, $\Delta\rho$, and $\Delta\kappa$, where $\Delta\theta$ determines the relative alignment of groove and interferometer in the reflection plane. The remote control of all important parameters of the diffractometer is necessary because the whole experiment is interlocked while the synchrotron radiation beam is turned on.

As will be discussed in §3 the polarization state and the higher collimation perpendicular to the orbital plane, which is not affected by the curvature of the orbit, require in many

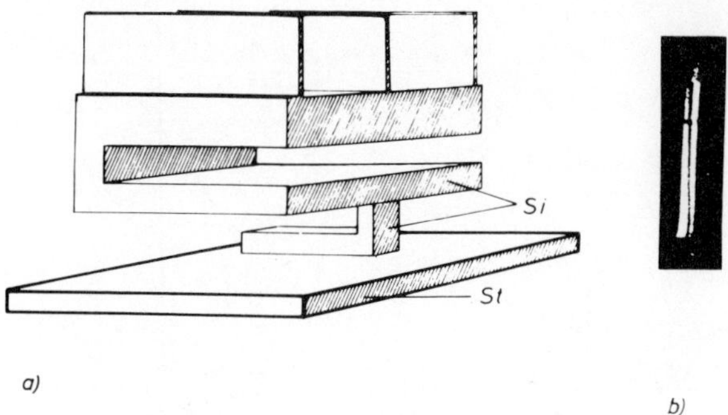


FIG. 3. (a) Interferometer mount for vertical reflection plane: St, steel; Si, silicon. (b) Build-in moiré.

applications a vertical reflection plane. This means that all X-ray optical components have to be rigidly mounted without inducing any strains. The interferometer designed for this purpose is shown in Fig. 3. The deep groove in the base serves as an obstacle for lattice deformations from the glue which connects the interferometer base with an L-shaped silicon block. This block buffers the different thermal expansions of the silicon and the steel plate.

Also shown in Fig. 3 is the built-in moiré pattern. The visible lattice dilatation with a fringe distance $\Lambda_D \approx 1$ mm is produced by a lattice constant variation $\Delta d/d = d/\Lambda_D \approx 2 \times 10^{-7}$. We have not yet investigated whether this dilatation is produced by differential heating of the interferometer mirrors when they are hit by the intensive synchrotron radiation beam.

## 3. Polarization and angular collimation

One peculiarity of synchrotron X-rays is their polarization state. The normalized intensities of radiation for the components of the electrical vector parallel ($P_{\parallel}$) and perpendicular ($P_{\perp}$) to the electron orbit are plotted in Fig. 4(a) as a function of the angle $\Psi$ to the horizontal plane. The width at half-maximum of $P_{\parallel}$ is about 0.16 mrad and the high collimation of the relativistically compressed radiation cone emitted by one

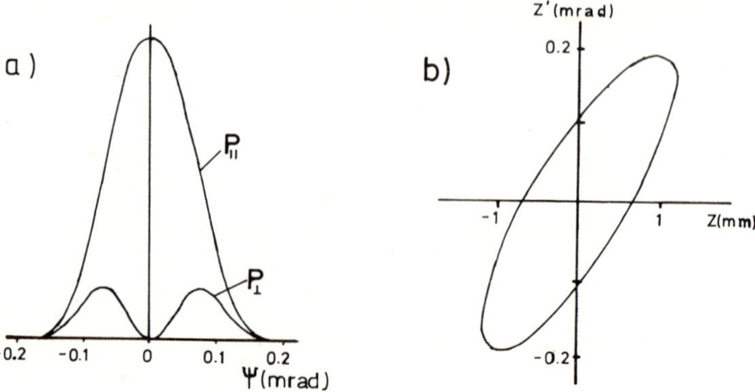


FIG. 4. (a) Intensity distribution of synchrotron radiation as a function of the inclination angle $\Psi$ to the orbit plane for $\lambda = 1.5$ Å. $P_{\|}$ and $P_{\perp}$ are the components of the electric vector parallel and perpendicular to the orbit. (b) Typical phase-space ellipse for DORIS at 3 GeV electron energy (for details see text).

electron is evident. $P_{\perp}$ exhibits two maxima with a separation of 0.14 mrad. Both are much smaller in height than the maximum of $P_{\|}$. A very high degree of linear polarization is reached close to the orbital plane, while the polarization is elliptical if the accepting slit moves out of the plane. It becomes more and more circular for increasing $\Psi$ (Kunz 1976) because the phase difference between $P_{\|}$ and $P_{\perp}$ is $\pm\ \pi/2$ and the amplitudes are then almost equal.

The ideal case in which all electrons are travelling along a central orbit is never fulfilled in an actual synchrotron or storage ring. In practice the electrons oscillate about a central orbit (betatron oscillations) in such a way that for each cross-section of the beam the distance of an electron to the centre of the bunch and the corresponding angular divergence are connected by an elliptical relation (Green 1975). Such a phase-space ellipse obtained with the storage ring DORIS in Hamburg for an electron energy of 3 GeV is shown in Fig. 4(b) for a vertical beam distribution in the **z** direction (**z** is perpendicular to the orbit) and a divergence $z' = dz/ds$ where **s** is the unit vector parallel to the tangent and perpendicular to **z**. The shape of the ellipse varies along the orbit and depends on the actual accelerator beam optics. However, if the particle is only radially accelerating, the area of the ellipse is a constant of motion and the vertical and horizontal betatron oscillations can usually be treated independently.

The curve representing the width at half-maximum of a normal density distribution for the electron beam is shown in Fig. 4(b). In this case the maximum divergence of the betatron oscillations is 0.38 mrad which is about twice the natural divergence of the synchrotron radiation at 1.5 Å photon wavelength. Of course this also changes the total degree of polarization reached within a receiving slit and a polarizer, as discussed at the end of this section, might become necessary.

The phase ellipse for the horizontal plane is usually larger than that for the vertical plane and the smearing by the curvature of the ring has also to be included. Both of these observations support the use of a vertical reflection plane if the highest possible wavelength resolution for one single crystal diffraction is to be reached. Fortunately machine parameters which determine the phase ellipses can be specified for each beam optics configuration and one can thus successfully treat the effect of the phase ellipse and of all slits placed into the beam to define which portion of the source is used and the corresponding divergence (Green 1975, Hastings 1976, Pianetta and Lindau 1977).

Proof of the high polarization and collimation of DESY and their dependence on beam optics is shown in Fig. 5. A wedge fringe system was recorded using a horizontal reflection plane with the arrangement shown in Fig. 1. The diagrams are densitometer profiles along the traces indicated by the arrows. The intensity distribution of the perpendicularly polarized component $P_{\perp}$ (Fig. 4(a)) is visible and comparison of Figs. 5(a) and 5(b) gives a contrast change which follows from an increase in source height. The separation of the two maxima in Fig. 5(a) is about 0.14 mrad. This shows that the divergence of the vertical betatron oscillations is very small at that particular section of the orbit.

The use of the Laue mirror as a Borrmann polarizer (Cole, Chambers, and Wood 1961) makes this illustration possible. It is well known that the Borrmann effect strongly favours wave-

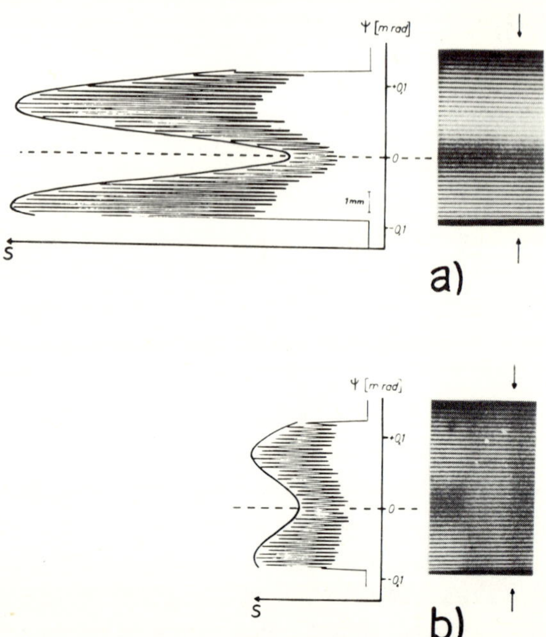


FIG. 5. Fringe moirés and densitometer traces along the directions indicated by arrows for two different electron beam optical configurations of DESY. S optical density. Note the contrast change and the similarity between S and $P_{\perp}$.

fields with an electrical vector perpendicular to the reflection plane. The anomalous small absorption coefficient for an energy flow parallel to the net planes in the two-beam case is given by (Borrmann 1955)

$$\mu_{min}^{c} = \frac{\mu_0}{\cos\theta_B}\left(1 - C\left|\frac{\chi_{i0}}{\chi_{ih}}\right|\right) \tag{1}$$

where $\mu_0$ is the normal photoelectric absorption coefficient and $\chi_{i0}$ and $\chi_{ih}$ are the imaginary parts of the O and H Fourier components of the electrical susceptibility. The polarization state ($C$ = 1 for perpendicular and $C = |\cos 2\theta|$ for parallel polarization to the reflection plane) introduces a strong difference in absorption. In the Laue case this difference can be used to construct a simple polarizer. Only by this means can full use of the polarization of synchrotron X-rays be made, especially if the influence of betatron oscillation is considered.

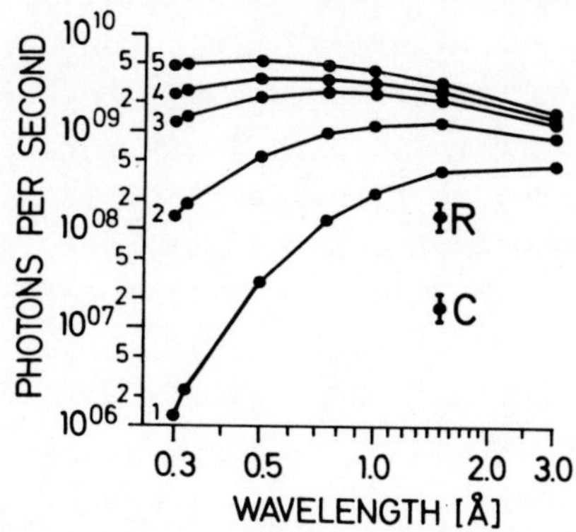


FIG. 6. Synchrotron radiation intensity of DESY at a slit 0.5 mm wide and 7 mm high situated at a distance of 29 m from the tangent point for different electron end-energies: curve 1, 4 GeV; curve 2, 5 GeV; curve 3, 6 GeV; curve 4, 6.5 GeV; curve 5, 7.2 GeV. The beam current was 10 mA and $\Delta\lambda = 10^{-3}$ Å. R and C denote the intensities from a 30 kW Rigaku Denki rotating anode and a conventional 1.5 kW X-ray tube respectively.

## 4. Continuous spectrum and higher orders

Easy wavelength tunability is another important advantage of synchrotron radiation. Figure 6 shows the intensity distribution from DESY in a given entrance slit 29 m away from the source as a function of wavelength in the region $\lambda$ = 0.3–3.0 Å. The available flux was compared with that from a Cu $K_\alpha$ line generated by a 30 kW Rigaku Denki rotating anode (R) and by a conventional 1.5 kW sealed X-ray tube (C). The comparison was made by using the same experimental arrangement (Fig. 1) in each case. The intensities were measured with films and normalized to the theoretical synchrotron radiation flux. As was pointed out in §3 the source divergence determines the actual energy passband. In our experiment this parameter was not controlled which led to uncertainties in $\Delta\lambda/\lambda$ (marked by the bars for R and C. A factor of about 10 can be gained when compared with R[†] and another factor of 10 when compared with C. However, this comparison is made for the intensities of a characteristic

---

[†] The intensity of the rotating anode could be doubled for short periods of time. However, the chances of vacuum and power failures then increased markedly for the model used.

X-ray line! When real tunability is wanted, as in dispersion spectrometry (Bonse and Materlik 1976), conventional X-ray generators only offer the bremsstrahlung spectrum and the comparable intensities are reduced by another factor of 500–1000. It should also be noted that the increase in intensity and thus the decrease in exposure or counting time is in many cases not only of quantitative but also of qualitative nature because the influence of thermal and vibrational disturbances is reduced.

The continuous spectrum extending far into the hard X-ray region offers the advantage of testing the interferometer for long and short wavelengths at the same Bragg angle. Thus one can carry out interferometry with varying absorption and extinction length up to the absorption-free neutron-like case (Bauspiess, Bonse, and Graeff 1976). This has been made possible by the development of a new method to eliminate the harmonics which are always present (with the exception of forbidden reflections) after one single-crystal reflection of a wide spectrum. The idea is to use the refractive index correction to the Bragg angle (Bonse and Materlik 1975a) which introduces a wavelength dependence of the angular position of the centre of the reflection range. The centre position is also coupled to the asymmetry factor $b = \sin(\phi+\theta)/\sin(\phi-\theta)$ where $\phi$ is the angle between the surface and the reflecting Bragg planes. Then (Zachariasen 1967)

$$\theta_0 - \theta_B = \left(1-\frac{1}{b}\right)\frac{|\chi_{r0}|}{2\sin 2\theta_B} + C\frac{|\chi_{rh}|}{|b|^{\frac{1}{2}}\sin 2\theta_B}y \tag{2}$$

and

$$\theta_h - \theta_B = -b(\theta_0 - \theta_B) \tag{3}$$

where

$$\chi_{r0} = -\frac{r_e}{\pi}\frac{F_0}{V}\lambda^2$$
$$\chi_{rh} = -\frac{r_e}{\pi}\frac{F_h}{V}\lambda^2 \tag{4}$$

Here $\theta_0$ and $\theta_h$ are the reflection angles for incidence and reflectance respectively, $\chi_{r0}$ and $\chi_{rh}$ are the real parts of the electrical susceptibility which are connected to the structure factors $F_0$ and $F_h$, $r_e$ is the classical radius of the electron, and $V$ is the volume of the unit cell. The parameter $y$ gives the deviation from the centre ($y$=0) of the reflection range. The principle of this method has been discussed in detail by Bonse, Materlik, and Schröder (1976) and its application to monolithic monochromators has been discussed by Materlik and Kostroun (1979).

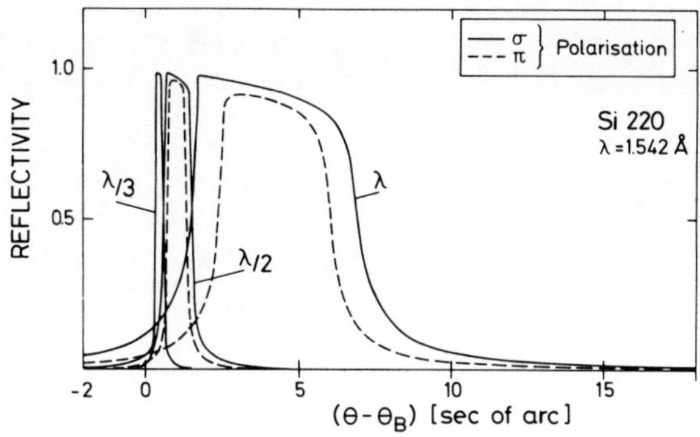


FIG. 7. Bragg case single-crystal reflection curves.

Silicon 220 reflection curves for λ = 1.54 Å and the higher orders λ/2 and λ/3 are shown in Fig. 7 for two polarization states assuming parallel incident plane waves. The centres of the λ and λ/2 reflection ranges are shifted by 3.1″. Since $b$ = 1 in the symmetrical Laue case all curves are then centred at $\theta_B$. The combination of a Bragg case groove and a Laue case interferometer as shown in Fig. 1 is a simple way of selecting harmonics by tuning the relative θ adjustment of the two optical components.

The wedge moirés in Fig. 8 which were photographed at each θ position simultaneously in the H and O beams exhibit different fringe spacings for λ, λ/2, and λ/3 and also show the remarkable change of the reflex structure caused by Pendellösung oscillations. For λ = 1.5 Å the interferometer is Borrmann

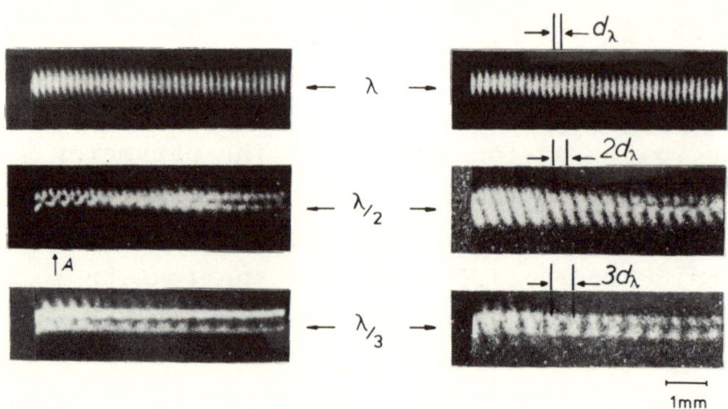


FIG. 8. Wedge moirés for $\lambda$ = 1.5 Å, $\lambda/2$, and $\lambda/3$ reflections: left-hand columne, H beam; right-hand column, O beam.

like ($\mu_0 t \approx 23$), but for $\lambda$ = 0.75 Å ($\mu_0 t \approx 2.9$) a further modulation is introduced (different in the H and the O beams) which is perpendicular to the interferometer base and becomes even more pronounced for $\lambda$ = 0.5 Å ($\mu_0 t \approx 0.9$).

## 5. Applications

As mentioned in §1 the two fields in X-ray interferometry which so far have profited most from synchrotron radiation are the measurement of the dispersion correction $f'$ close to the X-ray absorption edges and the development of dispersive X-ray interferometers. In this section we therefore explain which properties of synchrotron X-radiation and which of the methods of adaption to the source characteristics described earlier were crucial to these experiments.

### 5.1. *Dispersion spectroscopy*

The refractive index $n$ of a homogeneous medium with $N$ atoms per unit volume and the coherent forward scattering amplitude $f = f_0 + f' + if''$ are connected by

$$1 - n = \frac{r_e}{2\pi}\lambda^2 N(f_0+f') \tag{5}$$

in the case of X-rays. $n$ was measured on each side of the

nickel K absorption edge by Bonse and Materlik (1976) using the arrangement shown in Figs. 1–3. A lucite wedge produced a fringe system which was shifted when the sample was inserted into an interference beam. The fringe system was photographed both in the presence and the absence of the sample and the relative shifts were determined with a digitizing densitometer and successive computer analysis. A vertical reflection plane was chosen in order to achieve the highest possible wavelength resolution and intensity. The reflex position on the film placed in the H beam could be used to calibrate the wavelength, and harmonics were suppressed by the Bragg–Laue combination. The result of the measurement is plotted in Fig. 9. It shows a smooth wavelength dependence of $f'$ on the long-wavelength side of $\lambda_K$ and reflects the well-known extended X-ray absorption fine structure (EXAFS) of the absorption spectrum for $\lambda < \lambda_K$. Here the intermediate electron state is modulated close to the edge by the conduction bands of the solid and further out can be interpreted by taking the backscattering from neighbouring atoms into account. The arrows in Fig. 9 illustrate the correspondence of the $f'$ structure to the minima (below the curve) and maxima (above the curve) of the absorption spectrum measured by Cauchois and Manescu (1950).

The errors in this measurement of $f'$ were mostly due to

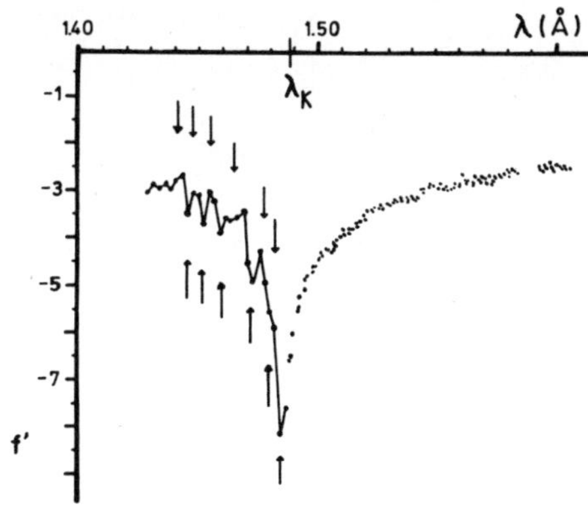


FIG. 9. Measured $f'$ dependence of nickel on each side of the K absorption edge. The arrows indicate the structure measured in absorption: the arrows above the curves give the positions of the maxima and the arrows below the curves those of the minima.

inhomogeneities of film shrinkage and sample thickness. Here the use of detectors and the calibration of the thickness with a wavelength far away from the absorption edge (Cusatis and Hart 1975), for example with $\lambda/2$, will establish a standard method for precise $f'$ determination.

High intensity, continuous wavelength spectrum, and high collimination are the decisive advantages of synchrotron radiation in this application.

### 5.2. *Three-beam interferometer*

The arrangement of the three-beam interferometer used by Graeff and Bonse (1977) at DESY is shown in Fig. 10. A 220 germanium groove was selected to avoid the spot hit by the primary beam and to suppress the higher orders. This was done by using the method described in §4 and combining two materials with different electronic densities (see equations (2) and (4)), in this case the germanium groove with the (440, 404) silicon interferometer.

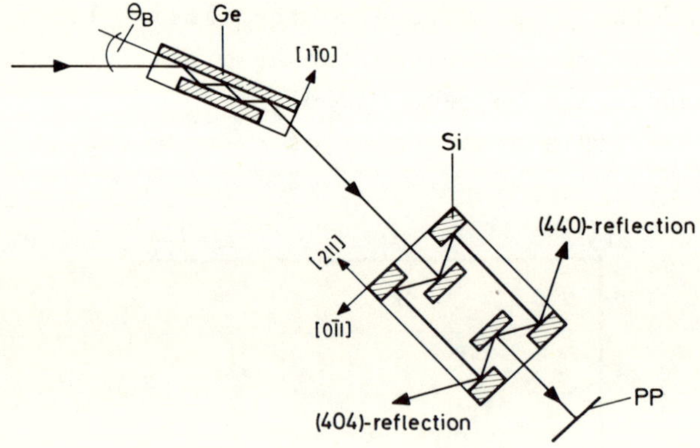


FIG. 10. Set-up and ray path for a three-beam interferometer: PP, photographic plate.

The path of rays in the interferometer is shown in Fig. 10. All rays are coplanar with the incident beam because the right wavelength can be chosen from the continuous spectrum. The choice of this coplanar case was crucial for the operation of the interferometer because the temporal coherence length

$l_c = \lambda^2/\Delta\lambda$ is determined by the wavelength resolution. If the incoming beam is tilted by an angle $\eta$ with respect to the ray plane shown in Fig. 10 it follows from Bragg's law and geometry that $\Delta\lambda/\lambda \approx \eta^2/2$ if the divergence is kept constant. The coherence length thus decreases. Of course this is also true if the divergence perpendicular to the reflection plane increases.

A large coherence length is necessary in this experiment because the surface roughness decreases the temporal coherence. Graeff (1976) estimated that the vertical divergence of synchrotron X-radiation corresponds for $\lambda = 1.66$ Å to a coherence length $l_c \approx 15$–30 μm which is enough to reflect consecutively from several Bragg mirrors each with a surface roughness of about 0.2 μm. A horizontal reflection plane was used and wavelength tunability and high collimation were the most important parameters of synchrotron X-radiation.

**Acknowledgment**

The author would like to thank Prof. Dr. U. Bonse for many discussions about the above topic.

## References

BAUSPIESS, W., BONSE, U., and GRAEFF, W. (1976). *J. appl. Crystallogr.* **9**, 68.

BONSE, U. and MATERLIK, G. (1975a). In *Anomalous scattering* (eds. S. Ramaseshan and S.C. Abrahams), pp. 107–109. Munksgaard, Copenhagen.

—— (1975b). *Acta Crystallogr. A* **31** (S3), 232.

—— (1976). *Z. Phys. B* **24**, 189.

BONSE, U., MATERLIK, G., and SCHRÖDER, W. (1976). *J. appl. Crystallogr.* **9**, 223.

BORRMANN, G. (1955). *Z. Kristallogr. Mineral. Petrogr.* **106**, 109.

COLE, H., CHAMBERS, F.W., and WOOD, C.G. (1961). *J. appl. Phys.* **32**, 1942.

COUCHOIS, Y. and MANESCU, I. (1950). *J. Chim. phys. Phys.-Chim. biol.* **47**, 892.

CUSATIS, C. and HART, M. (1975). In *Anomalous scattering* (eds. S. Ramaseshan and S.C. Abrahams), pp. 57–68. Munksgaard, Copenhagen.

DESLATTES, R.D. (1968). *Appl. Phys. Lett.* **12**, 133.

GRAEFF, W. (1976). Ph.D. Thesis, Univ. Dortmund.

GRAEFF, W. and BONSE, U. (1977). *Z. Phys. B* **27**, 19.

GREEN, K. (1975). *BNL Rep. 50595*, Brookhaven National Laboratory, N.Y.

HART, M. (1975). *Proc. R. Soc. A* **346**, 1.

HASTINGS, J.B. (1977). *J. appl. Phys.* **48**, 1576.

INDENBOM, V.L., SLOBODETSKII, I.S., and TRUNI, K.G. (1975). *Sov. Phys.—JETP* **39**, 542.

KUNZ, C. (1976). In *Optical properties of solids* (ed. B.O. Seraphin), pp. 477–553. North-Holland, Amsterdam.

MATERLIK, G. (1975). Ph.D. Thesis, Univ. Dortmund.

MATERLIK, G. and KOSTROUN, V.O. (1979). Accepted for publication in *Rev. Sci. Instrum.*

PIANETTA, P. and LINDAU, I. (1977). *J. electron. Spectrosc. relat. Phenom.* **11**, 13.

ZACHARIASEN, W.H. (1967). *Theory of X-ray diffraction in crystals*. Dover Publications, New York.

# 6. OPTICAL INTERFEROMETRY

M. FRANÇON

*Optical Institute of the P. and M. Curie University,*
*4 place Jussieu, tour 13, 75230 Paris Cêdex 05 (France)*

## 1. Luminous quasi-monochromatic vibration

A luminous quasi-monochromatic vibration can be represented by the expression

$$V(t) = a(t)\exp(j2\pi\nu_0 t) \qquad (1)$$

where $j = \sqrt{(-1)}$, $a(t)$ is a complex function of time $t$ and varies very slowly compared with the oscillatory term $\exp(j2\pi\nu_0 t)$, and the frequency $\nu_0$ represents an average frequency. The function $a(t)$ is the instantaneous complex amplitude of the vibration. $V(t)$ can be considered as a monochromatic vibration, the amplitude $a(t)$ of which is variable and the frequency of which is equal to the average frequency $\nu_0$. For thermal sources $a(t)$ still varies very quickly compared with our means of observation. If these phenomena are represented in the complex plane, the complex amplitude $a(t)$ is constant for a monochromatic vibration and can be represented by a point in the complex plane (Fig. 1). The real vibration is an infinite sinusoid and its spectrum contains only the frequency $\nu_0$. In the case of a limited sinusoidal vibration duration $\tau$ the amplitude is still represented by a point during the period of the vibration but the spectrum is of the type shown in Fig. 2 with the frequencies distributed about the mean frequency $\nu_0$. The

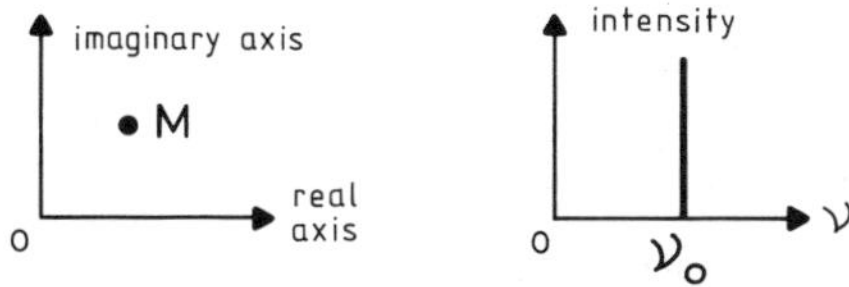


FIG. 1. Representation of monochromatic light.

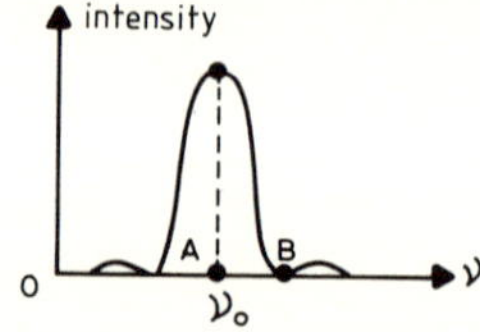


FIG. 2. Spectrum of quasi-monochromatic light (OA ≫ AB).

time $\tau$ is the coherence time. The length of the vibration $l = c\tau$ (where $c$ is the speed of light) is the coherence length and is related to the width AB of the spectrum by the equation

$$\Delta\nu = 1/\tau \; . \tag{2}$$

In the case of quasi-monochromatic light OA ≫ AB. In the general case the complex amplitude $a(t)$ varies in a very complicated manner and the curve is analogous to that shown in Fig. 3. If the light is quasi-monochromatic the variations of $a(t)$ are relatively slow compared with the variations of the term oscillating at the frequency $\nu_0$.

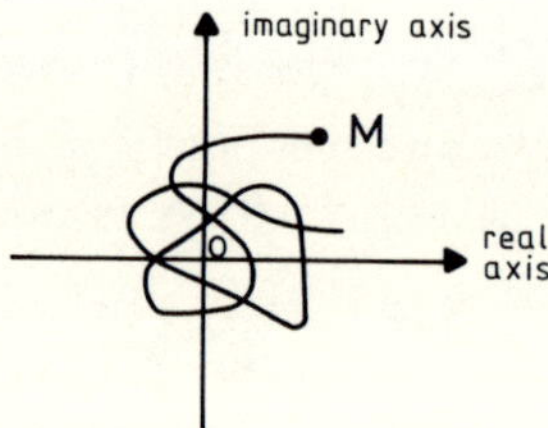


FIG. 3. Variation of the complex amplitude $a(t)$ in the general case.

## 2. Coherence of luminous vibrations

Let us consider two point sources $S_1$ and $S_2$ which emit quasi-monochromatic light (Fig. 4). They emit vibrations $V_1(t)$ and $V_2(t)$ which are superimposed at P. For the present we shall not

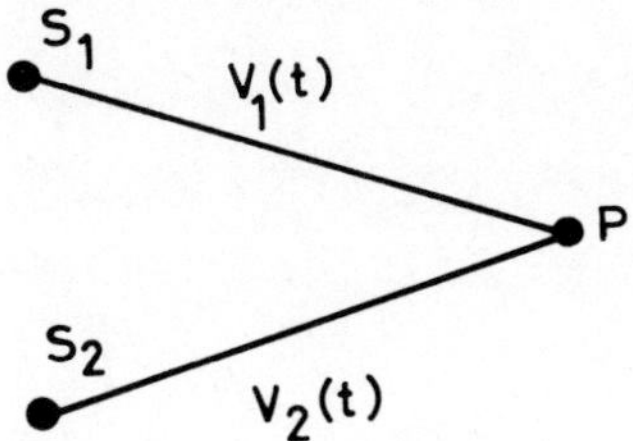


FIG. 4. Vibrations $V_1(t)$ and $V_2(t)$ emitted by two sources at P.

specify the nature of these two sources but we assume that the path difference $\Delta = S_1P - S_2P$ is much less than the coherence length $l$. During the time necessary to carry out an observation the point P receives a considerable number of vibrations and the luminous intensity observed at P can be expressed as

$$I = \langle V_1(t)V_2{}^*(t)\rangle = \langle a_1(t)a_2{}^*(t)\rangle \tag{3}$$

where the angular brackets indicate the mean value and the asterisk indicates that $V^*(t)$ is the complex conjugate of $V(t)$. It can be shown that if $I_1$ and $I_2$ are the intensities emitted at P by $S_1$ and $S_2$, the intensity at P is given by

$$I = I_1 + I_2 + 2I_1{}^{\frac{1}{2}}I_2{}^{\frac{1}{2}}|\gamma_{12}(0)|\cos 2\pi\nu_0\theta \tag{4}$$

where

$$\theta = \Delta/c \tag{5}$$

and

$$\gamma_{12}(0) = \frac{\langle a_1(t)a_2{}^*(t)\rangle}{I_1{}^{\frac{1}{2}}I_2{}^{\frac{1}{2}}} \tag{6}$$

$\gamma_{12}(0)$ is called the degree of complex coherence and its modulus $|\gamma_{12}(0)|$ is the degree of coherence of the vibrations emitted by $S_1$ and $S_2$ at P.

If $S_1$ and $S_2$ are incoherent the values $a_2(t)$ and $-a_2{}^*(t)$ are equally probable so that the mean value $\langle a_1(t)a_2^*(t)\rangle$ is zero. We then have

$$|\gamma_{12}(0)| = 0 \tag{7}$$

and

$$I = I_1 + I_2 \tag{8}$$

i.e. the intensity at P is the sum of the intensities emitted separately by $S_1$ and $S_2$. If $S_1$ and $S_2$ are coherent, for example synchronous, we then have

$$\langle a_1(t)a_2^*(t)\rangle = I_1^{\frac{1}{2}}I_2^{\frac{1}{2}} \tag{9}$$

and

$$|\gamma_{12}(0)| = 1. \tag{10}$$

If the sources are partially coherent $|\gamma_{12}(0)|$ is less than unity.

## 3. Spatial coherence and time coherence in interferometers

We consider Young's experiment (Fig. 5) where the source $S_0$, which is extended and quasi-monochromatic, illuminates the two holes $S_1$ and $S_2$ interference phenomena are observed on the

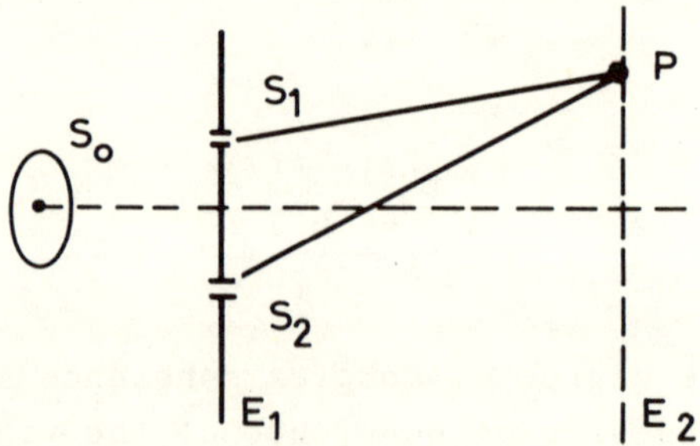


FIG. 5. Spatial coherence in Young's experiment.

screen $E_2$. The holes $S_1$ and $S_2$ are equivalent to the $S_1$ and $S_2$ of §2. Their degree of coherence depends on the extent of the source $S_0$. In this experiment the path difference $\Delta = S_2P - S_1P$ is always less than the coherence length $l$. It can be shown that the degree of coherence of $S_1$ and $S_2$ illuminated by $S_0$ is given by the normalized Fourier transform of the distribution of the energy emitted by the source. If we write

$$2\pi\nu_0\theta = \frac{2\pi\nu_0\Delta}{c} = \frac{2\pi\Delta}{\lambda} = \phi , \tag{11}$$

where $\lambda$ is the wavelength and $c$ the speed of light, the intensity at P is given by expression (4). The fringe contrast is defined as

$$\gamma = \frac{I_{max} - I_{min}}{I_{max} + I_{min}} \tag{12}$$

and we have

$$\gamma = \frac{2I_1^{\frac{1}{2}}I_2^{\frac{1}{2}}|\gamma_{12}(0)|}{I_1 + I_2} \tag{13}$$

Then if $I_1 = I_2$

$$\gamma = |\gamma_{12}(0)| . \tag{14}$$

In this form the degree of coherence $|\gamma_{12}(0)|$ intervenes as a degree of spatial coherence as the fringe contrast is related to the extent of the source and its structure.

Let us now consider another type of interferometer with well-separated beams, e.g. Michelson's interferometer. We assume that the source is a point but does not emit quasi-monochromatic vibrations (Fig. 6). The luminous vibration $V(t)$ can no longer be written in the form given by expression (1) but is composed of a continuous sum of complex components of different frequencies. This function $V(t)$ is called the 'analytic signal'. The degree of complex coherence is then written

$$\gamma_{12}(\theta) = \frac{\langle V_1(t+\theta)V_2^*(t)\rangle}{I_1^{\frac{1}{2}}I_2^{\frac{1}{2}}} \tag{15}$$

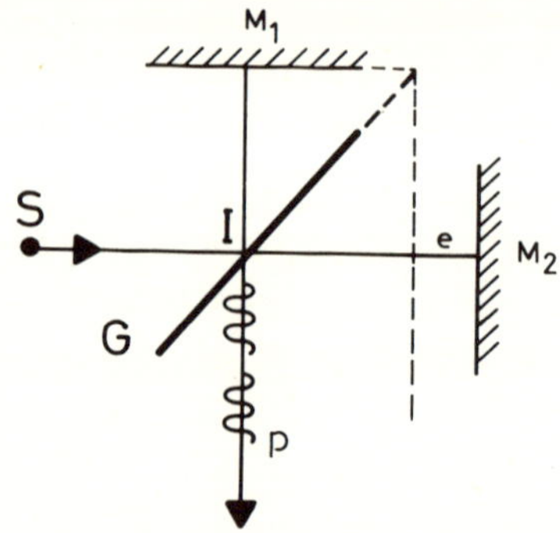


FIG. 6. Michelson's interferometer.

where θ is related to the path difference $\Delta = 2e$ by expression (5), i.e. $\theta = \Delta/c$. If $\theta \gg \tau$ we have

$$\gamma_{12}(\theta) = 0 \tag{16}$$

and there is incoherence from the point of view of time. The wavefronts which are superimposed at P do not come from the same initial wavefront. The path difference between these wavefronts varies arbitrarily and the interference phenomena disappear. The degree of coherence results here from the finite length of the wavefronts. If θ is of the same order of magnitude as τ there is partial coherence, and if $\theta \ll \tau$ there is coherence. In this latter case the interference phenomena are perfectly physical.

## 4. Interference of two waves with separate beams

In Young's experiment (Fig. 5) interference is caused by wavefront division, but in the majority of interferometers the interference is produced by amplitude division. The Michelson, Mach–Zehnder and Jamin interferometers are of the amplitude-division type. If Michelson's interferometer is illuminated by a wide source $S_0$ the visibility of the localized fringes on the mirrors depends on the path difference (Fig. 7): the smaller the path difference, the better the fringe contrast. If $M_2$ is parallel to $M'_1$ the fringes are observed at infinity and their

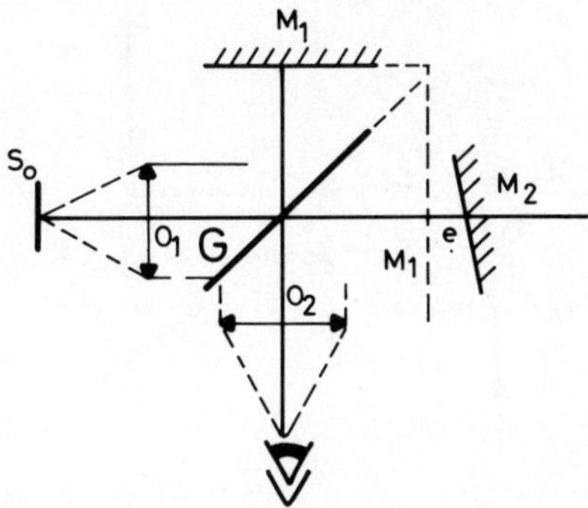


FIG. 7. Observation of the fringes of an air pocket with Michelson's interferometer.

contrast naturally does not depend on the extent of the source. It does, however, depend on the coherence time τ of the vibrations emitted by $S_0$ and the path difference. Figures 8 and 9 show the principles of the Mach–Zehnder and Jamin interferometers. If the source S is extended (Fig. 8) the fringes are localized in the area where the two rays which interfere arrive at the same incident ray. In Jamin's interferometer (Fig. 9) the path difference is zero, whatever the inclination of the rays, provided that the two plates $L_1$ and $L_2$ are parallel. If $L_1$ is inclined slightly relative to $L_2$, fringes are observed at infinity which are parallel to the angle formed by the two plates.

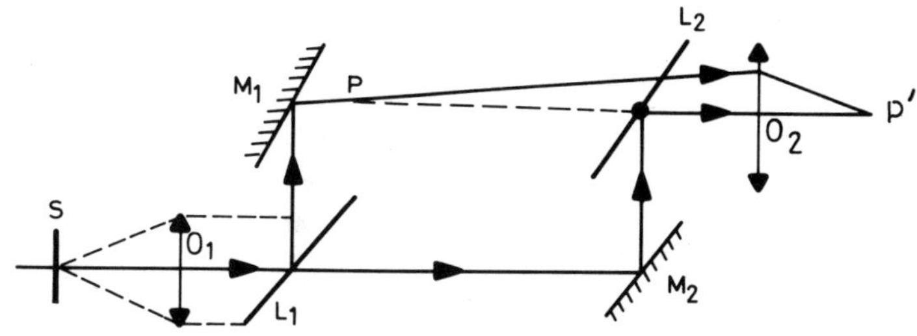


FIG. 8. Mach–Zehnder interferometer.

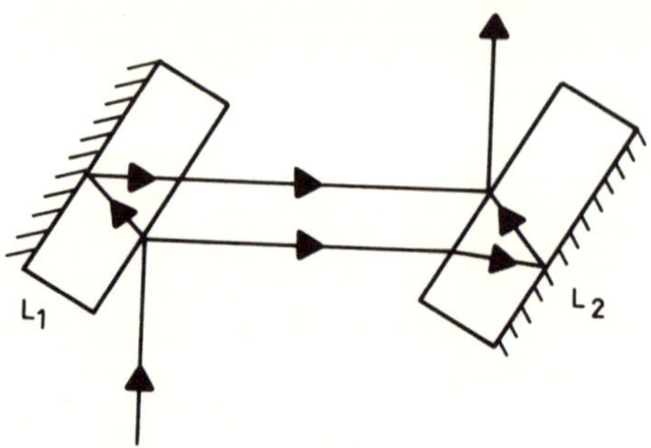


FIG. 9. Jamin interferometer.

## 5. Polarization interferometers

These interferometers use birefringent elements to produce the amplitude division. In general linear (Savart, Fig. 10) or angular (Wollaston, Fig. 11) doubling systems are used. Savart's polariscope consists of two identical quartz or spar plates cut at 45° to the axis and crossed. The incident ray is doubled in the first plate. The ordinary ray O becomes the extraordinary ray E in the second plate. The E ray in the first plate becomes

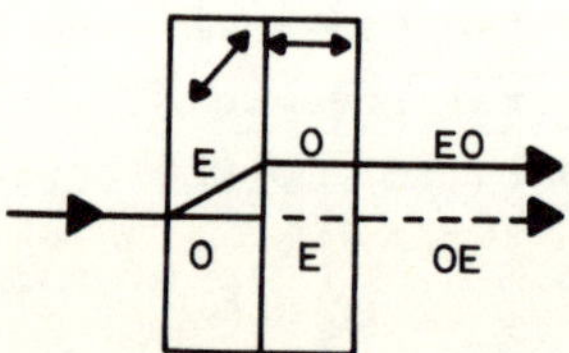


FIG. 10. Savart polariscope.

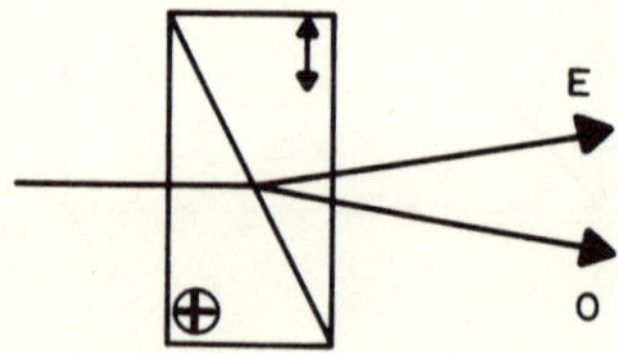


FIG. 11. Wollaston prism.

the O ray in the second sheet. The two emergent rays EO and OE are parallel but the ray OE is no longer in the plane of the figure. The principle is analogous in the Wollaston system but the doubling is angular.

These systems are generally used for shearing interferometry. The incident wave $\Sigma_0$ deformed by some transparent object A, for example a glass plate with deformed surfaces, traverses Savart's polariscope Q which doubles it. In the image A′ of A given by the objective O there are two waves identical to $\Sigma$ and slightly displaced. These two waves are coherent because of the polarizer $P_1$ and the vibrations are parallel because of the polarizer $P_2$. The interference of these two waves reveal the variations of the optical path and not the optical path itself. We have the derivative of $\Delta$ but not $\Delta$.

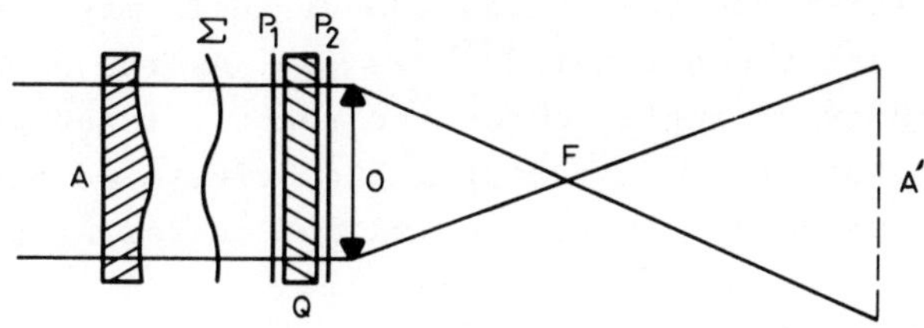


FIG. 12. Principle of a polarization interferometer.

It should be noted that shearing interferometers can also be made with classical interferometers, e.g. the Mach–Zehnder interferometer.

## 6. Holographic interferometry

One of the fundamentals of holography is the possibility of obtaining interference between waves recorded at different moments without of course contravening the basic principles of physics. Figure 13 shows the standard diagram of the recording of a hologram. The incident beam sends a wave $\Sigma_1$ onto the photographic plate H. The wave $\Sigma_1$ is the wave deformed by the transparent object to be studied. The beam R is the reference beam coherent with the previous beam. These two beams come

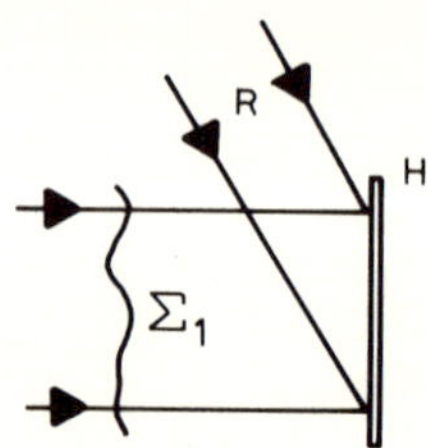


FIG. 13. Recording of a hologram of the wave surface $\Sigma_1$.

from the same source (laser) by means of a set-up which it is not necessary to show here. One exposure is made under these conditions. The object is then removed and a second exposure is made on the same plate which is now illuminated by a plane wave $\Sigma_{12}$ (Fig. 14). After development the negative H is illuminated by the reference beam R and two virtual images of the waves $\Sigma_1$ and $\Sigma_2$ are reconstituted (Fig. 15). As the negative H is illuminated by a single source the reconstituted waves $\Sigma_1$ and $\Sigma_2$ are coherent and interfere. The defects in the transparent object studied are thus observed and characterized by the wave $\Sigma_1$.

It should be noted that holography permits interference with ordinary diffusing objects which is practically impossible with conventional interferometers. In the first exposure the hologram of the diffusing object is recorded and after development the negative is returned to exactly the same place without the object being removed. Through the hologram one can see the object itself and its reconstruction by the hologram. These two 'objects' are perfectly superimposed. There is interference

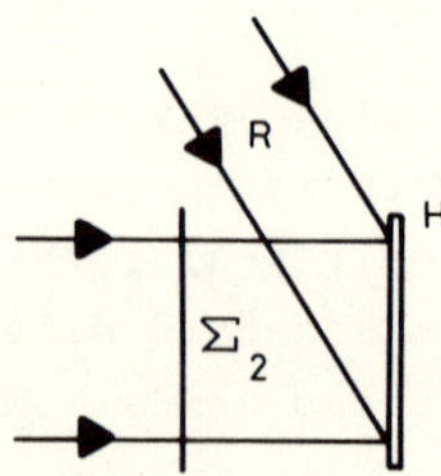


FIG. 14. Recording of a hologram of the plane wave $\Sigma_2$.

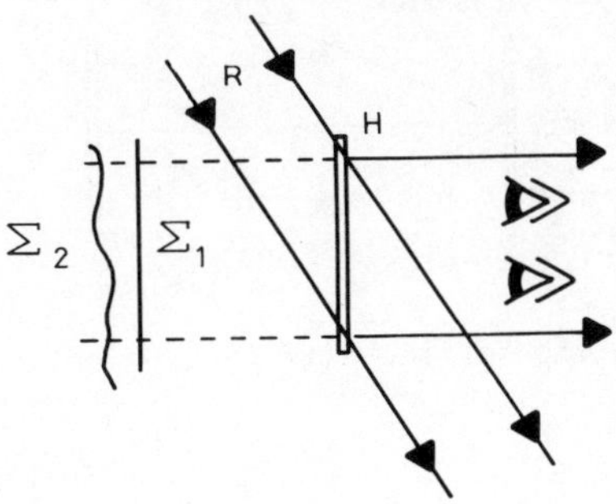


FIG. 15. The hologram H permits interference between the two waves $\Sigma_1$ and $\Sigma_2$ after reconstitution.

between the object itself and its image reconstituted by the hologram. If the object undergoes deformation this is immediately detected by the appearance of fringes.

## 7. Speckle interferometry

Let us consider any diffusing object, e.g. a ground glass screen G (Fig. 16). It is illuminated by a laser and consequently all the points on G are coherent. These diffracted vibrations are capable of interfering at any point in space. Any point P receives vibrations from all points on G. The interference results in the creation of a very fine granular structure on a plane H situated at any distance from G. The fineness, or speckle, of the grains in this structure depends on the angle at which one looks at G from a point on H. The diameter of the grains on H varies very little: the smallest have a diameter which is that of the diffraction spot of a virtual objective of diameter equal to G and with a focusing

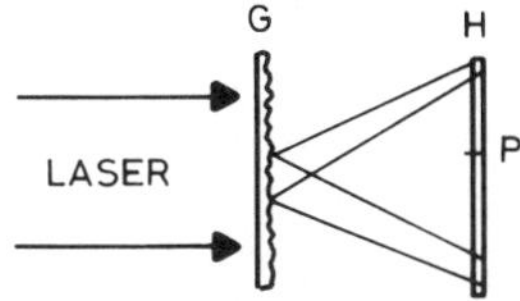


FIG. 16. Speckle produced by the diffusing object G (Fresnel speckle).

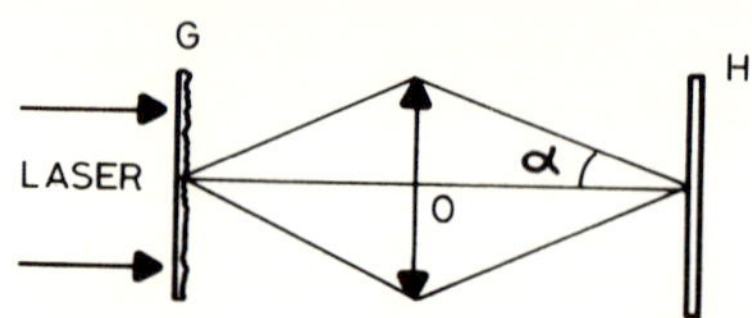


FIG. 17. Speckle in the image of the diffusing object G (Fraunhöfer speckle).

plane in the plane H.

The same experiment can be carried out by projecting an image of G on H with an objective O (Fig. 17). The diameter $\varepsilon$ of the grains of the speckle on the plane H is that of the diffraction spot of the objective O:

$$\varepsilon = \lambda/\alpha \ . \tag{17}$$

If we assume that a photographic plate is placed at H in Figs. 16 or 17 the experiment is the same. Let us make two successive exposures on the same plate H, translating the object G between the two exposures. The translation is effected in any direction but in the plane of G. The speckle is represented in H by an arbitrary function $D(\eta,\zeta)$ of two co-ordinates $\eta,\zeta$ in the plane H. The function $D(\eta,\zeta)$ represents the speckle intensity variations in H. The exposure can be shown as follows:

$$D(\eta,\zeta) \otimes (\delta_1+\delta_2) \tag{18}$$

since any translation is equivalent to a convolution by a delta function. The first exposure can also be represented by a function $\delta_1$ which does not change anything. After development the negative H is illuminated in a parallel beam and the focal beam of the objective O placed after H is observed (Fig. 18). In the focal plane the Fourier transform of the amplitude $t$ transmitted by H is observed. It is now known that the amplitude $t$ transmitted by the plate H is proportional to the intensity recorded. We therefore have

$$t' = a - b\{D(\eta,\zeta)\otimes(\delta_2+\delta_2)\} \tag{19}$$

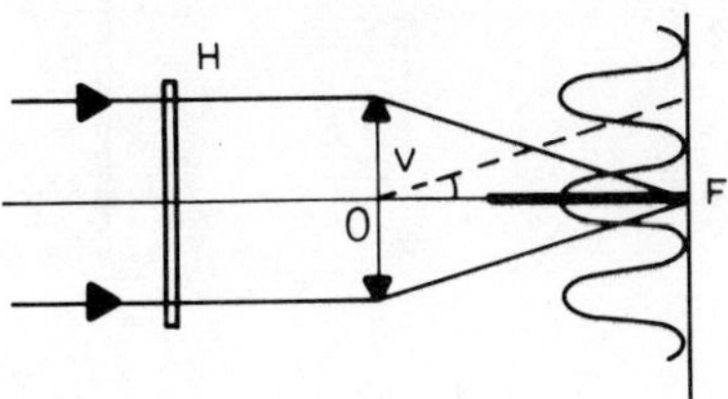


FIG. 18. Recording spectrum of two identical speckles translated with respect to each other.

where $a$ and $b$ are two constants. In the focal plane F the amplitude is given by the transform $\tilde{t}$ of $t$:

$$\tilde{t} = a\delta - b[\tilde{D}(u,v)\{1+\exp(jkv\zeta)\}] \quad (20)$$

where $k = 2\pi/\lambda$, $\zeta$ is the translation of G between the two exposures measured in the plane H, and $u$ and $v$ are the angular co-ordinates characterizing the position of the point where the phenomenon is observed in the focal plane F. The first term $a\delta$ corresponds to the geometrical image of the point source in F. This term is not of interest to us. In intensity the second term can be written, ignoring a constant factor

$$I = |\tilde{D}(u,v)|^2 \cos^2 \frac{\pi v \zeta}{\lambda} \quad (21)$$

which corresponds to a system of rectilinear parallel fringes with a separation equal to $\lambda/\zeta$ and modulated by $|\tilde{D}(u,v)|^2$. This last factor represents a very fine structure diffuser and it can be neglected at the first approximation. This experiment is the basis of the techniques known as speckle interferometry.

Let us consider a diffusing object consisting of two parts A and B. An image of A and B is formed on H. To simplify the description it is assumed that A is fixed and that B undergoes deformations which it is desired to investigate. Two successive exposures are made on H: one before and the other after the deformation of B, translating H between the two exposures. Two successive exposures can also be carried out during the

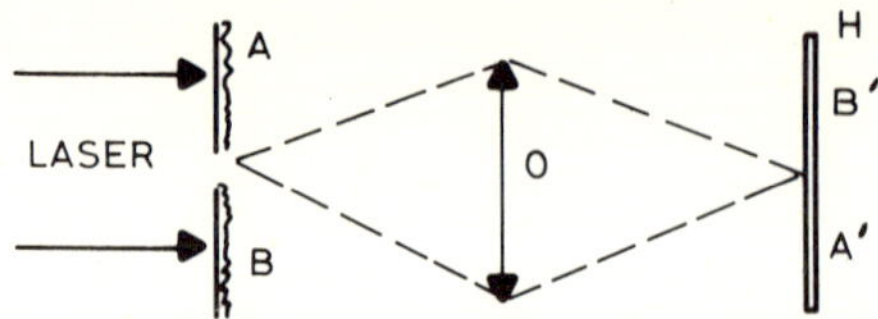


FIG. 19. Principle of a speckle interferometry experiment.

deformation. After development the negative H is illuminated as shown in Fig. 20. In the region A′ two identical and displaced speckles have been recorded because A has not moved and H has undergone a slight translation. The region A′ gives the fringes previously studied in the focal plane of the objective $O_1$. However, in the region B′ two speckles have been recorded which are not identical because the region B was deformed between the two exposures. If the deformation was so great that the two speckles recorded at B′ have no correlation, this region gives a practically uniform background in the focal plane of the objective $O_1$. In fact it is a background modulated by $|\tilde{D}(u,v)|^2$ but this modulation, which is very fine, can be neglected. If an opaque screen pierced by a slit T is placed in the focal plane F all the light coming from A′, i.e. from the fixed-A region, can be stopped. It is sufficient that the slit T coincides with a minimum zero of the fringes produced by A′. In the image H′ of H the parts of the object which were not modified between the two exposures will not be seen. However, as the area B′ does not produce fringes part of the light from this area will pass by T and appear in the image H′. Only

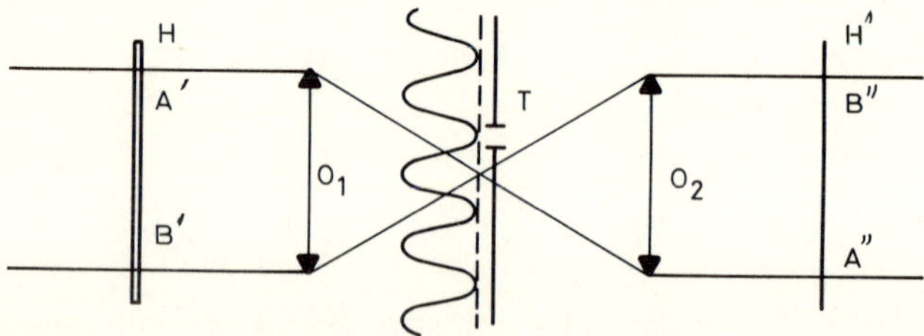


FIG. 20. Set-up for filtering the speckle spectrum recorded in the experiment shown in Fig. 19.

the areas modified between the two exposures will appear in the image.

We have only given the fundamental principles of the methods of speckle interferometry. There are numerous developments of these methods, which are both qualitative and quantitative.

## References

BORN, M. and WOLF, E. (1970). *Principles of optics*, Pergamon Press, Oxford.

BRYNGDAHL, O. (1965). Application of shearing interferometry. *Prog. Optics* **4**, 39.

DAINTY, J.C. (1975). Laser speckle and related phenomena. In *Topics in applied physics*. Springer-Verlag, Berlin.

DITCHBURN, R.W. (1963). *Light*. Blackie, London.

GOODMAN, J.W. (1966). *Introduction to Fourier optics*. McGraw Hill, New York.

LEITH, E.N. and UPATNIEKS, J. (1963). Reconstructed wavefronts and communication theory. *J. opt. Soc. Am.* **53**, 1377.

STEEL, W.H. (1966). Two-beam interferometry. *Prog. Optics* **5**, 147.

# 7. STUDY OF SOLIDS BY SUBMILLIMETRE-WAVE INTERFEROMETRY

J.F. ANGRESS and T.J. PARKER

*Department of Physics, Westfield College, Kidderpore Avenue, London, NW3 7ST, U.K.*

## 1. Introduction

Until comparatively recently optical spectroscopy in the range 1–200 $cm^{-1}$, now commonly referred to as the submillimetre region, was extremely difficult. The inefficiency of the grating systems used, the low fluxes from broad-band sources, and the poor performance of available detectors combined to give very low signal-to-noise ratios, and comparatively few laboratories had instruments operating at these frequencies. However, some 20 years ago it was realized that a Michelson interferometer used as a spectrometer has two advantages over a diffraction grating spectrometer: these are the light grasp (or Jacquinot) advantage and the multiplex (or Fellgett) advantage (see for example Loewenstein 1971). Used in the now conventional way, the output of a Michelson interferometer as a function of the relative time delay $\tau$ between the two arms is essentially the autocorrelation function $\gamma(\tau)$ of the input radiation. Thus, using the Wiener–Khintchine theorem (see for example Mandel and Wolf 1965) we can write

$$\gamma(\tau) = \int_{-\infty}^{\infty} P(f)\exp(-2\pi i f\tau)df \qquad (1)$$

where $P(f)$ is the power spectrum of the radiation and $P(f) = P(-f)$. The spectrum $P(f)$ can be obtained by Fourier transforming the measured 'interferogram' $\gamma(\tau)$, and this is the basis on which the instrument is used as a 'Fourier transform spectrometer'. (In practice the path difference $x = c\tau$ and wave number $\nu = f/c$ are usually used as the conjugate variables.) The transformations require the use of a fast computer, but with the general availability of such facilities, the advantages accruing to the interferometric method, and the high performance of modern detectors, broad-band spectroscopy in the submillimetre region is now being increasingly exploited in a wide variety of fields.

In the present paper we briefly review the use of submillimetre-wave interferometry in the study of solids and describe some recent advances in techniques made in our laboratory. Our own interests are in the study of photon–phonon interactions. The type of information concerning these which can be obtained from submillimetre-wave spectroscopy is discussed and illustrative examples from our recent results will be presented.

## 2. Experimental techniques

Spectroscopic measurements on solids at submillimetre wavelengths are now generally made with Michelson interferometers constructed in the symmetrical optical configuration illustrated schematically in Fig. 1(a). Radiation from a source L is divided by a beam splitter B into two partial beams which are reflected

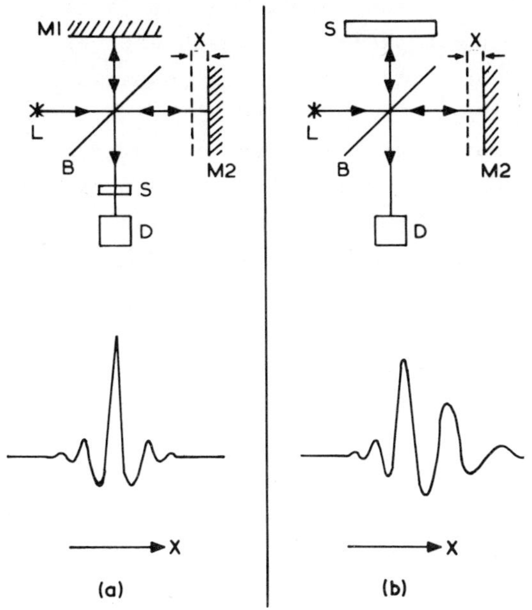


FIG. 1. Schematic diagram of a Michelson interferometer and a typical interferogram which would be recorded using (a) the symmetrical configuration used in power transmission spectroscopy and (b) the asymmetric configuration used in amplitude reflection spectroscopy.

from the two mirrors M1 and M2 and then recombined at the beam splitter before passing on to the detector D. A symmetrical background interferogram, with its maximum value at the position of zero optical path difference, is recorded when the optical path difference between the two arms of the interferometer is varied by scanning the mirror M2 as shown. The specimen is placed near the detector either in a transmission configuration, as shown, or in a reflection configuration, and as each partial beam interacts in an identical way with the specimen a symmetrical interferogram is again observed which can be analysed, together with the background interferogram, to determine either the power transmission spectrum $T(\nu)$ or the power reflection spectrum $R(\nu)$ of the specimen.

In general the exact forms of the measured spectra $T(\nu)$ and $R(\nu)$ depend on details of the experimental arrangement such as the specimen thickness and angle of incidence of the radiation as well as on the fundamental phenomenological optical parameters $n(\nu)$ and $k(\nu)$ or the real and imaginary parts $\varepsilon'(\nu)$ and $\varepsilon''(\nu)$ of the dielectric response. As it is the latter configuration-independent parameters which are conventionally calculated from theoretical models for comparison with experiment, it is usual to determine $n(\nu)$ and $k(\nu)$, or associated parameters, from the measured spectra. In some cases this is a relatively straightforward procedure. For example, if $k(\nu) \approx 0$, then $n(\nu)$ can be determined with reasonable accuracy from the measured $R(\nu)$, and if the channel spectrum is not resolved in $T(\nu)$, then $T(\nu)$ can be used with the known refractive index spectrum to calculate $k(\nu)$.

In the case of highly absorbing solids measurements on bulk samples are necessarily made in the reflection mode and power-reflection spectroscopy has for many years provided the basis for the conventional method of determining the optical constants or dielectric functions of such materials. The usual procedure is to measure the power reflectivity at near normal incidence over a wide spectral range and to calculate the associated phase angle $\phi(\nu)$ from the Kramers–Krönig relation (Robinson and Price 1953):

$$\phi(\nu) = \frac{2\nu}{\pi}\int_0^\infty \frac{\ln r(\nu')}{\nu'^2 - \nu^2}\mathrm{d}\nu'. \qquad (2)$$

In this expression $r(\nu)$ exp $i\phi(\nu)$ represents the complex amplitude reflection coefficient, and $r(\nu) = \{R(\nu)\}^{\frac{1}{2}}$. The Fresnel relations for normal incidence

$$\varepsilon'(\nu) + i\varepsilon''(\nu) = \{n(\nu)+ik(\nu)\}^2$$
$$= \left\{\frac{1 + r(\nu)\exp\, i\phi(\nu)}{1 - r(\nu)\exp\, i\phi(\nu)}\right\}^2 \qquad (3)$$

can then be used as a reasonable approximation to calculate the optical constants or dielectric functions. It should be noted that in crystals which are anisotropic more than two optical constants are required to describe the observed behaviour. The main disadvantage of this procedure is that, in addition to experimental errors which can easily be estimated, there is often some degree of uncertainty in the calculated values of $\phi(\nu)$ determined from equation (2) which results from the truncation of the measured spectrum at low and high frequencies and which is difficult to assess.

During the past 15 years a wide range of techniques, known collectively as dispersive Fourier transform spectroscopy (DFTS), have been developed for determining the optical constants of specimens in all three material phases by using Michelson interferometers in the asymmetric mode. The technique was pioneered by Chamberlain and co-workers at the National Physical Laboratory, Teddington (Chamberlain, Gibbs, and Gebbie 1963, 1969) and by Bell and co-workers at Ohio State University (Bell 1966, Russell and Bell 1966), and the principle is illustrated in Fig. 1(b) for the case of amplitude reflection spectroscopy of solids. A symmetrical background interferogram is first recorded as described above, and the specimen S is then substituted for the reference mirror M1 before recording the specimen interferogram. As the specimen now interacts with only one of the two partial beams, spectral components in that beam only are attenuated and delayed in time by amounts which are determined by the values of $r(\nu)$ and $\phi(\nu)$ respectively with the result that an asymmetric interferogram is observed as indicated in Fig. 1(b). The amplitude and phase reflection spectra of the specimen can then be recovered directly from the ratio of the complex Fourier transforms of the two interferograms. The information in the measured spectral range is thus complete and limited only by the experimental errors in the determination of $r(\nu)$ and $\phi(\nu)$, so that the optical constants can be calculated directly from equation (3). The uncertainties associated with the use of equation (2) are thus eliminated. However, it can easily be shown that if the phase spectrum is to be determined with an accuracy which is limited by the detector noise level, which is the usual criterion for assessing the quality of any spectral measurement, then for measurements in the submillimetre region the relative positions of the reference mirror and specimen surfaces must be known with a precision of better than 0.1 μm. As a result of this severe constraint the application of this technique was, until very recently, confined to measurements at room temperature (Russell and Bell 1966, Johnson and Bell 1969, Birch *et al.* 1975). However, following developments in our laboratory (Parker, Chambers, and Angress 1974) and elsewhere (Gast and Genzel 1973), measurements were later extended to liquid-nitrogen temperatures (Parker and Chambers 1975, Zwick, Irslinger, and Genzel 1976) and more recently to liquid-helium temperatures (Mead and Genzel 1978, Parker, Lowndes, and Mok 1978b).

Some applications of these various techniques in solid state spectroscopy are described in the following sections.

## 3. Photon–phonon interactions

In our laboratory we have been concerned primarily with the optical properties of solids associated with atomic vibrations. Lattice frequencies extend from the low-frequency acoustic modes to the maximum lattice frequency of the optic modes at typically 300 $cm^{-1}$, i.e. the frequencies span the submillimetre wave range and, provided the modes have an associated electric dipole moment, direct strong photon–phonon interactions can take place in this spectral region. However, in the case of pure crystalline materials of quasi-infinite extent and within the harmonic and other first approximations only a very restricted amount of information can be obtained from these interactions. The restrictions arise in the first place from the lattice translational and rotational-type symmetries specified by the crystal space group. The translational symmetry leads, via Bloch's theorem, to the characterization of the normal modes by their wavevector **q**. Since the inverse wavelength of the radiation is extremely small compared with the Brillouin zone dimensions, submillimetre waves, unlike neutrons, can only be used to probe modes of essentially zero **q**. Further, symmetry places restrictions on the number of these zone-centre modes which can interact with the radiation field in any particular case. However, these various restrictions can be relaxed in one way or another as outlined below.

Higher-order processes, such as anharmonicity and second-order dipole moments, can lead to the participation of more than one phonon in the interaction and, although energy conservation and symmetry-based selection rules still apply, these processes give rise to highly detailed spectra which can be interpreted to give a lot more information than is obtainable from first-order (single-phonon) processes. As an example, if a single phonon with $\mathbf{q} \approx 0$ is excited, anharmonic processes can cause it to decay by interaction with two or more other phonons having non-zero wavevectors (see for example Donovan and Angress 1971). This leads to a temperature-dependent frequency shift and broadening of the observed bands and structure associated with phonons from the entire Brillouin zone appears. Practical examples of the effects associated with both weak and strong anharmonicity are presented in §4.

The lattice symmetry, and hence the symmetry-based restrictions on the single-phonon (and multi-phonon) processes which occur, can be reduced in a number of ways, again leading to richer spectra and more information concerning the dynamics of the system. One method is to modify the point symmetry by an externally applied perturbation such as a homogeneous electric field: effects due to such perturbations have been observed in our laboratory although we shall not discuss them here (see for example Angress, Gledhill, and Maiden 1971). Another powerful method is to introduce impurities into the crystal; for a review see Barker and Sievers (1975). These destroy the translational symmetry thereby removing the highly restrictive selection rule requiring wave-vector conservation. The spectra of disordered systems such as mixed crystals and amorphous materials are, of course, free from symmetry-based selection rules. Some representative spectra of crystals containing impurities and the spectrum of a disordered material are presented and briefly discussed in §5.

## 4. The submillimetre-wave spectra of pure single crystals

In this section we shall describe measurements by DFTS on an alkali halide crystal, as an example of a weakly anharmonic system, and on a hydrogen-bonded ferroelectric crystal, which is an example of a strongly anharmonic system. In the former case structure has been revealed in the spectra by DFTS which has never been observed by other techniques, permitting a more critical comparison with theory than would otherwise have been possible. In the latter case the optical constants have been determined more accurately as a function of temperature in the region of the ferroelectric soft mode by DFTS than was previously possible, and the results are consistent with earlier coupled-mode descriptions of the dielectric response based on Raman scattering measurements.

The amplitude and phase reflection spectra of a single crystal of KBr measured at 100 and 300 K by Parker and Chambers (1975) by DFTS are shown in Fig. 2, and the optical constants calculated from these spectra together with values determined from dispersive transmission measurements at 90 and 300 K by Parker *et al.* (1978a) are shown in Fig. 3. Although the broad features of the reststrahlen band had previously been well established at both 90 and 290 K by power reflection spectroscopy (Wilkinson 1963) it was not until Johnson and Bell (1969) first used amplitude reflection spectroscopy to study KBr and KCl at room temperature that the detailed structure in the reflection spectrum was revealed. The reason for this is that most of the structure occurs in a spectral range where the power reflectivity $R(\nu) = r(\nu)^2$ is too small to be measured with the necessary precision by conventional spectroscopic techniques. The values of the optical constants shown in Fig. 3 on either side of the reststrahlen band were determined from

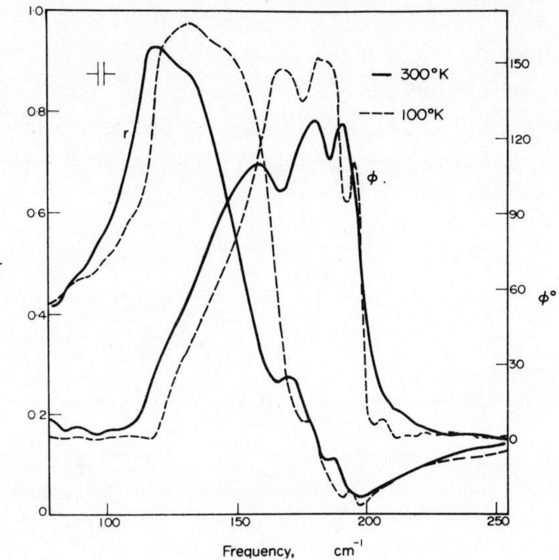


FIG. 2. The amplitude $r$ and phase $\phi$ reflection spectra of KBr at 100 and 300 K. (After Parker and Chambers 1975.)

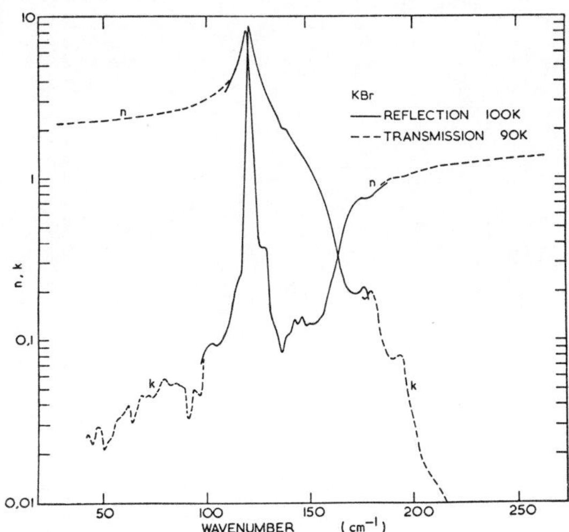


FIG. 3. The refractive index $n$ and extinction coefficient $k$ of KBr determined at 90 K by dispersive transmission spectroscopy (dotted curves) and at 100 K by dispersive reflection spectroscopy (solid curves). (After Parker *et al.* (1978a).)

dispersive transmission measurements because dispersive reflection spectroscopy is not suitable for determining $k(\nu)$ when the phase reflection coefficient of the specimen is close to that of the reference mirror. The results shown in Fig. 2 were found by Parker and Chambers (1976) to be in good agreement with calculations by Bruce (1973) who considered the effects of cubic and quartic anharmonicity on the dielectric response of KBr.

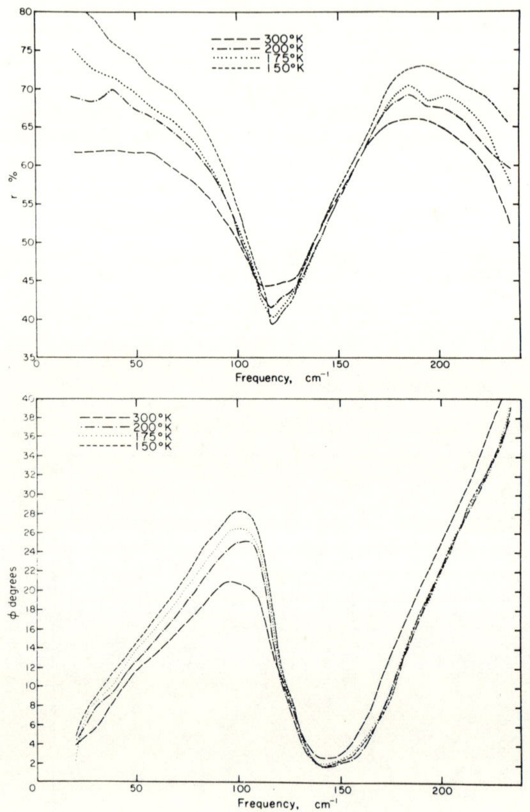


FIG. 4. The *c* axis amplitude *r* and phase ϕ reflection spectra of KDP measured at 5 $cm^{-1}$ resolution. (After Ledsham, Chambers, and Parker 1977.)

Similar techniques have also been used to study the dielectric response of strongly anharmonic systems, such as ferroelectric crystals, as a function of temperature, and Fig. 4 shows the *c* axis amplitude and phase reflection spectra of a single crystal of $KH_2PO_4$ (KDP) measured at a number of temperatures above the transition temperature by Ledsham, Chambers, and Parker (1977). The real and imaginary parts of the dielectric response calculated from these results are shown by the dotted lines in Fig. 5 and it should be noted that the strong low-frequency mode which is associated with the ferroelectric transition is clearly resolved in most of these curves.

The temperature dependence of the ferroelectric soft mode in KDP has been studied by a variety of techniques in a number of laboratories since it was first observed by Barker and Tinkham (1963). Raman scattering measurements by She *et al.* (1972) and Lagakos and Cummins (1974) produced results which could not be satisfactorily explained in terms of independent oscillators and revealed that the ferroelectric soft mode is strongly coupled to an optic phonon of the same symmetry near 180 $cm^{-1}$. The experimental results shown in Fig. 5 could similarly not be satisfactorily explained in terms of independent oscillators and were fitted with a model in which the ferroelectric soft mode is strongly coupled to the mode near 180 $cm^{-1}$ which is also clearly visible (Ledsham, Chambers, and Parker 1977). It can be seen that the general overall agreement between the experimental measurements and the theoretical results obtained using this model and indicated by crosses is very good.

## 5. The submillimetre-wave spectra of imperfect and mixed crystals

As we have already pointed out the presence of impurities in crystals removes the selection rule requiring wavevector conservation in photon–phonon interactions. This is most easily appreciated by considering an electric charge placed on one atom of an otherwise uncharged monatomic linear chain. It is clear that *any* mode in which the charged atom moves will have an associated dipole moment. In fact because the dynamics of the system are unchanged in this idealized case the absorption spectrum just maps the density of vibrational states of the chain. In practice, of course, even a simple isotopic substitutional defect will have an associated mass change and in general the force constants will also change. As a result the absorption spectrum can be very complicated, but it can be interpreted to give information about the dynamics of the host

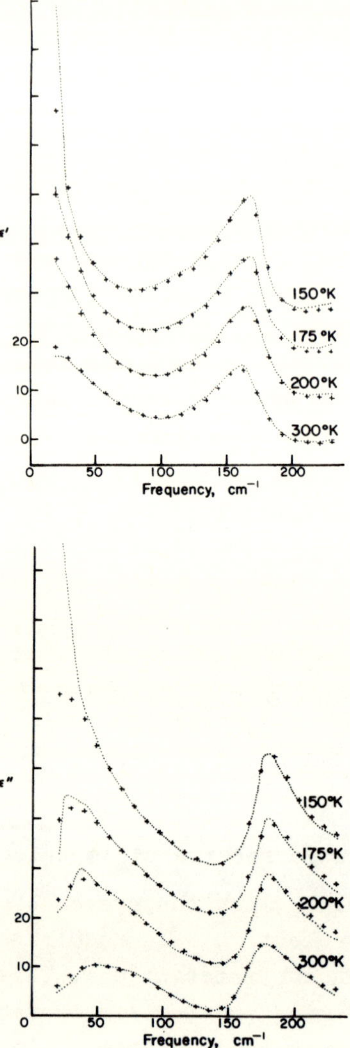


FIG. 5. Theoretical fits to the dielectric functions ε′ and ε″ of KDP. The dotted curves are the experimental curves calculated from the data of Fig. 4. The 200, 175, and 150 K curves are displaced by 10, 20, and 30 units respectively up the ε axes from the 300 K curve for clarity. The crosses are points from theoretical curves calculated using the coupled oscillator model described in the text. (After Ledsham, Chambers, and Parker 1977.)

crystal and about the structure and dynamics of the defects in favourable circumstances. The absorption features induced by an impurity can be classified as follows.

(i) A continuous spectrum associated with features in the density of vibrational states $\rho(\nu)$ of the host crystal.

(ii) Sharp lines due to modes localized around the defect (usually a light impurity) which occur at frequencies above the maximum frequency of the perfect crystal.

(iii) Sharp lines due to localized modes with frequencies in any gaps in the density of states. Such modes are usually referred to as gap modes to distinguish them from (ii).

(iv) Broader lines due to 'resonance modes' which occur in regions where $\rho(\nu)$ is low but not zero.

All these types have been studied in our laboratory over a number of years and in the following we present examples of types (i), (iii), and (iv) recently measured using a conventional Fourier transform spectrometer to obtain power transmission spectra. Localized modes (type (ii)) usually occur at high frequencies which are outside the submillimetre range.

In Fig. 6 the absorption spectra of unirradiated GaAs and

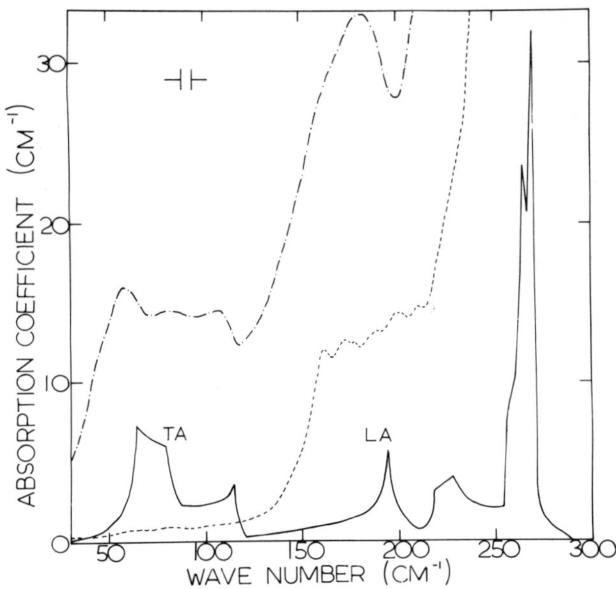


FIG. 6. The absorption spectra at 100 K of unirradiated GaAs (dotted curve) and of GaAs irradiated with $2.5 \times 10^{18}$ fast neutrons $cm^{-2}$ (chain curve). The solid curve is the density of states for GaAs calculated by Dolling and Cowley (1966). (After Gledhill *et al.* 1977.)

of GaAs irradiated with $2.5 \times 10^{18}$ fast neutrons $cm^{-2}$ obtained by Gledhill *et al.* (1977) are shown. The density of vibrational states for GaAs calculated by Dolling and Cowley (1966) is also shown. This consists of two main bands in the acoustic mode region which can be assigned in a rough way to TA-like and LA-like vibrations. It can be seen that in this region the induced one-phonon structure bears some resemblance to $\rho(\nu)$; in particular there is a pronounced minimum in the absorption at 117 $cm^{-1}$ which is undoubtedly associated with the sharp cut-off of the TA band near this frequency. We associate the absorption bands on each side of this dip with the TA and LA bands in $\rho(\nu)$. The spectrum is an example of type (i) above. The induced absorption is clearly associated with the structural defects induced by the irradiation. However, the precise nature of the defects responsible is not known and the reason why the absorption is so similar to $\rho(\nu)$ in this case is not fully understood.

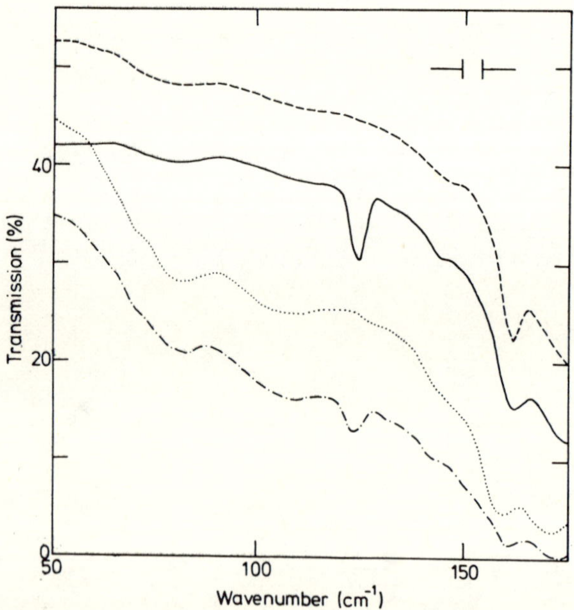


FIG. 7. The transmission spectra of pure GaAs at 100 K (broken curve) and 300 K (dotted curve) and of GaAs containing B and Si at 100 K (solid curve) and 300 K (chain curve). (After Laithwaite *et al.* 1977.)

An example of an absorption band associated with a defect resonance can be seen in Fig. 7 (Laithwaite *et al.* 1977). This figure shows the transmission spectra at 300 and 100 K of an essentially pure GaAs specimen and of a GaAs specimen containing the impurities Si and B (among others). The additional sharp band in the doped specimen occurs at 125 $cm^{-1}$ (100 K value) which is in a region where $\rho(\nu)$ is low and featureless just above the broad band associated with TA-like modes (see Fig. 6). It is thought that the defect responsible for the resonance mode is B on a Ga site.

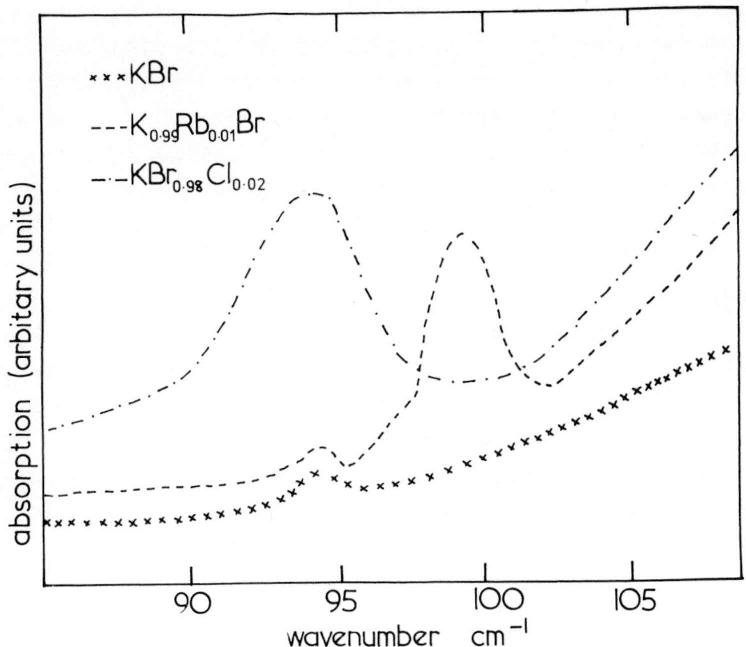


FIG. 8. The absorption spectra of pure KBr (crosses), $K_{0.99}Rb_{0.01}Br$ (broken curve), $KBr_{0.98}Cl_{0.02}$ (chain curve) at 100 K. (After Gledhill, Syms, and Angress 1978.)

In Fig. 8 we show as examples the absorption due to gap modes arising from Rb and Cl impurities in KBr (Gledhill, Syms, and Angress 1978). The weak band at 94.5 $cm^{-1}$ in nominally pure KBr is thought to be due to residual impurities — possibly Cl. The gap in the density of states of KBr extends from about

94 to 100 $cm^{-1}$, and the Rb-induced mode can be thought of as having fallen into this gap from the optic band while the Cl-induced mode has risen out of the acoustic band. Strangely enough, calculations using a simple mass-defect model indicate that the Cl mode should occur at a higher frequency than the Rb mode, and comparison with the experimental results suggests that considerable force-constant changes are associated with the ion substitution.

As the concentration of defects increases they can no longer be considered as isolated, and the spectra become more complicated owing to defect aggregation. In the limit the system is no longer a crystal containing defects but must be thought of as a disordered solid. In the case of structural defects the material becomes amorphous. In the case of substitutional impurities, as exemplified by $K_xRb_{1-x}Br$ described above for $1 - x = 0.01$, the system becomes what is called a mixed crystal (random alloy) in which the K and Rb ions are randomly arranged on the cation sublattice. This presupposes, of course, that KBr and RbBr are completely miscible and that superlattices are not formed. A large number of mixed crystals have been examined in our laboratory and in Fig. 9 we show the reflection spectra of some $K_xRb_{1-x}Br$ alloys at 100 K obtained by Angress *et al.* (1977). The weak band near 95 $cm^{-1}$ in $K_{0.75}Rb_{0.25}Br$ thought to be associated with the gap mode due to Rb impurities in KBr (Fig. 8) should be noted. Of particular interest is the band splitting which is clearly observable in the spectrum of $K_{0.3}Rb_{0.7}Br$. Such splitting occurs in some of the spectra of some alloy systems but not in others. Various criteria for the formation of split bands have been proposed but in our opinion none of these is entirely satisfactory and further work remains to be done.

## 6. Conclusion

Spectroscopy in the submillimetre-wave region has developed rapidly over the past few years owing to advances in interferometric and other techniques. Interest in these techniques and their applications is still growing. In this paper we have reviewed some of the recent advances in this field and have

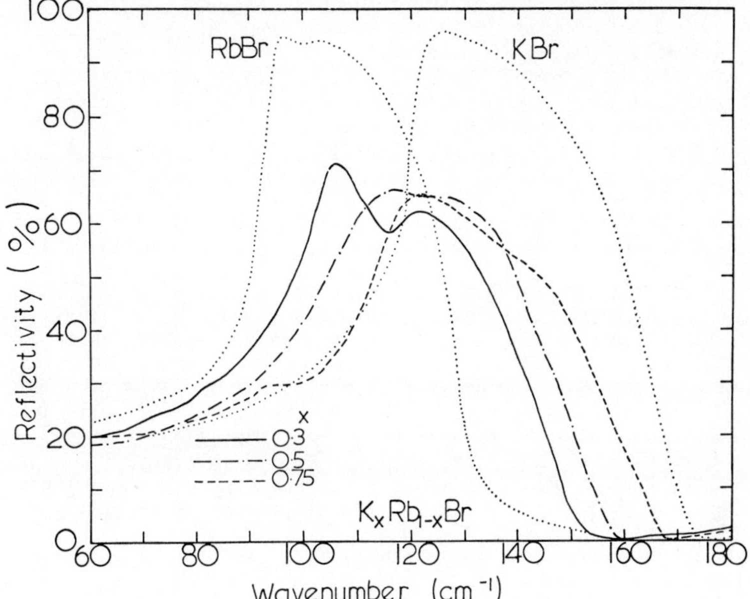


FIG. 9. Reflection spectra of some $K_xRb_{1-x}Cl$ alloys at 100 K: solid curve, $x$ = 0.3; chain curve, $x$ = 0.5; broken curve, $x$ = 0.75. (After Angress *et al.* 1977.)

described the application of interferometry in the submillimetre range to problems in lattice dynamics.

## Acknowledgments

The authors are members of a team which includes Dr. G.A. Gledhill, Dr. W. Chambers, and latterly Dr. M. Syms. We thank these colleagues and past research students for many helpful discussions and for permission to present some of their results in this paper. We also thank Professor R.C. Newman for his interest and for permission to present the spectra of his GaAs samples.

## References

ANGRESS, J.F., GLEDHILL, G.A., CLARK, J.D., CHAMBERS, W.G., and SMITH, W. (1977). *Proc. Int. Conf. on Lattice Dynamics*, p. 422. Flammarion Sciences, Paris.

ANGRESS, J.F., GLEDHILL, G.A., and Maiden, A.J. (1971). *Int. Conf. on Phonons Rennes*, p. 459. Flammarion Sciences, Paris.

BARKER, A.S. and SIEVERS, A.J. (1975). *Rev. mod. Phys.* (Suppl. 2) **47**, 1.
BARKER, A.S. and TINKHAM, M. (1963). *J. chem. Phys.* **38**, 2257.
BELL, E.E. (1966). *Infrared Phys.* **6**, 57.
BIRCH, J.R., COOK, R.J., HARDING, A.F., JONES, R.G., and PRICE, G.D. (1975). *J. Phys. D* **8**, 1353.
BRUCE, A.D. (1973). *J. Phys. C* **6**, 174.
CHAMBERLAIN, J.E., GIBBS, J.E., and GEBBIE, H.A. (1963). *Nature (Lond.)* **198**, 874.
—— (1969). *Infrared Phys.* **9**, 185.
DOLLING, G. and COWLEY, R.A. (1966). *Proc. phys. Soc.* **88**, 463.
DONOVAN, B. and ANGRESS, J.F. (1971). *Lattice vibrations*. Chapman and Hall, London.
GAST, J. and GENZEL, L. (1973). *Opt. Commun.* **8**, 26.
GLEDHILL, G.A., ANGRESS, J.F., BROZEL, M., and NEWMAN, R.C. (1977). *Proc. Int. Conf. on Lattice Dynamics, Paris*, p. 388. Flammarion Sciences, Paris.
GLEDHILL, G.A., SYMS, M., and ANGRESS, J.F. (1978). *Infrared Phys.* **18**, 915.
JOHNSON, K.W. and BELL, E.E. (1969). *Phys. Rev.* **187**, 1044.
LAGAKOS, N. and CUMMINS, H.Z. (1974). *Phys. Rev. B* **10**, 1063.
LAITHWAITE, K., NEWMAN, R.C., ANGRESS, J.F., and GLEDHILL, G.A. (1977). *Inst. Phys. Conf. Ser . No. 33a*, p. 133. Institute of Physics and Physical Society, London.
LEDSHAM, D.A., CHAMBERS, W.G., and PARKER, T.J. (1977). *Infrared Phys.* **17**, 165.
LOEWENSTEIN, E.V. (1971). *Aspen Int. Conf. on Fourier Spectroscopy, U.S. Air Force Rep. No. 114, AFCRL-71-0019*, p. 3.
MANDEL, L. and WOLF, E. (1965). *Rev. mod. Phys.* **37**, 231.
MEAD, D. and GENZEL, L. (1978). *Infrared Phys.* **18**, 555.
PARKER, T.J. and CHAMBERS, W.G. (1975). *IEE Conf. Publ. No. 129*, p. 169. IEE, London.
—— (1976). *Infrared Phys.* **16**, 349.
PARKER, T.J., CHAMBERS, W.G., and ANGRESS, J.F. (1974). *Infrared Phys.* **14**, 207.
PARKER, T.J., CHAMBERS, W.G., FORD, J.E., and MOK, C.L. (1978a). *Infrared Phys.* **18**, 571.
PARKER, T.J., LOWNDES, R.P., and MOK, C.L. (1978b). *Infrared Phys.* **18**, 565.
ROBINSON, T.S. and PRICE, W.C. (1953). *Proc. R. Soc. B* **66**, 969.
RUSSELL, E.E. and BELL, E.E. (1966). *Infrared Phys.* **6**, 75.
SHE, C.Y., BROBERG, T.W., WALL, L.S., and EDWARDS, D.F. (1972). *Phys. Rev. B* **6**, 1847.
WILKINSON, G.R. (1963). In *Infrared spectroscopy and molecular structure* (ed. M. Davies), p. 85. Elsevier, Amsterdam.
ZWICK, U., IRSLINGER, C., and GENZEL, L. (1976). *Infrared Phys.* **16**, 263.

# 8. SPIN INTERFEROMETRY AND PHASE RELATIONS IN THREE-LEVEL SYSTEMS

M. MEHRING, M.E. STOLL and E.K. WOLFF

*Institut für Physik, University of Dortmund, 4600 Dortmund, Federal Republic of Germany*

Although the sign of the wave function of a fermion under a $2\pi$ rotation ('spinor character') has been a well-established theoretical fact since the time of Pauli, it has aroused much interest recently. After the proposal of a *Gedanken* experiment by Bernstein (1967) and Aharanov and Susskind (1967) an experimental demonstration of spinor character was given in 1975 by neutron interferometry experiments (Rauch *et al.* 1975, Werner *et al.* 1975) and magnetic resonance experiments (Klempt 1975). More recently the ideas of nuclear magnetic resonance (n.m.r.) interferometry were introduced by Stoll, Vega, and Vaughan (1977) to demonstrate the observability of the phase dependence of the wave function in an n.m.r. double-resonance experiment in the case of two coupled spin-½ particles.

In all the above cases four different states are involved in Hilbert space either given by a spin-½ particle with two different **k** states (for wave vector **k**) as in neutron interferometry or by two coupled spin-½ particles in the n.m.r. case. Recently we have shown that the same phenomenon can be observed even with a spin-1 particle, namely deuterium (Stoll, Wolff, and Mehring 1978). In fact this is the simplest possible case for demonstrating this kind of Hilbert space behaviour since at least two states are involved in the 'rotation' and at least one 'reference' state is needed for the detection of the phase change of the other two wave functions, i.e. any three-level system in which at least two different transitions can be excited separately will show spinor character if irradiated suitably.

In order to show spinor character in such a system we chose a spin-1 nucleus, deuterium. In a large magnetic field the three energy levels of the Zeeman Hamiltonian have equal spacing $\omega_0$, whereas inequivalent transition frequencies $\omega_0 + \omega_Q$

and $\omega_0 - \omega_Q$ will occur if a quadrupolar interaction of strength $\hbar\omega_Q$ is included. The corresponding three-level diagram is shown in Fig. 1.

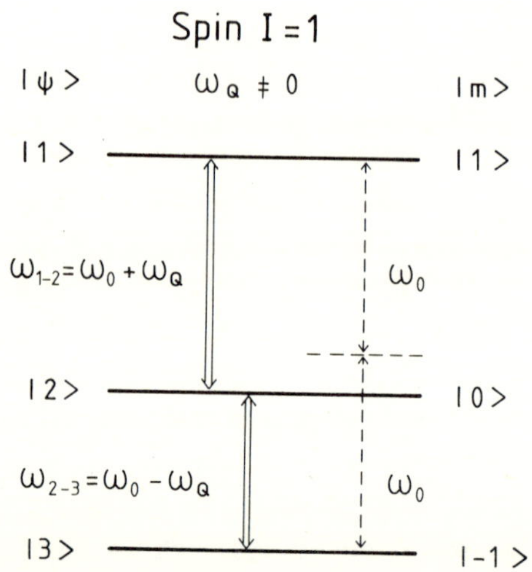


FIG. 1. Energy level diagram of a spin-1 particle with quadrupole interaction $\omega_Q$ in a magnetic field $H_0 = \omega_0/\gamma$. The eigenstates of the three-level system are designated $|1\rangle$, $|2\rangle$ and $|3\rangle$. The allowed transitions 1–2 and 2–3 as well as the double-quantum transition at $\omega_{1-3}/2 = \omega_0$ are indicated.

The chemical system selected was a single crystal of 98% deuterated hexamethylbenzene (HMB, $C_6(CD_3)_6$).

The transitions $\omega_{1-2}$ and $\omega_{2-3}$ are separated by $2\nu_Q = 32.5$ kHz, where $\nu_Q = \omega_Q/2\pi$ is the quadrupolar frequency of the deuterium nuclei, if the magnetic field is oriented perpendicular to the benzene ring. Two different pulsed radiofrequency (r.f.) magnetic fields were supplied to the separate transitions 1–2 and 2–3, i.e. 'rotations' represented by $\exp(-i\beta_{12}I_x^{1-2})$ and $\exp(-i\beta_{23}I_x^{2-3})$ respectively were performed on each transition separately. Here the rotation angle $\beta_{ij} = \sqrt{2}\gamma H_1 t_p$ is actually given by a time evolution of the transition $i$–$j$ under the influence of the r.f. magnetic field $H_1$ applied for a pulse duration $t_p$ where $\gamma$ is the gyromagnetic ratio of the spins involved. The fictitious spin-½ operators $I_\alpha^{k-j}$ $(\alpha=x,y,z)$ for the transition $i$–$j$ will be introduced later.

Spin echoes could be excited at time $2\tau$ by the application of the two pulse sequences $(\pi/2)_x-\tau-(\pi/2)_y$ and $(\pi/2)_x-\tau-(\pi)_y$ on either transition 1–2 or 2–3. In Fig. 2 the echo formation at time $2\tau$ is shown if the first pulse sequence is applied to the 1–2 transition. The echo signal is positive if no irradiation is applied at the 2–3 transition. If, however, a $\beta_{23}$ 'rotation' is applied to the 2–3 transition after the first $(\pi/2)_x$ pulse at the 1–2 transition as explained above, the echo signal at the 1–2 transition reverses sign under a '$2\pi$ rotation of the 2–3 transition, as seen in Fig. 2. In Fig. 3 we have plotted the echo height *versus* $\beta_{23}$ up to $8\pi$ and the expected $\cos(\beta_{23}/2)$ behaviour is observed. The depletion of the echo signal for large values of $\beta_{23}$ is due to relaxation and r.f. field inhomogeneity.

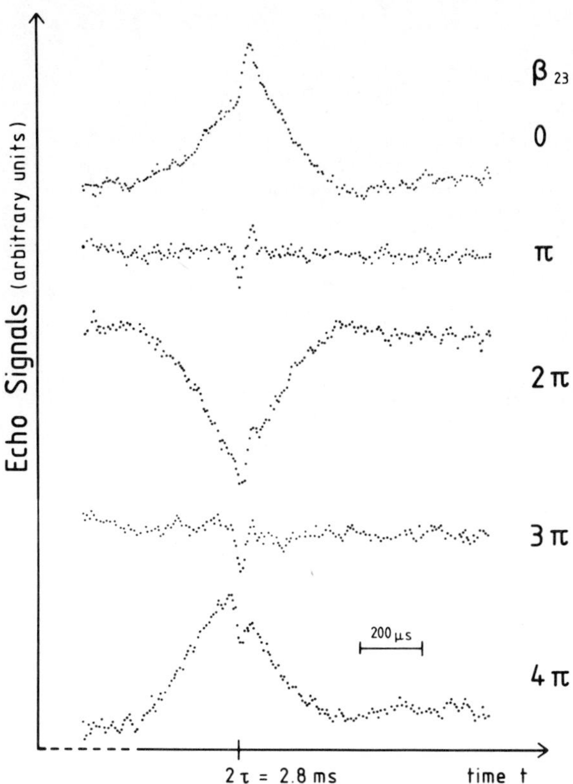


FIG. 2. Echo amplitude at the 1–2 transition as influenced by the rotation $\beta_{23}$ at the 2–3 transition in the case of deuterium (spin-1) in solid HMB. The sign reversal for a $2\pi$ rotation ($\beta_{23}=2\pi$) is evident.

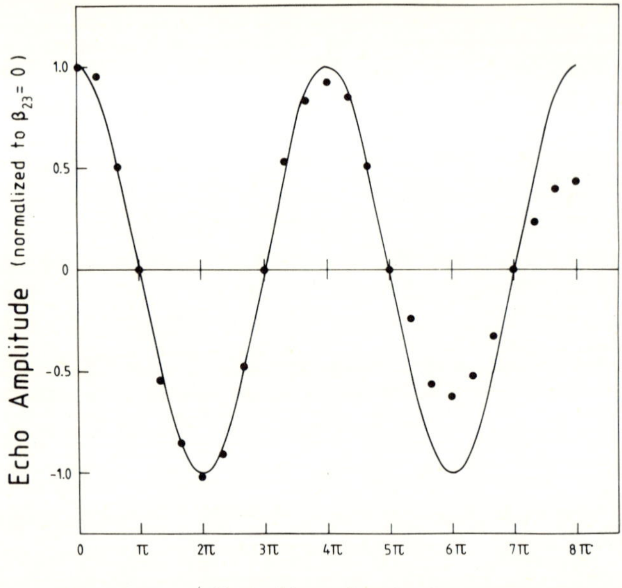


FIG. 3. Normalized echo amplitude at the 1–2 transition *versus* the rotation angle $\beta_{23}$ at the 2–3 transition obtained from data as shown in Fig. 2. The theoretical behaviour (solid curve, $\cos(\beta_{23}/2)$) and the experimental data demonstrate the $4\pi$ symmetry.

In order to discuss the other experiments we have performed in the nine-dimensional Hilbert space, we refer to the following nine fictitious spin-½ operators (Vega and Pines 1977, Wokaun and Ernst 1977, Vega 1978) in the Zeeman eigenstates:

$$
\begin{aligned}
I_x^{r-s} &= \tfrac{1}{2}(|r\rangle\langle s| + |s\rangle\langle r|) \\
I_y^{r-s} &= -\tfrac{i}{2}(|r\rangle\langle s| - |s\rangle\langle r|) \\
I_z^{r-s} &= \tfrac{1}{2}(|r\rangle\langle r| - |s\rangle\langle s|) \; .
\end{aligned}
\tag{1}
$$

In the following experiments we have always started with a density matrix prepared in the state $\rho(0) = I_x^{1-2}$ or $I_y^{1-2}$ by the application of an inversion prepulse ($\pi$ pulse) at the 2–3 transition a long time before the actual experiment begins and a $(\pi/2)_{y,x}$ initiating pulse at the 1–2 transition at time $t = 0$.

The following experiments have been performed.

(i) Simultaneous irradiation of both transitions with field strength $\omega_1$ in the $x$ direction, i.e. $\beta I_x^{1-2}$ and $\beta I_x^{2-3}$ where $\omega_1 t_p = \beta$.

(a) Spin-locking condition (Stoll, Wolff, and Mehring 1978).
Initiating pulse at 1–2: $(\pi/2)_y$, i.e. $\rho(0) = I_x^{1-2}$.
It follows that

$$\rho(\beta) = I_x^{1-2}\tfrac{1}{2}(1+\cos\beta) + I_x^{2-3}\tfrac{1}{2}(1-\cos\beta) - I_y^{1-3}\frac{1}{\sqrt{2}}\sin\beta \ . \quad (2)$$

This behaviour has been observed experimentally by looking at the magnetizations $\langle I_x^{1-2}\rangle$ and $\langle I_x^{2-3}\rangle$ respectively.

(b) Quadrature conditions.
Initiating pulse at 1–2: $(\pi/2)_x$, i.e. $-\rho(0) = I_y^{1-2}$.
It follows that

$$-\rho(\beta) = I_y^{1-2}\tfrac{1}{2}(\cos\beta+\cos 2\beta) + I_y^{2-3}\tfrac{1}{2}(\cos\beta-\cos 2\beta)$$
$$+ I_z^{1-3}\frac{1}{\sqrt{2}}\sin 2\beta + (I_z^{1-2}-I_z^{2-3}+I_x^{1-3})\frac{1}{2\sqrt{2}}\sin\beta \ . \quad (3)$$

The magnetizations $\langle I_y^{1-2}\rangle$ and $\langle I_y^{2-3}\rangle$ have been observed experimentally and show the theoretical behaviour.

(c) Phase variation $\phi_{1-2}$.
Initiating pulse at 1–2: $(\pi/2)_x$, i.e. $-\rho(0) = I_y^{1-2}$.
Irradiation: $\pi(I_x^{1-2}\cos\phi_{1-2} + I_y^{1-2}\sin\phi_{1-2})$; $\pi I_y^{2-3}$.
As follows from cases (a) and (b), only 2–3 magnetization $\langle I_\alpha^{2-3}\rangle$ ($\alpha=x,y$) is expected in this case. The experimental observations verify the theoretical expression

$$-\rho(\phi_{1-2}) = I_x^{2-3}\cos\phi_{1-2} + I_y^{2-3}\sin\phi_{1-2} \ . \quad (4)$$

(ii) Consecutive irradiation at both transitions with $\pi$-pulses.
Initiating pulse at 1–2: $(\pi/2)_x$, i.e. $-\rho(0) = I_y^{1-2}$, followed by a $(\pi)_y$ pulse at the 2–3 transition, transfers coherence from transition 1–2 to the forbidden transition 1–3 as has already been shown by Hatanaka, Terao, and Hashi (1975). Coherence decay of the 1–3 transition was observed by us in HMB by applying a readout $(\pi)_x$ pulse at the 1–2 transition a time $\tau$ later. This readout pulse at the 1–2 transition creates 2–3

magnetization as can readily be seen by inspecting the density matrix. The phase of the 2–3 magnetization was observed to vary with the phase $\phi_{1-2}$ of the $\pi$-pulse $\pi(I_x^{1-2}\cos\phi_{1-2}+I_y^{1-2}\sin\phi_{1-2})$ as follows:

$$\rho(\phi_{1-2}) = I_x^{2-3}\cos\phi_{1-2} - I_y^{2-3}\sin\phi_{1-2} \,. \tag{5}$$

This theoretical expression was experimentally verified.

(iii) Double quantum spinor behaviour was demonstrated by the application of r.f. fields of strength $\omega_1 = \gamma H_1$ at the 'double-quantum transition' frequency $\omega_0$. The corresponding double-quantum rotation angle can be defined as $\beta = (\omega_1{}^2/\omega_Q)t_p$ as shown by Vega and Pines (1977). By applying a double-quantum $\beta$ pulse after an initiating $\pi/2$ pulse at the 1–2 transition we have observed a $\cos(\beta/2)$ variation of the 1–2 magnetization, which demonstrates the spinor behaviour of this forbidden transition.

We finally remark that double-quantum coherence transfer has also been observed in the case $\beta = \pi$ in the above sequence (Hatanaka, Terao, and Hashi 1975). Magnetization was created at the 2–3 transition in this case and was observed to be phase coherent to the frequency $2\nu_0 - \nu_{1-2}$ as generated by the two-frequency source $\nu_0$ and $\nu_{1-2}$.

More details of these and other experiments are published elsewhere (Wolff and Mehring (1979), Mehring, Stoll and Wolff (197

**Acknowledgment**

Financial support from the Deutsche Forschungsgemeinschaft is gratefully acknowledged.

## References

AHARANOV, Y. and SUSSKIND, L. (1967). *Phys. Rev.* **158**, 1237.
BERNSTEIN, H.J. (1967). *Phys. Rev. Lett.* **18**, 1102.
HATANAKA, H., TERAO, T., and HASHI, T. (1975). *J. phys. Soc. Japan* **39**, 835.
KLEMPT, E. (1975). *Phys. Rev. D* **13**, 3125.
MEHRING, M., STOLL, M.E. and WOLFF, E.K. (1979), *J. Magn. Res.* (in press).
RAUCH, H., ZEILINGER, A., BADUREK, G., WILFING, A., BAUSPIESS, W., and BONSE, U. (1975). *Phys. Lett. A* **54**, 425.
STOLL, M.E., VEGA, A.J., and VAUGHAN, R.W. (1977). *Phys. Rev. A* **16**, 1521.
STOLL, M.E., WOLFF, E.K., and MEHRING, M. (1978). *Phys. Rev. A* **17**, 1561.

VEGA, S. (1978). *J. chem. Phys.* **68**, 5518.
VEGA, S. and PINES, A. (1977). *J. chem. Phys.* **66**, 5624.
WERNER, S.A., COLELLA, R., OVERHAUSER, A.W., and EAGAN, C.F. (1975). *Phys. Rev. Lett.* **35**, 1053.
WOKAUN, A. and ERNST, R.R. (1977). *J. chem. Phys.* **67**, 1752.
WOLFF, E.K. and MEHRING, M. (1979). *Phys. Lett.* **20A**, 125.

# AUTHOR INDEX

# SUBJECT INDEX